PACKED COLUMN SFC

RSC Chromatography Monographs

Series Editor: Roger M. Smith, *University of Technology, Loughborough, UK*

Advisory Panel: J. C. Berridge, *Sandwich, UK*.
G. B. Cox, *Indiana, USA*, I. S. Lurie, *Virginia, USA*.
P. J. Schoenmaker, *Eindhoven, The Netherlands*,
C. F. Simpson, *London, UK*, G. G. Wallace, *Wollongong, Australia*.

This series is designed for the individual practising chromatographer, providing guidance and advice on a wide range of chromatographic techniques with the emphasis on important practical aspects of the subject.

Supercritical Fluid Chromatography
Edited by Roger M. Smith, University of Technology,
Loughborough, UK.

Chromatographic Integration Methods
by N. Dyson, Dyson Instruments Ltd.,
Houghton le Spring, UK.

Packed Column SFC
by T. A. Berger, Berger Instruments,
Newark, Delaware, USA.

How to obtain future titles on publication
A standing order plan is available for this series. A standing order will bring delivery of each new volume immediately upon publication, at a substantial discount price. For further information, please write to:
The Royal Society of Chemistry
Distribution Centre
Blackhorse Road
Letchworth
Herts. SG6 1HN

Telephone: Letchworth (01462)672555

Packed Column SFC

T. A. Berger
Berger Instruments,
Newark, Delaware, USA

THE ROYAL
SOCIETY OF
CHEMISTRY

ISBN 0-85404-500-7

A catalogue record of this book is available from the British Library.

© The Royal Society of Chemistry 1995

Published by The Royal Society of Chemistry,
Thomas Graham House, The Science Park, Cambridge
CB4 4WF

Typeset by Computape (Pickering) Limited, North Yorkshire
Printed by Bookcraft (Bath) Ltd., Bath

Preface

Supercritical Fluid Chromatography (SFC) is more than 30 years old and is now reaching a critical mass in terms of the number of workers in the field and the number of journal articles published describing its uses. Hundreds of supercritical fluid chromatographs exist and, *ca.* 1992, the number of journal artucles on the topic exceeded 1000.

The most dramatic SFC event in the last three years has been the re-emergence of packed column instrumentation, and a switch in emphasis to more polar solutes. These include pharmaceuticals, especially in chiral separations, and agrochemicals. This has amounted to nothing less than a redefinition of the technique and has generated the need for this book. There is no practical guide dedicated to packed column SFC method development that prospective users can purchase to familiarize themselves with the subtle aspects of the technique. New users must largely rely on concepts developed in either LC or GC which are often inappropriate or misleading. Just as importantly, packed column SFC has little in common with capillary SFC. In fact, many of the approaches used in capillary SFC are among the worst things to try with packed columns (and *vice versa*).

A secondary reason for this book is to try to clear up some of the confusion that surrounds the analytical use of supercritical fluids. The advantages of SFC over LC are practical, not fundamental. There are differences between super-critical fluids, gases, and liquids but they are not as dramatic as often supposed. There is no fundamental characteristic of supercritical fluids which differentiates them from gases or liquids other than their definition. Packed column SFC can actually be thought of as an odd form of LC. The greatest difference is simply the need to hold the outlet pressure above ambient to prevent expansion of the fluid.

It is hoped that the information presented in this book will allow the potential user to understand the power and utility of packed column SFC. Chapter 1 attempts to compare packed column SFC with the other separation techniques. Chapter 2 tries to point out the difference in hardware between packed and capillary instruments and also help the user avoid common practical problems. Chapters 3, 4, and 5 try to dispel many of the myths surrounding supercritical fluids and packed columns. In addition, it is hoped that the systematic method development strategy outlined in Chapters 6 and 7 will help new users be more effective in rapidly achieving their analysis goals. The remaining chapters deal with specific application areas.

Following the strategies suggested in this book, a reader should be able to determine in a few days whether packed column SFC will solve a particular analysis problem, or not. In addition, a few days work should produce a viable method for all but the most difficult applications.

Contents

Chapter 1   **Putting Packed Column Supercritical Fluid**   1
            Chromatography into Perspective

        1   Introduction   1
            What Is Packed Column SFC?   1
            What is a Supercritical Fluid?   5
            Do Packed and Capillary SFC Compete with Each
                Other?   5
            Are Capillaries Inherently Superior?   7
            Is There a Need for SFC?   8

        2   Chromatographic Attributes or Figures of Merit   9
            Resolution   9
            Speed   10
            Sample Capacity   11
            Sensitivity   11
            Classes of Detectors   12
            Sensitivity of Optical Methods   12
            Column Impedance   13

        3   Comparing Separation Techniques Based on Figures of
                Merit   13
            The Speed–Resolution–Sensitivity Triangle   13
            Spidergrams   14

        4   Summary   20

        5   References   21

Chapter 2   **Practical Aspects of SFC Hardware**   22

        1   Introduction   22

        2   Instrumental Overview   22
            Other Configurations   23

        3   Practical Aspects of Fluid Supplies   26
            Supply Line Characteristics to Avoid   26
            Padded Tanks   27
            Pre-mixed Binary Fluids   27
            Empty?   27
            Inertness   28
            Safety   28
            Gas Quality   29

4   Pumps   29
     Reciprocating Pumps   29
     Compressibility Adjustment   30
     Adiabatic Heating   31
     Leak Compensation   32
     Mixing Binary Fluids   32

5   Pressure Control   34
     Fixed Restrictors   34

6   Injection   35
     Aqueous Injection – Solvent Polarity   36

7   Detection   37
     UV Detectors   37
     Optimizing UV Sensitivity   38
     GC Detectors   39

8   Columns   40

9   References   41

Chapter 3   Physical Chemistry of Mobile Phases Used in Packed
               Column SFC   43

1   Introduction   43

2   Why SFC? Mobile Phase Considerations   43
     Diffusion Coefficents   44
     Viscosity   45

3   What's in a Name? or 'There is NO Such Thing
       as SFC!!'   47
     Diffusion Coefficients in Near Critical Fluids   47
     Viscosity of Near Critical Fluids   47
     Effect of Physical Parameters on Retention   49
     Summary   51

4   Solvent Strength   51
     Solvatochromic Dyes   52

5   Elution Strength   55
     Effect of Density on Retention   56
     Modifier Effects Change with Column Polarity   57
     Relative Effect of Modifiers and Density on Retention 58
     Solubility   59
     Ultra-High Pressures   59
     Does Chromatography occur at 'Infinite Dilution'?   60
     Explaining Non-linear log *k*' *vs.* Modifier
     Concentration Plots   61
     Limited Density Range in Binary Fluids   62

	6	Phase Behaviour	64
		What's Wrong with Phase Diagrams?	65
		Modified Fluid Phase Behaviour	65
		Phase Behaviour of Methanol/Carbon Dioxide Mixtures	66
	7	Conclusions	69
	8	References	70
Chapter 4		**Physical Chemistry of the Stationary Phase**	72
	1	Introduction	72
	2	The Nature of the Packing Materials	72
		Surface Area	73
		Phase Ratio	73
		Active Sites	75
	3	Adsorption of Mobile Phase Componentrs	77
		Adsorbed Carbon Dioxide Acts as a Stationary Phase	77
		Phase Ratio of an Adsorbed Film	78
		Volumes Adsorbed	78
		SFC with Pure Carbon Dioxide is Normal Phase	79
		Swelling	79
		Chromatographic Evidence of Adsorbed Carbon Dioxide Acting as Part of the Stationary Phase	79
	4	Adsorption from Binary Mixture	80
		Adsorbed Binary Fluids Sometimes Increase Retention	84
	5	Polar Additives	84
		The Roles of Additives	86
	6	Choosing an Additive	93
		Solutes Can Act as Additives	93
		Support Deactivation – Polymer Based Columns	94
	7	Conclusion	95
	8	References	96
Chapter 5		**The Effect of Instrumental Parameters on Retention, Selectivity, and Efficiency**	97
	1	Introduction	97
	2	Summarizing the Effects of Physical Parameters	97
	3	The Effect of Physical Parameters in Controlling Retention of Polar Solutes	98
		Modifier Concentration–Retention Effects	98

Pressure and Retention   99
Temperature and Retention   100
Mobile Phase Flow and Retention   102

4   The Effect of Physical Parameters on Selectivity
     between Polar Solutes   102
     Representing Selectivity Changes   103
     Modifier Concentration and Selectivity   103
     Pressure and Selectivity   103
     Temperature and Selectivity   104
     Flow Rate and Selectivity   106

5   The Effect of Physical Parameters on Efficiency with
     Polar Solutes   106
     Pressure and Efficiency   106
     Temperature and Efficiency   107
     Mobile Phase Composition and Efficiency   107
     Flow Rate and Efficiency   108

6   The Effect of Physical Parameters Using Pure Fluids
     on Non-polar Solutes   108
     Pressure Control   109
     Temperature Control   111

7   References   111

Chapter 6   **Concepts That Simplify Phase Selection**   113

1   Introduction   113

2   'Polarity Windows' Concept Helps Choice of
     Phases   113
     SFC Is Always Normal Phase   114
     Contrary Views   114
     Active Sites   114
     Avoiding Unwanted Interactions   115
     Subtle *vs.* Gross Separations   115
     The Most Polar Entities Present Dictate Retention   115
     Examples of Polarity Windows   116

3   Previous Polarity Scales in GC, LC, and SFC   119
     The ϵ^0 Elution Scale from Normal Phase LC   119
     Gas Chromatographic Retention   120
     The '*P*' Scale of Solvent Polarity and Selectivity   120
     Solvatochromic Dyes   121
     Previous Polarity Scales in SFC   122

4   A Solvent Strength Scale for SFC   123
     Hydrocarbons, Ethers, Esters, Aldehydes, and Ketones   124

	Alcohols	124
	Acids	125
	Bases	125
5	Factors Affecting Retention	126
	Functional Group Polarity	127
	Number of Functional Groups	127
	Steric Hindrance	127
	Molecular Size	128
	Stationary Phase Polarity	129
6	A Summary of Retention in packed Column SFC	130
7	References	131

Chapter 7   **Systematic Method Development**   133

1	Introduction	133
2	Initial Non-chromatographic Tests	133
	Screening Question No. 1: Are the Solutes Soluble in a Desirable Mobile Phase?	133
	Screening Question No. 2: What is the Concentration of the Solute in the Real Sample?	134
	Screening Question No. 3: Is the Samplew Matrix Compatible with Supercritical Fluids?	134
	Screening Question No. 4: Is a Universal, Near Constant Response Factor Critical?	135
3	Choosing between Packed and Capillary Columns	135
	Capillary Columns	135
	Packed Columns	136
4	Guidelines for Simplifying Method Development	136
	Solute Characteristics That Affect Retention	137
	Phase Selection Guidelines	137
	Instrumental Strategy	137
5	Mobile and Stationary Phase Selection Guide for Packed Column SFC	138
6	Step-by-Step Speed–Resolution Optimization Scheme for Molecules That Elute with Pure CO_2	140
	Initial Conditions	140
	Non-polar Solute, Asymmetric Peaks	140
	Non-polar Solute, Symmetric Peaks	140
7	Step-by-Step Speed–Resolution Optimization Scheme for Molecules That Require Modifier	141
	Initial Conditions for Polar Solutes	141

    Observe Peak Shape 141
    Enhancing Resolution 143

8 Alternative Approaches to Method Development 144
    Systematic Study of Each Variable 144
    Additional Approaches to Method Development 147

9 Optimizing the Detection of Real Samples 148
    Choosing a Detector 148

10 Other Considerations. Injection Volume/Sample
    Solvent Polarity 149

Chapter 8 Pharmaceutical Analysis by Packed Column SFC 151

1 Introduction 151

2 Separations of Some Specific Pharmaceuticals 151
    Miscellaneous 152
    Steroids 154
    Bile Acids 155
    Ecdysteroids 156
    Carbohydrates – Oligosaccharides 157
    Barbiturates 158
    Opium Alkaloids 159
    Imidazole Derivatives 159
    Crotamitron in Creams and Lotions 159
    Taxol 160
    Caffine, Theophylline, Theobromine 160
    Benzodiazepines 163
    Phenothiazine Anticonvulsants 164
    Tricyclic Antidepressants 167
    Stimulants 169
    Sulfonamides 171
    Diuretics 173

3 References 174

Chapter 9 Chiral Analysis of Drugs 176

1 Introduction 176

2 Characteristics of Chiral Separations 177

3 Developing a Chiral Method 178
    The Effect of Physical Parameters on Chiral Separations 178
    Initial Conditions 179
    Optimization after the Initial Experiment 179
    Other Figures of Merit of Chiral Analysis 181

4 Some Examples   182
   Isolation and Detection of Ibuprofen from a Biological
      Fluid   182
   Mixed 'Universal' Phase?   183
   Benzadiazepans and Metabolities   184
   Long Column as Easiest Way to Increase Resolution   186
   LC Worked, SFC Didn't   186
   Other Chiral Separations   186

5 References   190

**Chapter 10   Separation of Agricultural Chemicals by Packed
               Column SFC**   192

1 Introduction   192

2 Trace Contaminants/Quality Control   193

3 Residue Analysis   193
   Carbamate Pesticides   193
   Phenylurea Herbicides   198
   Sulfonylureas   201
   Triazines   204

4 Screening Methods for Multiple Pesticide Residues   204
   Programming Selectivity and Retention   206
   Multiple Detectors   208

5 Summary   210

6 References   210

Chapter 11   SFC and the Petroleum Industry   212

1 Group Separations   212
   Why SFC?   216
   Other Aspects   219
   Historical Development   219

2 High Temperature GC *vs.* SFC   221

3 High Efficiency Packed Column SFC   221

4 Several Instrumental Concerns Seldom Mentioned in
      Print   224
   Characteristics of the FID   224
   The Problem of Changing Split Ratio with Pressure
      Programming   225

5 References   226

Chapter 12   Miscellaneous Applications of Packed Column SFC   227

    1   Introduction   227

    2   Specific Separations   227
       Polystyrenes   227
       Silicone Oils   228
       Methyl Vinyl Silicone Stationary Phases – Peroxide
         Reaction Mixtures   232
       Methyl Methacrylates   232
       Epoxy Resins   233
       Polymer Additives   235
       High Speed Chromatograms   236
       Triton X-100   236
       Natural Products   239
       Ubiquinones in *Legionella*   240
       Underivatized Fatty Acids   242

    3   References   244

       Subject Index   245

CHAPTER 1

Putting Packed Column Supercritical Fluid Chromatography into Perspective

1 Introduction

What Is Packed Column SFC?

Packed column supercritical fluid chromatography (SFC or pSFC) is an analysis technique similar to liquid chromatography (LC) that uses supercritical fluids (SFs), instead of liquids, as the mobile phase (MP) (supercritical fluids are defined in the next section). The MP solvates the solutes. The stationary phase (SP) consists of a bed of very small particles packed in a tube capable of withstanding high pressures. Some SPs are the surfaces of uncoated particles. Some are organic films bonded to the surface of the particles. Solutes are separated by differential attraction to the SP.

Compared with LC, packed column SFC is faster, more efficient, has a wider range of selectivity, and detection options, and produces less toxic waste. Not surprisingly, the fields most likely to be affected by packed column SFC in the future are traditional LC application areas. In particular, pharmaceutical and agricultural chemical development will derive significant benefits. Chiral separations will likely be a major application area for packed column SFC. This list is likely to surprise many readers since the application areas most often associated with capillary SFC have involved less polar but perhaps more complex solute mixtures such as homologous series, of surfactants, polymers, and the like.

In reality, the characteristics of interest in 'SFC' have more to do with intermolecular interactions in the MP than the name of the fluid. Many of the characteristics that make SFs interesting to chromatographers (*e.g.*, high diffusivity, low viscosity) are also available from some fluids **defined** as gases or liquids. Unfortunately, the name SFC is somewhat misleading. SFC differs from LC in that the MP is a dense compressed fluid which will dramatically expand if external pressure is removed.

The most widely used supercritical fluids (like carbon dioxide, nitrous oxide, or CHF_3) are inorganic and do not produce a response in some GC detectors,

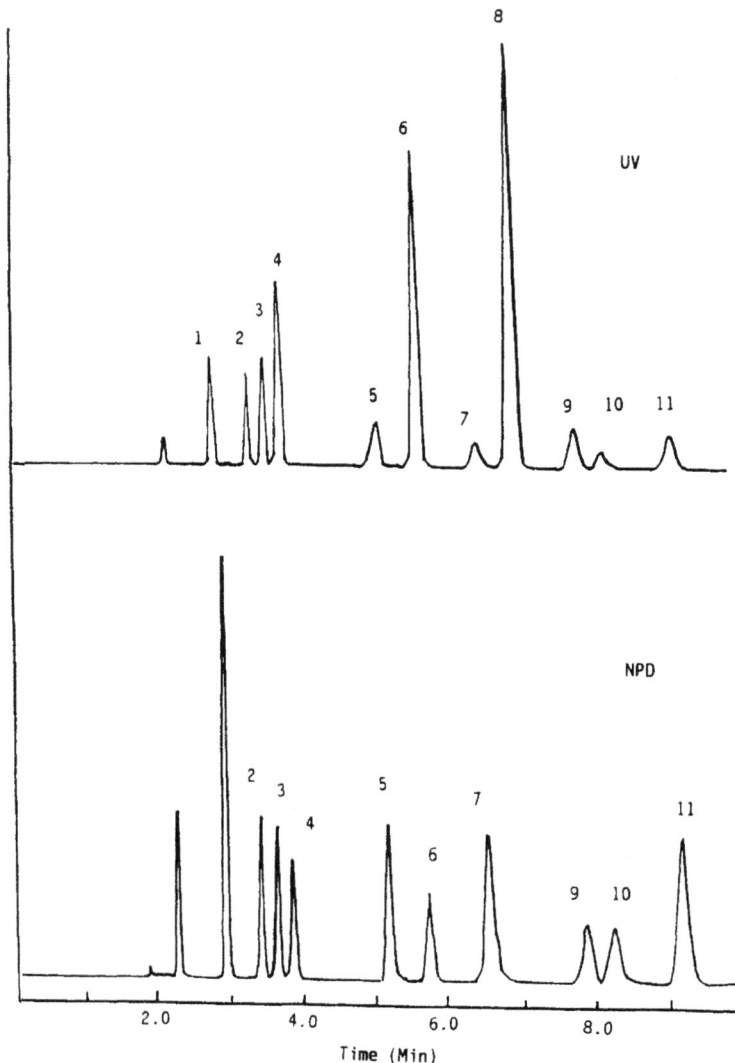

Figure 1.1 *Separation of carbamate pesticides of EPA Method 531.1 by packed column SFC with simultaneous UV and NPD detection. 1. aldicarb, 2. Baygon, 3. carbofuran, 4. methiocarb, 5. aldicarb sulfone, 6. carbaryl, 7. methomyl, 8. 1-naphthol, 9. 3-hydroxycarbofuran, 10. aldicarb sulfoxide, 11. oxamyl*

like the FID. This combination of characteristics allows some LC-like separations with more GC-like figures of merit, such as high speed, high resolution, and multiple detection options.

Several modern packed column SFC chromatograms may help convey the features that make the technique desirable. In Figure 1.1, the 11 carbamates of EPA Method 531.1 are separated[1] in 9 minutes and directly detected using both a UV and an NPD (Nitrogen–Phosphorus Detector).

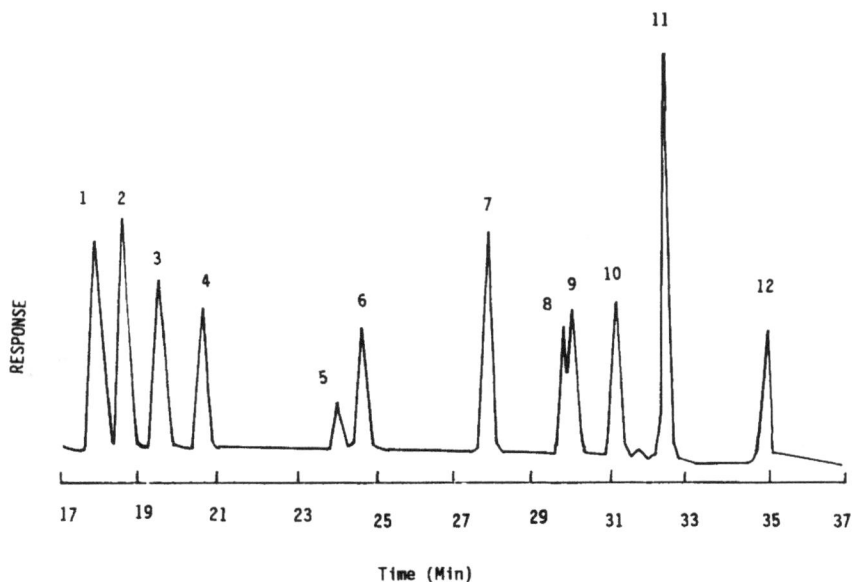

Figure 1.2 *Separation of carbamate pesticides by AOAC HPLC standard method.*
1. aldicarb sulfoxide, 2. aldicarb sulfone, 3. oxamyl, 4. methomyl, 5. 3-hydro-
xycarbofuran, 7. aldicarb, 8. carbofuran, 9. carbaryl, 10. methiocarb
(Reproduced by permission from ref. 4)

The methanol/carbon dioxide ($MeOH/CO_2$) MP flow rate is 2.5 ml min^{-1}, producing near optimum chromatographic efficiency. This is approximately 3.5 times the optimum flow rate in LC (on this column) and illustrates the superior diffusion rate in supercritical fluids.

The standard method uses gradient elution LC[2] followed by two postcolumn reactions to yield fluorescent products. Although the separation takes *ca.* 40 minutes, the column must then be re-equilibrated. The whole process requires *ca.* 1 hour between injections. A representative chromatogram is shown in Figure 1.2.

Cumulatively, the SFC separation and detection options produce a throughput approximately six times that of the LC standard method, and avoid the complexity of the postcolumn reactions.

An alternative example of the unusual characteristics of SFC is shown in Figure 1.3. The separation in Figure 1.3 was developed to suggest the feasibility of using SFC for screening pesticides not amenable to GC analysis. A 10 ml water sample containing 6–22.5 p.p.b. of 31 carbamate, sulfonylurea, pheny-lurea, and triazine pesticides was injected into a precolumn mounted in place of an external loop on a six port valve. The water was blown off with helium, and then the precolumn was switched into the flowing stream. The solutes were eluted by a gradient of 1–16% MeOH in CO_2, 90–140 bar, from a 1.6 m long LC column packed with 5 μm particles. At 2 ml min^{-1} of 20% MeOH, the pressure drop was 150 bar. After the column, the flow was split, diverting a

Current Chromatogram(s)

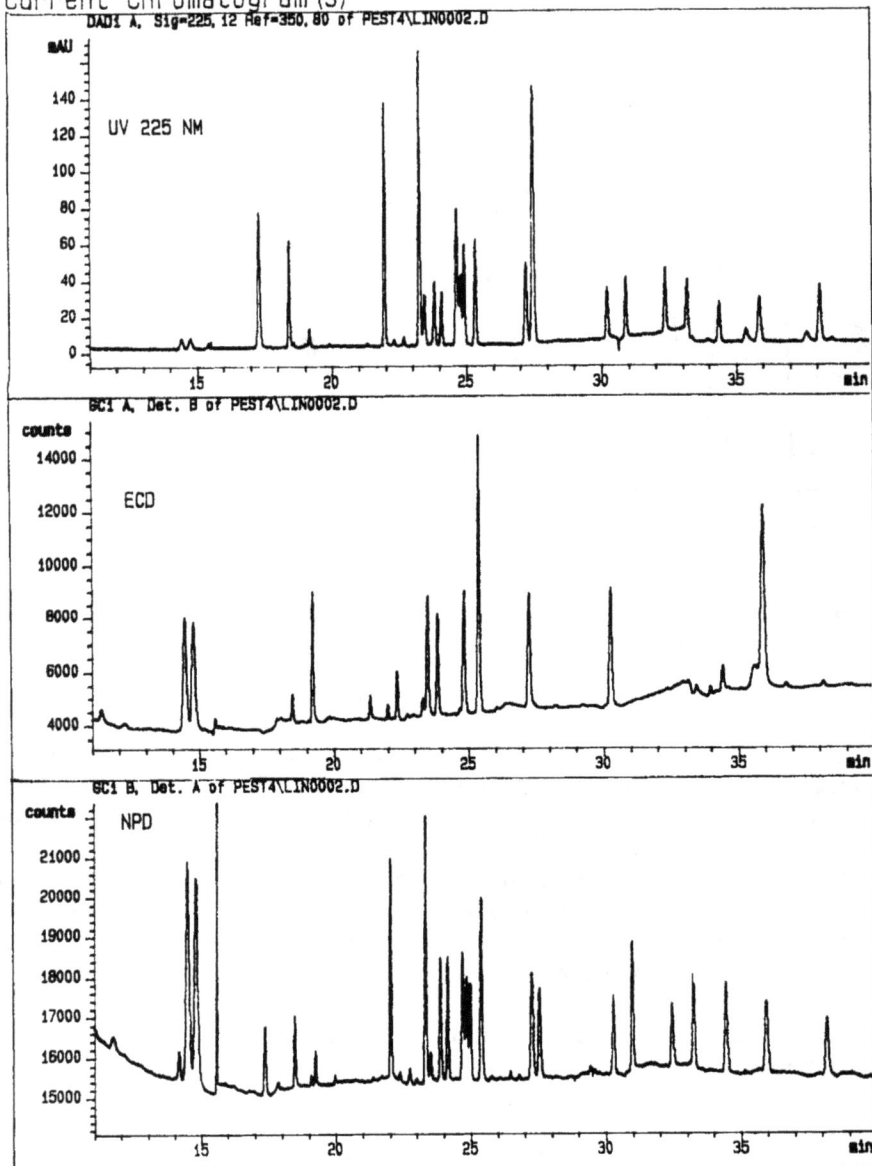

Figure 1.3 *High efficiency, high sensitivity separation of 32 pesticides at 6–22.5* p.p.b. *in 10 ml water. Solid phase extraction by replacing the loop in the 6-port injection valve with a C$_{18}$ cartridge guard column. Simultaneous detection by UV, NPD, and ECD*

fraction to an ECD and an NPD while most passed through the UV diode array detector. The detection limits for some solutes were a few tens of parts per trillion ($1/10^{12}$).

One trend in LC is toward the use of smaller diameter packed columns, requiring less MP. Major reasons are a desire to reduce solvent cost and minimize toxic waste generation. In some locations, it is already more expensive to dispose of solvents than purchase them. Unfortunately, smaller columns require more stringent instrumental design. In general, it is more difficult to achieve the same high efficiencies, and high sensitivity on a small column as on a large column. Packed column SFC offers an attractive alternative. Inert CO_2 replaces most of the liquid solvent. Modifiers typically represent 2–20% of the mobile phase. An SFC method on a 4.6 mm column creates the same or less liquid waste as a 2 or even 1 mm LC column. By retaining the larger column format, SFC allows relaxed constraints on extra column effects, while often providing higher capacity, better detection, and reproducibility.

What is a Supercritical Fluid?

It is important to understand that 'Supercritical' is only a **defined** state. Supercritical fluids are not a separate state of matter (there are only gases, liquids, and solids). To be 'supercritical', a fluid must be above BOTH its critical temperature, T_c, and critical pressure, P_c. The combination of T_c and P_c is known as the 'critical point'. Above its critical point, a fluid cannot be liquified, no matter how high the pressure is raised.[3] Note that the definition only deals with $T > T_c$ and $P > P_c$. The definition ignores what happens at conditions BELOW the critical point.

There has been a great deal of confusion about transitions from subcritical to supercritical conditions. Such transitions are NOT phase transitions. They are only transitions from one DEFINED state to another. This ambiguity is dealt with in depth in Chapter 3.

Supercritical fluids lack adequate intermolecular interactions which would otherwise condense them to liquids. This low intermolecular energy gives the fluids certain advantageous characteristics compared with normal liquids familiar as mobile phases in LC.

With SFs (and some similar fluids), the pressure can be increased until the molecules are as close to each other as the molecules in a condensed liquid. This molecular closeness and resulting high collision frequency between molecules makes the fluids reasonable solvents for many solutes. Simultaneously, less intermolecular interaction results in lower viscosity and high diffusivity of solutes in the fluid (molecules do not 'stick' to each other). Both will be discussed in detail in Chapter 3.

Do Packed and Capillary SFC Compete with Each Other?

Many readers may recall the controversy a decade ago over whether capillary or packed columns were the 'best' column type for SFC. Reopening that controversy is counter productive and has some of the characteristics of an old beer commercial on American television. Two retired athletes argue about WHY their (same) beer is the 'best' (Figure 1.4a). Each sees different aspects of

Figure 1.4 (a) *Tastes great–less filling argument about which is the best attribute of the same beer.* **(b)** *The old controversy about the 'best' form of SFC has many of the same attributes of the beer commercial*

the same product as its most important attribute. The arguments over column type in SFC are much the same (Figure 1.4b). Individual users concentrate on different aspects of the same technique. In reality, the two column types are best suited for different kinds of samples and compound classes, producing different figures of merit using different fluids and detectors.

A few major attributes of packed SFC are: independent dynamic pressure and flow controls,[4] common use of binary and tertiary[5] MPs, composition programming preferred over pressure programming, elution of much more polar solutes[6,7], trace *vs.* major–minor component analysis, and UV, electron capture (ECD), nitrogen–phosphorus (NPD), and sometimes FID (when no modifier is present) detection.

Capillary SFC should be characterized as an extension of GC to larger, low volatility, but mostly thermally stable molecules. However, either packed or capillary SFC can perform many of the separations for which the other is nominally superior. For example, capillary SFC can also produce high speed separations of small polar molecules like agricultural chemicals[8], as shown in Figure 1.5. The capillary method is fast but lacks easy selectivity adjustment and sensitivity and reproducibility are likely to be poorer than with a packed column.

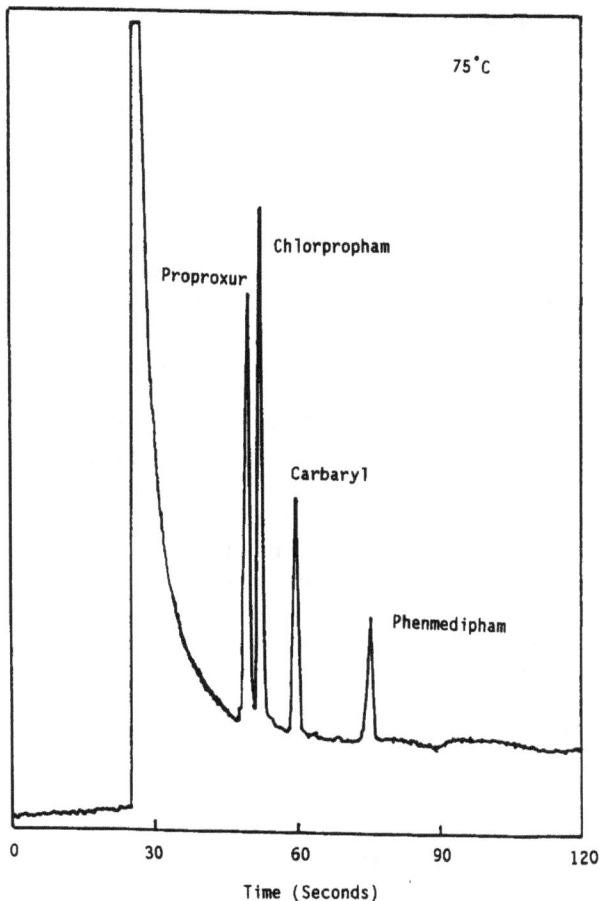

Figure 1.5   *Rapid separation of agrochemicals, including some carbamates by capillary SFC with pressure programming and FID detection* (Reproduced by permission from ref. 8)

Are Capillaries Inherently Superior?

The last 20 years has seen dramatic improvements in capillary chromatography. There has, subsequently, been a tendency among chromatographers to assume capillary columns are inherently superior to packed columns in all cases. However, there are situations where either capillary or packed columns produce superior results. It is the analysts job to match the **best** technique and column type to **each** application.

In GC, the mix of column applications has shifted from > 70% packed to > 70% capillary since 1979. However, this change was NOT primarily due to speed, resolution, or sensitivity considerations. Packed columns remain faster and allow higher sensitivity even today. Capillary columns were used in GC for nearly 20 years without becoming the columns of choice. These columns had many attributes of modern capillaries. However, they were also relatively active, causing tailing of polar solutes. Fused silica capillaries introduced a new level of inertness to GC, producing symmetrical peaks for even quite polar compounds. This inertness was the primary reason for the switch to capillaries. The overall figures of merit of capillary GC are so dramatically superior that the technique can easily compromise on sample capacity and speed to achieve inertness and efficiency.

In LC, open tubular columns were demonstrated over 15 years age.[9,10] Some have predicted[11] a switch from packed to capillary columns similar to the switch that occurred in GC. However, capillaries have still made almost no inroads into LC applications. LC has fewer or at least different problems than GC. The trade-off between speed–sensitivity and resolution in LC is much less favorable than in GC. LC on packings is slow compared with GC. Capillaries would be even slower unless $d_c < 5$–10μm. With such small d_c, injection volumes < 1 nl are required, with reproducibility $< \pm 1$–10 pl. There would be 10^{-15} grams of a solute in a p.p.b. injection. Extra column effects are extremely difficult to eliminate at the tiny dimensions required.

The liquid solvents and additives used in LC decrease the tailing problems associated with packing activity. Capillaries offer little improvement in LC inertness.

Supercritical fluids are intermediate between gases and liquids in terms of passivation and tailing problems. Higher diffusivity allows the use of larger d_c or higher speed on traditional LC packings. Modifiers suppress activity on packings and capillaries. Capillaries make it easier to achieve high efficiency. Packed columns inherently win both speed and sensitivity comparisons. While GC is primarily a capillary technique, and LC is a packed column technique, SFC is intermediate. From the author's perspective, packed columns have an edge in SFC since SFs tend to be more like liquids than gases.

Is There a Need for SFC?

SFC has not been an instant success. The technique was first demonstrated more than 30 years ago.[12] In the 1960s, LC was correctly recognized as the more general of the two techniques and most subsequent development effort has been spent on LC instead of SFC. This was undoubtedly the right choice at the time. Today, LC is reaching maturity and its strengths and weaknesses are well understood. To improve on LC, both SFC and electrophoresis are undergoing a renaissance.

Some have questioned the need for an additional separation technique. Many problems can be solved by either GC or LC, yet no one is surprised when one is arbitrarily chosen over the other for a specific application. Yet, it has often been suggested that SFC should be considered ONLY for separation problems that CANNOT possibly be solved by either GC or LC. To limit SFC (or any technique) only to cases where no other technique works dooms it at the start and is not representative of how real laboratories work.

A significant fraction of analytical methods are at the margins of techniques.

Such analyses are often 'expensive' in terms of time, uncertainty in the result, or in the level of operator intervention. Alternatives that offer enhanced performance at lower cost should be attractive. There is, in fact, a continual shift in many applications back and forth between GC and LC due to subtle changes in technology favouring one then the other.

Packed column SFC has characteristics which logically make it superior to LC for most molecules that can be solvated by SFs. As will be shown later, SFC should actually be the second technique of choice, after GC and before LC in terms of such performance factors as speed, efficiency, and detection options. Of course, LC is unlikely to be displaced from applications that exist. However, in the future, chromatographers are likely at least to consider SFC before LC, although the two are closely enough related technologically that they might eventually merge. Laboratories will probably eventually have 1/5th as many SFCs as LCs.

2 Chromatographic Attributes or Figures of Merit

It is difficult to compare separation techniques in any general way. However, some basis of comparison is needed to give the chromatographer a means of understanding why a new technique is worth considering. In this section common attributes of separation techniques are described. These attributes, such as speed, resolution and sensitivity, are often called figures of merit.

For each figure of merit, the common separation techniques can be ranked. However, care must be exercised in comparing techniques based solely on figures of merit. For any specific separation, one attribute may be absolutely critical. A technique vastly superior to others in all but the critical attribute may be useless in the specific application.

Resolution

Resolution (R_s) is a fundamental measure of separation between two solutes. The universal resolution equation, $R_s = \text{constant} (N^{0.5})[(\alpha - 1)/\alpha][k'/(1 + k')]$, indicates three aspects of chromatographic systems that lead to the physical separation of solutes: efficiency (N or plates), selectivity ($\alpha = k_2/k_1$), and retention ($k' = (t_R - t_0)/t_0$, where k' is called a partition ratio, t_R is the solute retention time, t_0 is the column transit time of an unretained peak).

Efficiency

Column efficiency [$N = L/H = 5.54(t_R/W_h)^2$, where W_h is the peak width at half height] is based on physical dimensions, like the size of particles (d_p) or the diameter of a capillary (d_c), the length (L) of the column, plus mobile phase flow rate, and sometimes retention (k'). Generating large N is not, in itself, desirable, since it is extremely expensive in terms of speed and sensitivity. A 10-fold increase in L increases N by 10 times but R_s by only 3.1 times ($10^{0.5}$), while t_R increases 10 times, and sensitivity degrades (peaks get broad and more dilute).

Selectivity

Selectivity ($\alpha = k'_2/k'_1$) is a measure of relative retention. If retention is very different for two solutes, they are easily separated with low N and modest t_R. In GC, α is primarily a function of the SP. If two peaks cannot be resolved by brute force (large N, or large k') then the only option is to change the column.

In LC, both the SP and the composition of the MP are important in determining α. Most liquid chromatographers would suggest that LC is more powerful than GC because the MP is not inert and can change α. The universal resolution equation shows that α is more powerful than N in resolving specific pairs of overlapped peaks.

SFC actually provides a wider range of α adjustment than normally available in LC. Besides the identity and composition of the MP and the SP, both temperature and pressure also play a major role in α adjustment in SFC.

Retention

With very little retention, resolution of two solutes requires either large N or large α. Resolution increases with k', up to $k' \approx 10$. With larger k' no additional resolution is achieved. Increasing k' directly trades speed for resolution. A k' of 10 roughly doubles R_s over $k' = 1$ but increases t_R by 5 times.

Peak Capacity

Peak capacity, $n_c = (N^{0.5})/4$, is a measure of the level of sample complexity a column can resolve. The standard deviation, sigma (σ), is a measure of peak width: $\pm 1\sigma$ represents *ca.* 68% of peak area; 2σ represents 95% of peak area. The 4 in the peak capacity equation represents 4σ ($\pm 2\sigma$). The result, n, indicates how many ideally spaced peaks could be resolved to 4σ per order of magnitude of retention. Real solutes are almost never ideally spaced on the baseline.

A column with 20 000 plates is considered adequate for separating samples containing no more than 35 peaks. Another definition of peak capacity [$n_c = (1 + 0.2\sqrt{N})$] suggests a maximum of 29 peaks. For both measures of n, a 10-fold increase in N (*i.e.* 10 × longer column) allows sample complexity to increase $(10)^{0.5} = 3.1$ times.

In GC, an upper limit of 1000 peaks have been partially resolved. In LC, only a few chromatograms of > 100 resolved peaks exist. In SFC, there has been little exploration of the limits of sample complexity but more than 100 compounds have been at least partially resolved.

Speed

Chromatographic 'speed' is an indication of how fast two solutes can be separated. Speed is a function of both solute diffusion coefficients (D_M) and the distance solutes diffuse before hitting a surface. Solute D_M is not constant but changes with temperature and the density of the fluid. In gases, D_M is *ca.* 0.1–1.0 cm^2 s^{-1}; in liquids, *ca.* 10^{-5}–10^{-6} cm^2 s^{-1}, and in supercritical fluids,[13–16] *ca.* 10^{-4}–10^{-5} cm^2 s^{-1}.

Using the same column and solute for GC, LC, or SFC, one can generate the same N, but GC is 1000 times faster than SFC, and 10 000 times faster than LC.

To achieve similar analysis times in LC and GC, the analyst typically uses both much smaller d_p or d_c and shorter columns in LC. LC is usually performed on 3–10 μm particles in 10–25 cm long columns whereas GC is performed on 200–530 μm i.d. open tubular columns, 10–60 m long. N is also usually much lower in LC than in GC. Smaller diameters dramatically increase N/L, allowing much shorter columns for the same N. A further decrease in L decreases total N.

Comparing observed speeds of analysis: GC[17] > 24 000 plates s^{-1}; 2000 plates s^{-1} routine; LC, with $d_p = 5$ μm, speed < 170 plates s^{-1}; SFC > 450 plates s^{-1}. In capillary electrophoresis, > 10 000 plates s^{-1} have been generated.

Sample Capacity

Sample capacity is the maximum weight or volume that can be injected on a column. Sample capacity is related to detection limits and dynamic range. With packed columns, sample capacity is nominally unrelated to d_p but is related to the square of column diameter. Smaller pore sizes and larger column diameters produce higher sample capacity. Packed columns have about the same sample capacity in either LC or SFC.

Sample capacity in capillaries is proportional to d_c^3. The physical length of solute bands corresponds to a characteristic number of capillary diameters. Thus, the dimension most responsible in creating high speed and N (d_c) dramatically degrades sensitivity. In both LC, and SFC, small D_M values dictate the use of very small d_c to produce reasonable analysis times. This in turn dictates extremely low sample capacity.

Sensitivity

Sensitivity claims must be carefully evaluated. It is sometimes stated that smaller diameter packed columns produce enhanced sensitivity in LC. This is only true when the **same** sample volume (V_i) is injected onto both large and small diameter columns using concentration sensitive detectors. This is an artificial comparison. The maximum concentration of a solute in a peak (c_{max}) is a fraction of the initial concentration of the solute (c_0) in V_i. When V_i is chosen to be the same fraction of the retention volume (V_i/V_r = constant), both large and small columns dilute peaks equally and there is no difference in sensitivity.[15]

$$c_{max}/c_0 = [V_i(N)^{0.5}]/[V_r(2\pi)^{0.5}]$$

Spectacular detection limits are often quoted in terms of some very small number of molecules detected. However, such masses are often present in only a few **nanolitres**. The **concentration** of the analyte may be orders of magnitude greater than in more common measurements using large volumes. The ONLY time that measurements using minute samples are realistic is when only a small sample is available.

One of the benefits of capillary SFC is the use of the FID with samples not eluted in GC. However, the detection limits of capillary SFC and GC are NOT similar. A five-fold decrease in d_c (250 μm in GC to 50 μm in SFC) results in a 125-fold decrease in the volume injected. Considering both the lower D_M **and** smaller d_c, the SFC peak is 40 times wider (in time) (N/time) at the same N. If 1/125th as much is injected, which emerges over a 40 times longer period, the **flux** into the FID (g s^{-1}) and the current output is 5000 times lower in SFC than in GC. Increasing flow rate in the column improves sensitivity, but degrades N. Ten $\times \mu_{opt}$ in SFC produces only 1/500th of the flux as in GC, but degrades efficiency more than five times.

Techniques using very small d_c are increasingly used with on-column pre-concentration steps to overcome at least partially inherently poor detection limits and dynamic range. On-column preconcentration is no different from off-line, preconcentration, except that it adds to the **instrumental** analysis time.

Classes of Detectors

There are two basic kinds of chromatographic detectors: universal and selective. In the extreme, a universal detector responds to everything except the carrier fluid. Selective detectors respond to only some solutes. In the extreme, selective detectors can respond to a single compound or class. Real detectors tend to be somewhere between ideal universal and ideal selective detectors.

Both universal and selective detectors can have constant or variable response factors. The flame ionization detector (FID) is highly valued, in part, because it responds nearly equally ($\pm$ 1–2%) to all organic compounds.

There are also concentration and mass detectors. Many detectors, like the FID, respond to the number of molecules of solute per unit time. Other detectors respond to the concentration of the solute in the MP. Concentration detectors are usually more compatible than mass detectors with very small columns.

Sensitivity of Optical Methods

UV detection limits are similar in both LC and SFC using large diameter packed columns. With a good chromophore, some solutes can be detected in the low part per (American) billion (1/10^9)(p.p.b.) range, but 1 p.p.m. is more common.

Optical methods of detection (UV or laser induced fluorescence) are also used with capillary columns. The detector pathlength is usually the d_c (50–150 μm). Beer's law ($A = \varepsilon l c$, where ε = molar absorptivity, l = pathlength, c = solute concentration) shows that a 0.005 cm pathlength has serious disadvantages compared with the 1 cm pathlength used with 4.6 mm packed columns.

To try to get around this problem, 'Z' shaped cells (a bent capillary) are sometimes used. However, the physical length of solute bands in a capillary is a small multiple of d_c. The volume of the detector must be limited to *ca.* 1/6th of the volume containing the band to avoid band broadening. Using a long Z shaped cell with a cell diameter the same as d_c produces enhanced sensitivity but severely degrades N. With a 50 μm i.d. capillary column and 'Z' cell, a path length of 500 μm (0.5 mm) may contain the entire peak and degrade N (at optimum velocity). In Z shaped cells, $l = 10 - 20$ times longer, making them even less appropriate for use with small capillaries operated at near optimum velocity.

Column Impedance

Although smaller dimensions improve speed, pressure requirements increase dramatically. A figure of merit introduced[17] to indicate a technological cost of pressure drops is column impedance. However, altogether too much emphasis has been placed on column 'impedance' as a figure of merit. While the pressure drop per plate is an indicator of the total N that can be achieved with a particular column and pumping system, there is little difference in the technological price between a 10 bar and a 400 bar or even 600 bar pressure drop. The maximum pressure capability of pumps has not been the primary limitation in developing smaller column dimensions. If small particles or capillaries could be fabricated into high quality columns, and all other requirements could be met (reproducibility, sensitivity) very high pressure and pressure drops would undoubtedly be used. Switching valves, tubing, fittings, and other components that operate at > 600 bar are not widely available or are difficult to use owing to their massive construction.

3 Comparing Separation Techniques Based on Figures of Merit

As stated before, it is difficult to compare separation techniques. The strengths and weaknesses of one technique may not even apply to another. Distortions arise particularly where qualitative statements are substituted for quantitative comparisons (*e.g.*, 'some' loss in efficiency produces 'dramatically higher' sensitivity), and when one technique using a more sensitive detector is compared to another separation mechanism but with a less sensitive detector.

The Speed–Resolution–Sensitivity Triangle

Speed, resolution, and sensitivity are each desirable in a separation. Unfortunately, an analyst cannot simultaneously have the maximum values of all these attributes. A traditional way to indicate the trade-offs between them is to draw an equilateral triangle and assign each to a different corner. Each analysis can be represented as a point inside the triangle. The distance from each corner indicates how much of the theoretical maximum of that attribute is achieved in the specific analysis. The areas of the triangles representing different techniques could be used to compare the relative importance of different techniques.

Spidergrams

Comparing different separation techniques solely in terms of the classic triangle can be misleading. **Although LC is less efficient, slower, and has many fewer detector options than GC, it is NOT a less powerful technique.** Additional attributes should also be considered, such as selectivity, cost, ease of use, and the range of solutes that can be separated. Using polygons with more sides than a triangle allows a wider range of attributes to be compared visually. In the following section, common separation techniques are compared using twelve attributes.

Each technique has its own polygon where each apex represents a specific attribute. The distance from the centre of the polygon to each apex represents the relative strength of the technique being evaluated compared to the best of the other techniques in that attribute. Drawing lines between the values of neighbouring attributes encloses an area. The larger the area, the more power-ful and generally applicable is the technique. Because the figures can look like a spider's web, they are called spidergrams.

The top half of the spidergrams represent performance characteristics, such as speed, efficiency, *etc.* The lower half represents the breadth of applications, such as molecular weight range, polarity, *etc.*

Gas chromatography

The attributes of GC are evident in the spidergram, presented in Figure 1.6. GC possesses by far the best speed–resolution–detection trade-off of any separation technique. It is fast, and efficient, with a tremendous range of detector options. In addition it is inexpensive. Reflecting these attributes, there are perhaps 200 000 GCs in use throughout the world. The first rule of chromatography ought to be:

If you *can* perform a separation by GC, you *should* use GC.

Unfortunately, GC has one major problem. It is limited to the separation of between 10 and 20% of all known compounds (*i.e.* modest molecular weight, thermally stable, volatile and semi-volatile compounds). The top part of the spidergram, describing figures of merit, encloses a large area. Conversely, the lower part, which represents the breadth of applications, encloses a relatively small area. Specifically, GC is a poor choice for thermally labile molecules, molecular weights more than a few thousand, and multi-functional polar molecules.

The μ_{opt} in GC is typically very high (10–100 cm s^{-1}). Up to 24 000

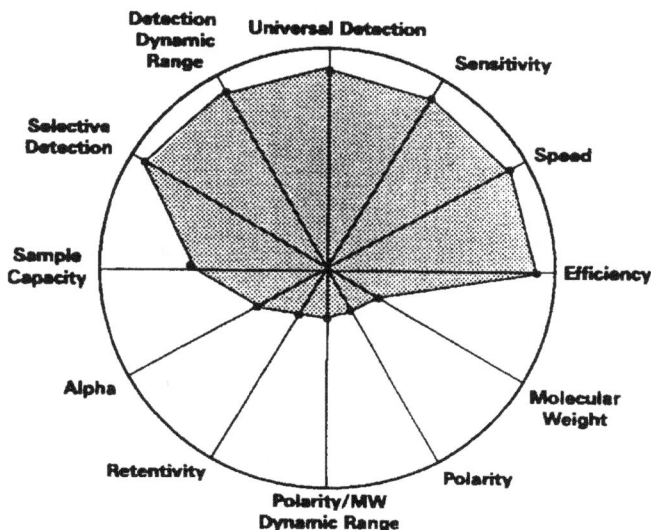

Figure 1.6   *Spidergram of GC. The upper part lists figures of merit, such as speed, sensitivity, and efficiency. The lower part attempts to characterize the breadth and depth of application areas*

plates s^{-1} [17] and more than a million plates[18] have been generated. Even in routine operation, 250 000 plates, generated in a few minutes are common.

The D_M values in the specific fluids and the present state of technology dictate commercially important column dimensions in the various techniques. In GC, the inherent high D_M of solutes allows large d_c, yet produces large N and high sensitivity, in modest times. A typical GC application might use a capillary with $d_c = 250$ μm, $L = 50$ m, with which it takes two to three minutes to generate 200 000 plates.

There is a wider range of detection options in GC than in any other chromatographic method. These include **universal** [*e.g.* FID, thermal conductivity (TCD), mass spectrometry (MS), atomic emission (AED)], and **selective** detectors [such as nitrogen–phosphorus (NPD), electron capture (ECD), electrolytic conductivity (ELCD), photo-ionization (PID), flame photometric (FPD), O-FID, sulfur chemiluminesence, *etc.*]. These detectors exploit practically every physical and chemical characteristic of molecules and atoms. The FID has near constant response factors and is linear over nearly seven decades of concentration. Other detectors tend to be less linear and cover a narrower dynamic range. The ECD is semi-selective and can directly detect some compounds down to the sub-part per trillion level. MS, IR, and AED detectors give structural, and elemental composition information.

Liquid Chromatography

Liquid chromatography is in many ways an opposite to GC. Its range of applicability is the widest of any chromatographic technique. Liquids can

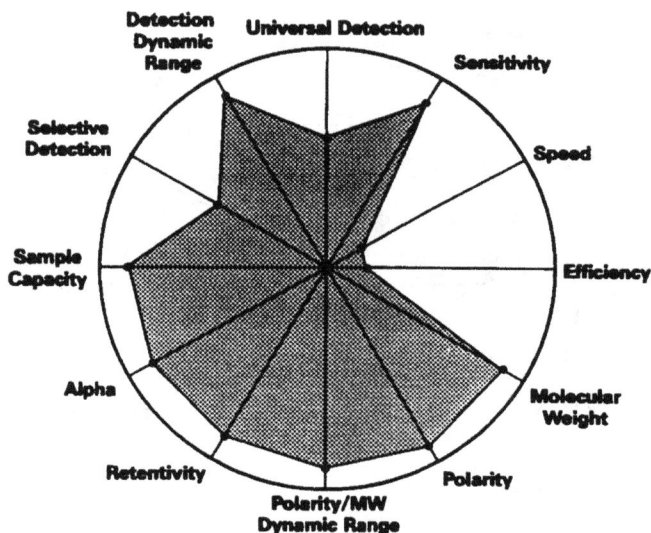

Figure 1.7   *Spidergram for LC*

solvate and elute perhaps 90% of all compounds so LC is, thus, a nearly universal separation technique. Unfortunately, its figures of merit are relatively poor. A spidergram for LC is shown in Figure 1.7. Physical attributes like speed–resolution–detection enclose a small area. On the other hand, the areas of application, at the bottom of the spidergram, enclose a large area. Applications include most of the compounds for which GC is a poor choice, including thermally labile, high molecular weight, and multi-functional polar molecules.

In LC, low D_M and poor quality of small capillary columns dictate the use of packed columns, with low N, modest speed, and high sample capacity. Columns can be built with some of the attributes of GC. A micropacked LC column produced 220 000 plates. However, $t_0 = 33$ minutes, and the pressure drop was 360 bar.[19] The high pressure drop and long t_R mean few practical chromatographers choose to use such an approach. Instead, a typical LC application might use a 4.6 × 200 mm column packed with 5 μm particles. The small d_p increases N/L by *ca.* 25 times compared with the 250 μm × 50 m GC column described above. For the LC column $L = 1/400$th, $N = 1/10$th. Because N in LC is expensive, the LC chromatographer typically chooses to use α instead of N to resolve solutes. The trend in LC continues toward smaller d_p to increase speed, while maintaining modest N.

The most serious weakness of LC is its limited choice of detectors. The most widely used is UV, which is only semi-universal. Response factors vary by > 3 orders of magnitude. The linear dynamic range seldom exceeds 10^5 (0.01–1000 mAU). Fluorescence detectors are selective and sensitive but with highly variable response factors. Few compounds naturally fluoresce, so most applications require solute derivatization. Dielectric constant, and refractive index (RI) detectors are more or less universal but exhibit low sensitivity. Electrochemical (amperometric or coulometric) detectors are selective and sensitive, but response factors are variable and difficult to predict.

There is no universal, sensitive, constant response factor LC detector equivalent to the FID. LC selective detectors tend to respond to structural, surface, or chemical characteristics like oxidation potential. There are no widely available, easy to use, selective detectors, like the NPD, ECD, or FPD, although each of these has been used at least superficially in LC. LC–MS is an important field but is much more expensive, and less widely used than GC–MS.

The areas enclosed in the GC and LC spidergrams are approximately the same, with that for LC being perhaps 50% larger. Not surprisingly, the worldwide markets for GC and LC are roughly similar in size. Approximately 50% more is spent on LC but, owing to their lower price and greater simplicity, more GC units are sold.

Supercritical Fluid Chromatography

Since supercritical fluids have properties between those of gases and liquids, the figures of merit of SFC are intermediate between those of GC and LC. A spidergram for packed column SFC is presented in Figure 1.8. SFC possesses inferior figures of merit compared with GC but is more widely applicable. On the other hand, SFC possesses superior figures of merit compared with LC but is less widely applicable. SFC may be usable with 30% of all molecules.

SFC allows a wider range of α adjustment than the other techniques (see Chapter 5). Like LC, MP composition and SP identity produce large k' and α changes. Both temperature and pressure are routinely used to produce additional k' and α changes.

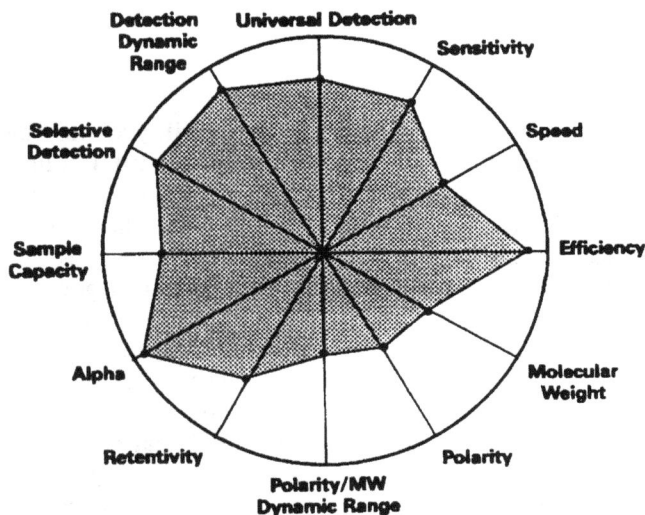

Figure 1.8 *Spidergram for packed column SFC*

The low viscosity of supercritical fluids allows use of high flow rates to scout out the effects of instrumental parameters on retention and selectivity. Flows of 5 ml min^{-1} on a 4.6 mm i.d. column are reasonable. Low viscosity also allows longer columns and higher N than typical in LC.

Pure fluids like CO_2 are not very polar and are limited to use with modestly polar solutes. Binary fluids will elute a wider range of polar molecules than pure liquids. Strong or multifunctional acids and bases often need tertiary MPs.

Molecules containing a strong base with a weak acid, or a weak base with a strong acid, can generally be separated by SFC. Many drugs that are salts can be eluted. SFC is not a good candidate for biological molecules such as peptides and proteins. Molecules containing both strongly acidic and strongly basic functional groups also tend to be difficult to separate.

Packed column SFC is a normal phase technique. For those familiar with normal phase LC this has both positive and negative connotations. Normal phase techniques are effective in separating compounds with only subtle differences, including isomers. On the other hand, normal phase LC is often difficult to use, requiring long periods of equilibration. Traces of water can cause drift and irreproducibility. SFC offers normal phase attributes without the problems. Small aqueous samples can be directly injected, equilibration is extremely rapid, and reproducibility is similar to that of reversed phase LC.

Most of the detectors used in both LC or GC can at least sometimes be used in SFC. This is moderated by the fact that many detectors perform less than optimally with supercritical fluids.

The main niches packed column SFC is likely to expand into involve high speed, trace analysis and complex separation of moderately polar compounds. We have often characterized several niches as 'GC like detection on LC like separations', and 'GC like separations (high N) of fairly polar substances normally separated by LC'.

Capillary SFC

The spidergram for capillary SFC in Figure 1.9 will probably be considered controversial since it indicates both poor figures of merit and limited application areas. However, its applications tend to be **unique**. They include solutes that can be solvated with pure CO_2 and quantified with the FID. Linear density programs typical in capillary SFC are ideal for homologous series found in surfactants, many pre-polymers, *etc*. Capillary SFC is a good substitute for high temperature GC.

In the traditional figures of merit, speed, resolution, and sensitivity, capillary SFC does not compete well with GC or LC. In speed, the much higher D_{MS} in gases allow the use of much larger d_c. GC generates much greater N in less time, plus much higher sample capacity (sensitivity and linear range). LC is inherently slower but that problem was largely solved by using very small d_p. In LC, d_p is *ca.* $1/10$ d_c in capillary SFC. This shorter diffusion distance in LC compensates for the higher D_{MS} in SFC. The use of packed columns in LC also provides extreme sample capacity for repeatability, and good detection limits. Capillary

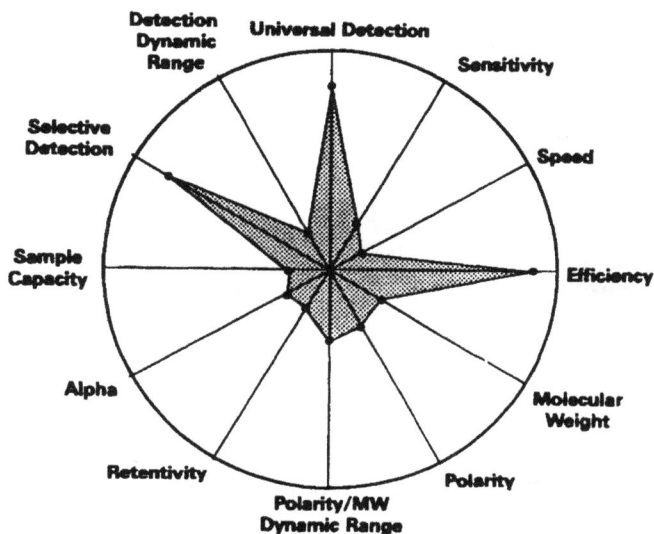

Figure 1.9   *Spidergram for capillary SFC*

SFC can be superior to LC in R_s partly because it is much easier to generate high N.

The most important attribute of capillary SFC compared with LC is the use of the FID. The FID represents the universal, constant response factor detector missing in LC. LC has always been more or less limited to molecules with chromophores.

Unfortunately, the fluids most compatible with the FID can only solvate modest molecular weight solutes up to perhaps 7000 MW. The pure fluids also have difficulty solvating molecules with multiple polar functional groups.

In SFC, intermediate D_Ms allow either a capillary or a packed column. Packed columns produce high efficiency in less time, higher speed (shorter diffusion path with the same D_M), and the highest sample capacity (20 μl direct injection). However, packed columns often require modifiers, negating the use of some detectors like the FID.

Capillary Electrophoresis

Capillary electrophoresis (CE) is another relatively new technique competing with LC.[20] A spidergram for CE, in Figure 1.10, illustrates its strengths and weaknesses. CE is similar to GC in terms of speed and efficiency, yet is most applicable to ionic and highly polar compounds. It can also be used for the separation of small, moderately polar solutes. Problem areas include quantitative and qualitative reproducibility, modest sensitivity, and limited detection options. At present, CE is a poor choice for trace analysis or situations requiring a high degree of quantitative precision.

The characteristics that make the technique fast and efficient also create

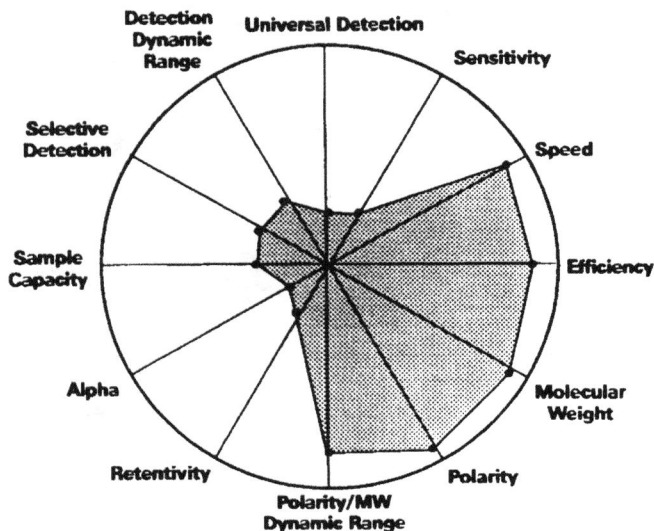

Figure 1.10 *Spidergram for capillary electrophoresis*

difficulties. High fields are desirable but can result in significant I^2R heating. In the extreme, this heat boils the buffer. Smaller d_c minimizes heating and allows higher fields and faster separations but makes injection and detection more difficult by decreasing sample capacity.

For high N, the bands must be no more than a few mm long. With $d_c = 50\ \mu$m, L = 2 mm corresponds to 4 nl. To get better than 1% RSD in area counts, the injected volume must be controlled to better than *ca.* ± 40 pl.

There is also some difficulty in changing α. In a given buffer, a solute has a characteristic migration velocity. If several solutes have similar velocities, they are difficult to resolve. To change solute velocity, its mass or charge must be altered. This can be done by changing ionic strength or buffer pH through electro-osmotic flow. Multi-dimensional micro-LC, CE offers exciting possibilities.

Contaminants on the capillary surface cause varying degrees of 'electro-osmotic' flow, resulting in variable t_R. Internal standards are required to allow the time base to be determined. The most modern instruments automatically wash the capillary before or after each run to try to minimize this problem, but this decreases through-put.

Summary

SFC brings GC-like figures of merit to the separation of many compounds that cannot be easily separated by GC. SFC is capable of making LC like separations with GC-like detection. Packed column SFC is superior in speed, N, and detector options compared to LC. If GC is the technique of choice for all molecules it is capable of separating, then packed column SFC should be the second technique of choice.

Packed column SFC and CE are both likely to make inroads into the application area served by LC, but from opposite extremes of polarity. There is unlikely to be much overlap between the two. CE is likely to be more efficient, perhaps faster, but mostly applicable to very polar molecules and ions. SFC is likely to be a more reproducible, trace technique, with greater selectivity and multiple detection options. However, neither is likely to displace LC methods that are well established and straightforward. Marginal LC methods are likely to be rapidly replaced by one or the other, depending on specific analytical goals.

References

1. T.A.Berger, W.H. Wilson, and J.F. Deye, *J. Chromatogr. Sci.*, 1994, **32**, 179.
2. Method and Fig. 991.06, 'AOAC Official Methods of Analysis' (1990) 15th Edn., 2nd Supplement (1991), Association of Official Analytical Chemists, p. 71.
3. R.M. Smith, *Pure App. Chem.*, 1993, **65**, 2397.
4. T.A. Berger and J.F. Deye, *J. Chromatogr. Sci.*, 1991, **29**, 390.
5. M. Ashraf-Korassani, M.G. Fessahaie, L.T. Taylor, T.A. Berger, and J.F. Deye, *J. High Resolut. Chromatogr.*, 1988, **11**, 352.
6. T.A. Berger and J.F. Deye, *J. Chromatogr. Sci.*, 1991, **29**, 141.
7. T.A. Berger and W.H. Wilson, *J. Chromatogr. Sci.*, 1993, **31**, 127.
8. B.W. Wright and R.D. Smith, *J. High Resolut. Chromatogr.*, 1985, **8**, 8–11.
9. C.G. Horvath, B.A. Preiss, and S.R. Lipsky, *Anal.Chem.*, 1967, **39**, 1422.
10. G. Nota, G. Marino, V. Buonocore, and A. Ballio, *J. Chromatogr.*, 1970, **46**, 103.
11. M. Verzele and C. Dewaele, *J. High Resolut. Chromatogr.*, 1987, **10**, 280.
12. E. Klesper, A.H. Corwin, and D.A. Turner, *J.Org. Chem.*, 1960, **27**, 700
13. Z. Balenovic, M.N. Myers, and J.C. Giddings, *J. Chem. Phys.*, 1960, **52**, 915.
14. I. Swaid and G.M. Schneider, *Ber. Bunsenges. Phys. Chem.*, 1979, **83**, 969.
15. R. Feist and G.M. Schneider, *Sep. Sci. Tech.*, 1982, **17**, 261.
16. H.H. Lauer, D. McManigill, and R.D. Bored, *Anal. Chem.*, 1983, **55**, 1370.
17. A. van Es, J. Janssen, C. Cramers, and J. Rijks, *J.High Resolut. Chromatogr.*, 1988, **11**, 852.
18. D.H. Desty, *J. Chromatogr. Sci.*, 1987, **25**, 552.
19. K.E. Karlsson and M. Novotny, *Anal. Chem.*, 1988, **60**, 1662.
20. J.W. Jorgenson and K.D. Lukacs, *Anal. Chem.*, 1981, **53**, 1298.

CHAPTER 2

Practical Aspects of SFC Hardware

1 Introduction

Although there have been many 'home-built' packed column SFCs in use, capillary SFC systems dominated the commercial marketplace until 1992. In the last few years the commercial emphasis has shifted to packed column systems. Many potential users are only vaguely aware of the fundamental differences between packed and capillary instruments. There are also unrecognized problems with 'home built' packed column instruments.

There is an understandable desire to use equipment already paid for. Managers are confused when asked to buy a new SFC when they already have paid for a capillary system. They assume that if the name is the same (SFC), the instruments must be similar. However, attempts to use capillary systems with packed columns are the most common reason research efforts fail.

In this chapter, packed column instrumentation is described. In addition many practical pitfalls are discussed in the hope that new users can avoid them. Capillary equipment is also discussed but primarily to highlight differences.

2 Instrumental Overview

Packed column SFC should be thought of as an extension of LC because of similarities in the equipment and approach. These include pumps, injection valves, columns, and some detectors. Equipment manufacturers may eventually modify standard LCs to include SFC.

Like LC, the packed columns used in SFC generally contain small particles (3–10 μm) packed in short (6–25 cm) stainless steel tubes. High pressure pumps are required to push the MP through the packing. In SFC, the column pressure drop (dP) is as little as 1/5th of that in LC, even though the optimum flow rate is 2–5 times higher. However, in SFC, $p_0 > 75$ bar (for CO_2) to keep the fluid *from expanding*. Columns, and pumps capable of high pressure are still required.

The most widely used pure SFs are simply too non-polar to solvate polar molecules. To increase the polarity range, binary mixtures and gradients are used, similar to LC. The most common mixture in SFC is methanol in CO_2.

With binary SFC MPs, the modifier concentration is the primary variable[1] for controlling retention. Pressure and temperature are important but secondary in retention control. They are more important in adjusting α. The **pumps** control the **flow** of each MP component and, subsequently, fluid **composition** in both LC and packed column SFC.

High pressure injection valves from LC allow the introduction of the sample into the column. The most common detector in SFC is the same UV absorbance detector used in LC.

The **differences** between LC and SFC are also informative. Mechanically, the pumps in LC and SFC are similar with a few exceptions. Commonly used fluids (*e.g.* CO_2) are purchased as a liquified gas. The pressure of CO_2 in a cylinder (P_{cyl}) is *ca.* 65 bar, depending on temperature. The pump must be capable of handling such a pressurized source.

Pumping gases to high pressure while accurately controlling flow is difficult. It is much easier to control the flow of a liquid accurately. The pump head used in SFC is generally chilled to ensure that the liquid from the supply cylinder remains a liquid during pump filling.

All modern LC pumps compensate for the compressibility of the fluids pumped. Although the pump delivers **liquid** CO_2, its compressibility factor (Z) is quite different from those of normal liquids. The pump must be capable of an extended Z range. Both pump head chilling and Z are dealt with in greater detail later in this chapter.

High pressure pumps are expensive. Most modern gradient LCs mix solvents at low pressure and use one high pressure booster pump. In SFC, such approaches will not work because one of the fluids (*i.e.* CO_2) is already at high pressure. Consequently, SFC must use at least two high pressure pumps to create binary or ternary mixtures.

The biggest difference from LC is the use of a back pressure regulator (BPR) in SFC. The BPR dynamically controls the **outlet pressure** (p_o) of the system and can be used to control pressure programs. Without the BPR, the fluid would expand to a low pressure, low density gas with little solvating power.

UV detectors are mounted between the column and the BPR. Since pressure in the detector can be 400 bar, special windows are required.

GC detectors can also be interfaced to packed columns through a 'tee' between the column and the BPR. A fixed restrictor diverts a small fraction of flow from the tee to the detector. Most flow passes through the BPR. Detectors like the ECD or the NPD remain selective but can be less sensitive than in GC.

Other Configurations

The above description, outlined in Figure 2.1, is the most useful for packed columns of i.d. ≥ 2 mm. It is called 'downstream' control because pressure is controlled **downstream** of the column. Capillaries and micropacked columns require low flow rates and have such low dead volumes that it is presently impossible to perform downstream control with existing equipment.

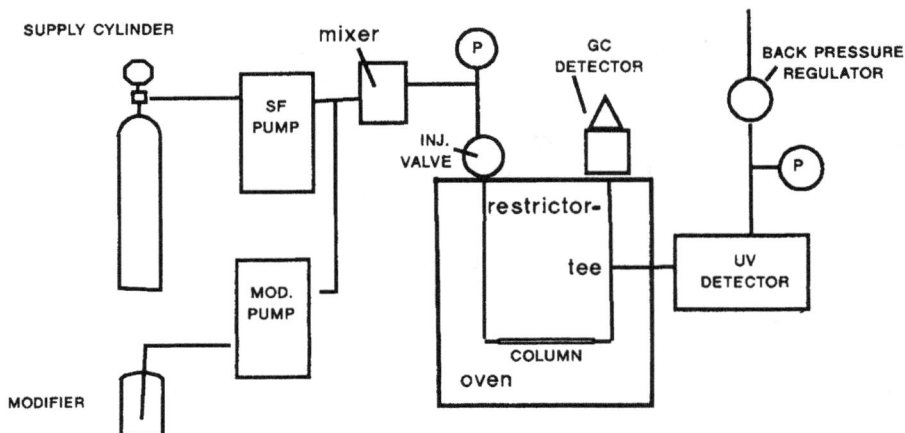

Figure 2.1 *Schematic diagram of 'Downstream' pressure control. Typical operation uses two high pressure pumps operating as flow sources. Outlet pressure is dynamically controlled using a back pressure regulator. Both inlet and outlet pressure is monitored. Detection is either via UV or a GC detector, like the FID, or both. A column oven is required. Columns are generally the same as used in LC*

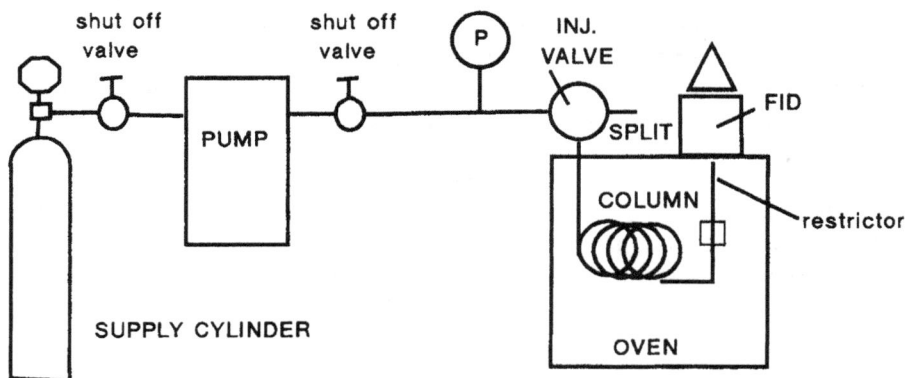

Figure 2.2 *Schematic diagram of 'Upstream' pressure control. Fluid from the supply cylinder is periodically introduced into the pump. The pump acts as a pressure source. A fixed restrictor in the base of the flame ionization detector (FID) acts as the passive flow control. A GC is typically used as the column oven. The injection valve is mounted on top of the oven. The column shown is a capillary but micropacked or standard columns could also be used*

Another instrumental configuration, called 'upstream' control, outlined in Figure 2.2 can be used with virtually any column size, but the quality of control is degraded compared with downstream control. Consequently, upstream control is most often used in situations requiring very low flow rates of pure fluids (*e.g.* capillary SFC). With larger columns, upstream control is much less desirable. Flow control is poor and irreproducible. In addition, binary or ternary fluids are nearly impossible to use.

In the upstream mode, the **pump** acts as a pressure source and the column head pressure, not p_o, is controlled. **Flow** is controlled by a passive **restrictor** located at the column outlet. The pump delivers whatever flow is required to achieve the pressure set-point. Commercial capillary systems have used syringe pumps, but properly designed reciprocating pumps[2] provide equal performance.

With pure MPs, pressure programming in SFC provides retention control analogous to temperature programming in GC.[3,4]. Retention is a nearly linear function of density.[5] Negative temperature programming with either fixed[6-9] or programmed[10] pressure can also change density. However, retention is a function of both density and temperature, so the two programming modes (positive pressure ramps and negative temperature ramps) are not equivalent.

As will be shown below, pre-mixed binary fluids purchased from gas vendors can be delivered by a single pump to create pressure programs of binary fluids. However, the fluid composition actually changes as the cylinder is used up,[11,12]. Attempts to mix fluids dynamically or form gradients with two pumps in the pressure control mode are also less than ideal. It is difficult to vary the mass delivery rate of two separate pumps simultaneously when one or both are used as the pressure source for the system. The composition accuracy and precision of binary fluids are, thus, less than ideal in the upstream mode.

The dP across **packed** columns varies depending on the composition of binary fluids. Pressure transducers are seldom placed between the column outlet and a fixed restrictor. Small (< 1 mm) or even intermediate i.d. (1–2 mm) columns cannot tolerate much extracolumn band broadening. With upstream control on small diameter columns, the density at the column outlet is, therefore, generally unknown because the dP is unknown. The location with the lowest density is also the position with the poorest solvent strength. In the worst case, unmeasured large dPs can cause low densities at the column outlet, creating numerous problems of which the operator may be unaware.

Binary fluids often produce inadequate retention[13,14] on capillaries, and also prevent the use of the FID. Finally, the density range available for density programming is narrower with binary fluids (typically 0.5–0.95 g cm^{-3}) than with pure fluids (0.02–0.95 g cm^{-3}) (see Chapter 3).

In the upstream mode, **flow** through the column is controlled by a fixed restrictor mounted on the column exit. The characteristics of such restrictors cause major problems for the upstream approach. When pressure increases, flow through fixed restrictors dramatically increases,[15] changing the t_0, fluid linear velocity, k', and N. It becomes extremely difficult to deconvolute the individual effects of all the parameters that change. Fixed restrictors are discussed in greater detail later in this chapter.

All the problems with binary fluids, pumping, unknown dP, and *restrictor* characteristics make upstream control a poor choice for most applications using packed column.

3 Practical Aspects of Fluid Supplies

Carbon dioxide (and other common fluids) is usually purchased in high pressure gas cylinders, but the fluid actually exists mostly as a liquid, with a gas headspace. Tank pressure, P (actually the fluid vapour pressure, P_v), depends on cylinder temperature. For CO_2, $P \approx 65-70$ bar, near 'room' temperature. As the liquid is used, P only changes slightly since some of the remaining liquid boils off to maintain a constant P_v.

It is easier to increase P by pumping the liquid from the bottom of the cylinder, rather than the gas in the headspace. Although some users invert standard cylinders, most order cylinders equipped with 'dip tubes' extending from the valve to the bottom of the cylinder. The head pressure of the gas forces liquid up through the dip tube and valve, through plumbing, and into the chromatographic pump.

In the cylinder, liquid is in equilibrium with gas: $P_{cyl} = P_v$ at the cylinder temperature. Storing cylinders in direct sunlight on a hot day can raise P_{cyl} to a point where the safety 'blow-out' disc ruptures and all the fluid is vented to the atmosphere. On very cold days, the P_v in the cylinder can be very low. If cylinders are brought inside when the outside temperature is very low, P_{cyl} can remain very low (unusable) for hours.

Supply Line Characteristics to Avoid

Safety officers sometimes require installation of supply cylinders outside the laboratory, perhaps even outside the building. If fluid is transported through plumbing lines to locations with lower pressure or higher temperature than the cylinder, the liquid in the plumbing will 'boil' and separate into two phases.

In winter with outside storage, low temperature depresses P_v. If the cylinder remains outside but cold fluid enters a heated building through plumbing lines, the P_v of the fluid *in the plumbing* increases but P_{cyl} remains low. The fluid is likely to separate into two phases. Gas bubbles must be reliquified, or the pump will cavitate, resulting in erratic flow.

When cylinders are changed, the pressure in connecting plumbing drops to ambient. On re-pressurization, the plumbing lines remain largely filled with gas, unless the gas is vented. At one new installation with initially very long supply lines, more than 30 minutes were required to achieve stable pump operation when a cylinder was replaced. Venting long delivery lines wastes substantial expensive fluid.

To avoid problems, the cylinder should be as close to the instrument as can be reasonably accommodated and kept at or above lab or supply line temperature. Multiple cylinders can be manifolded together, to prevent depressurization when changing cylinders. Supply lines should be short and with an internal diameter adequate to avoid pressure drops in the tubing at the flow rates used.

'Normally closed' electric shut-off valves are sometimes desirable or mandated to shut off the fluid supply in the event of a power failure. Since such valves must be held 'open' during operation, a current must flow through a solenoid. Some of this current will be converted into heat, warming the fluid. Raising the fluid temperature, $P_v > P_{cyl}$, causing phase separation. The power required to open such valves should be minimized. Valves in the supply lines should be mounted so that their internal volume is well swept.

Padded Tanks

It has been common practice to add a helium 'pad' to the supply cylinders, so that $P_{cyl} > 1200$–1500 psi (83–103 bar). Thus, $P_{cyl} \gg P_v$. With slightly lower pressure or higher temperature in the supply lines, the fluid remains a liquid. No phase separation can occur.

The helium is intended to act as an inert headspace gas, pushing the liquid at the bottom of the tank up and out the dip tube. However, helium does dissolve in the fluid,[11] gradually changing its composition slightly. With properly designed pumping systems, the recurring, added expense of the pad is not necessary.

Pre-mixed Binary Fluids

Several gas suppliers have offered cylinders of pre-mixed binary fluids, such as low (*i.e.*, 5–20%) concentrations of MeOH in CO_2. However, the headspace gas has a substantially different composition from the liquid (the P_v values of the two components are very different). As fluid is used, the modifier concentration in the delivered fluid can change by more than two times[11,12] between the initial and final use of the cylinder. Although pre-mixed cylinders allow some exploratory research for workers with a single pump, the results cannot be reproduced. Pre-mixed binary fluids should be avoided for SFC.

Empty?

When the liquid in the cylinder is nearly gone, the pump can suck a mixture of gas and liquid, resulting in progressive loss of flow control. This loss in control can be so subtle that it is not noticed until the pressure becomes unstable. If such behaviour occurs, the operator should check the data from the previous few chromatograms to see whether they need to be repeated. Such situations should be avoided.

There is no obvious warning when cylinders are nearly empty, since P_{cyl} doesn't change until the liquid is gone. There are two ways to avoid such problems. The simplest is to buy a scale and weigh the cylinder. The cylinders are provided with a tare weight. An aluminium cylinder completely filled with gas (no liquid remaining) still contains nearly 4 kg (8–9 lb) of CO_2. We generally change cylinders when there is *ca.* 4.5 kg (10 lb) of fluid left.

If a cylinder is left standing on a scale for long periods, the calibration changes and needs to be updated. The best practice is periodically to lift the cylinder onto the scale, weigh it, and then remove the scale, instead of leaving the scale permanently under the cylinder.

One manufacture is supplying cylinders with a built-in level sensor, which indicates whether the level is adequate for overnight operation. They have also been developing an electronic version to provide a warning light next to the instrument, for those who store the cylinder in a remote location.

Typical flow in a 4 or 4.6 mm i.d. column is 2–3 mls min^{-1} which is approximately 2–3 g min^{-1}, 120–180 g h^{-1} of CO_2. A CO_2 cylinder contains 9.5–10 kg (*ca.* 20 lb) of usable fluid (14–15 kg, 30–31 lb, total). Consequently, a cylinder should provide 55–83 hours of operation with such columns. A smaller diameter column increases cylinder lifetime in proportion to column cross sectional area. Thus, a cylinder should last up to 330 hours with a 2 mm i.d. column.

Inertness

Many users continue to search for the 'ultimate' mobile phase. However, much more polar fluids will undoubtedly have much more intense intermolecular forces and exist as liquids at STP. In addition, some fluids can damage hardware. Pump manufacturers generally list fluids that are compatible with the various seal materials present in check valves, pump seals, and switching valves. Care should be exercised before using new fluids.

Even CO_2 can cause dramatic swelling of some elastomers. In the late 1970s Gere and co-workers made a $\frac{1}{2}$ inch diameter 'O' ring swell to > 3 inches in diameter. When removed from a pressure vessel and placed on a lab bench, the 'O' ring outgassed plumes of CO_2 (it looked like a steaming doughnut) and randomly vibrated. When all the gas had escaped, it returned to its original dimensions. However, it was brittle and disintegrated when squeezed. All the light oligomers, plasticizers, *etc.* had been extracted by the CO_2, making it worthless as a seal material. This explained why these 'O' rings kept failing in a PBR. It is a good idea to test all elastomers for such problems before use in SFC.

Safety

Some of the early work in SFC used pentane at > 200 °C. A few workers still use such fluids. Pentane and other light hydrocarbons are extremely flammable. Many modern SFCs use 'hot wire' GC ovens in which a wire is resistively heated. If the column should suddenly leak while the wire is hot, a fire or explosion would result. Such fluids should not be used in such ovens.

There has been at least one fatal accident reported from an explosion involving nitrous oxide (N_2O). N_2O is a strong oxidizer. Mixing it with alcohols or other organic modifiers is equivalent to making a fuel/oxidizer mixture. Even large solute or solvent injections could cause problems. There is usually little to recommend N_2O over CO_2 (other than detection options) since they have similar solvent properties.

Nitrous oxide is 'laughing gas' sometimes used by dentists in place of pain killers. Employees have been known to sniff the gas to become intoxicated. The use and storage of N_2O is avoided in the author's laboratory.

There have been periodic reports of the use of ammonia as a supercritical fluid. However, initial findings of high polarity have never been reproduced. Ammonia attacks a number of polymers which might sometimes be used as sealing materials in LC equipment. It can be extremely dangerous if even a dilute mixture in air is inhaled. There is a report of workers absentmindedly mixing CO_2 with wet NH_3 to try to make polar mixtures. They spent days rebuilding their chromatograph after solid ammonium carbonate formed in much of their plumbing.

Gas Quality

SFC grade CO_2 costs five times more than some other grades. Although it is still cheaper than LC grade liquid solvents, some users specify less expensive grades to save money. Most instrument manufacturers recommend only the more costly, **better defined** supercritical grade, primarily to avoid warranty problems. In the early 1980s we used many different inexpensive grades. However, two cylinders containing a great deal of foreign material including a large amount of iron rust caused such extensive contamination and subsequent clean-up problems that we avoid using such undefined gases.

For UV detection, cheaper fluids will probably cause no detection problem. However, for use with some GC detectors, like the ECD, lower quality gas may not be acceptable.

There is an SFE grade with extremely low concentrations of halogenated compounds. SFE grade costs two or three times more than the SFC grade, but is seldom required in SFC.

4 Pumps

Most capillary SFCs use syringe pumps, which are best used as pressure sources, that can be programmed as high as 600 bar. Rapidly filling the pump can cause phase separation. If heat from the motor reaches the fluid, its temperature increases, causing $P_v > P_{cyl}$ and more gas bubbles form. Chilling the pump with a refrigerated bath, or using padded tanks, or both, keeps $P_{pump} \approx P_v$ and ensures nearly complete fills with liquid MP. A completely filled pump allows operation with capillaries for many days, even weeks. With a 4.6 mm i.d. packed column, requiring 2–5 ml min^{-1} of flow, the operating time between refills could be as short as 20 minutes. A refilling operation could take 10 minutes.

Reciprocating Pumps

Reciprocating pumps are more accurate and can also deliver essentially an 'infinite' volume before refilling. With packed columns, binary fluids are commonly used. The best way to produce binary fluids is to use two independent high pressure pumps; one designed to pump normal liquids, and the other specifically designed to pump highly compressible fluids.

Reciprocating pumps should also be chilled to ensure that the fluid remains a liquid during pumping. The temperature chosen is fairly arbitrary. At $T \ll T_c$ and P_{cyl}, the fluids are usually liquids, with liquid like densities. Small changes in temperature or pressure generally have little effect on this density.

Although padded tanks are superficially an alternative to pump head chillers, the recurring expense, and variable results obtained, mean that pump head chilling offers cheaper (in the long run) and more precise control of flow.

The better the temperature control the more accurate and precise the flow. In a pump built specifically for SFC, the pump head should be thermally isolated from the drive mechanism. An elegant solution on the new Hewlett Packard SFC uses a Peltier element with a fluid pre-cooler (to control temperature precisely), insulating stand-offs, and a low thermal conduction sapphire piston.

In humid regions, some cooled pump heads have caused condensation of substantial amounts of water from the atmosphere on the chilled surfaces. Such condensation should be nothing more than an annoyance as long as the water is channeled away from electronics or parts that rust. A few users have placed pumps in an isolation box and purged the box with dry air or nitrogen to avoid excessive condensation. Newer pumps use better insulation and have minimized this problem.

Compressibility Adjustment

Reciprocating pumps have inlet and outlet check valves. On the fill stroke, the pump piston moves backwards and the outlet check valve closes, isolating the pump from the higher pressure in the column (P_{col}). Eventually the pressure in the pump (P_{pump}) < P_{cyl}. The higher P_{cyl} forces open the inlet valve and fluid flows from the supply cylinder into the pump. After the pump is completely filled at P_{cyl}, the piston reverses direction, compressing the fluid in the pump until $P_{pump} > P_{cyl}$. This closes the inlet valve. At this point both check valves are closed.

On the compression stroke, the piston moves very rapidly until $P_{pump} > P_{col}$. The outlet check valve then opens and fluid begins to enter the column. The piston only slows down to the delivery speed when enough extra fluid has been pushed into the column to make up for the lack of flow during the fill stroke.

The distance the piston must rapidly travel to just compress the fluid to P_{col} is calculated, based on the known volume of the components and a characteristic of the specific fluid, called its compressibility factor, Z. With the right Z a pump can nearly eliminate flow or pressure ripple. With an inappropriate Z, the pump either under- or over-compresses the fluid causing characteristic ripples in flow and pressure.

If the fluid is under-compressed, the outlet check valve cannot open on time. Part of the delivery stroke is used up further compressing the fluid in the pump, not delivering flow to the column. The net result is that the delivered fluid, in ml min^{-1}, is less than the set point.

Over compensation results in the piston travelling too fast, too long. After $P_{pump} = P_{col}$, the piston continues to travel faster than required to deliver the desired flow. There is a momentary increase in pressure (and flow) at the head of the column. The pump actually delivers more than the desired amount of fluid, so the real flow rate is momentarily higher than the set flow rate.

Either under- or over-compensation results in periodic variations in both pressure and flow with the characteristic frequency of the pump (flow in ml min^{-1}, divided by the pump stroke volume in ml). This can result in noisy baselines and irreproducibility. Better LC pumps have compressibility adjustments to account for differences in fluid compressibilities.

The compressibility factors of fluids used in SFC are much higher than the Zs of normal liquids. Z also changes with temperature and pressure. Standard LC pumps modified only with a chilled pump head, but no extended Z range, are likely to under-compress the fluid dramatically. During a pressure program, the compressible fluid delivery rate drops, but the modifier flow remains unchanged (flow drops, modifier concentration increases). There are a number of reports in the literature using modified LCs, where such programs were inadvertently produced and went unrecognized. Attempts to reproduce the work show that much higher modifier concentrations are required than initially reported using under compensated pumps.

Pumps designed specifically for SFs require extended Z ranges and the ability to change Z dynamically during programs. Z can be dynamically calculated as a function of temperature and pressure. Alternately, look-up tables can provide the same information. However, neither pressure or temperature measurements are likely to be accurate enough to produce ideal compensation. The proper value, producing the best compensation, is likely to be somewhat different from the calculated value. There are techniques to overcome this problem.

Z can be continuously changed between slightly higher and slightly lower than the calculated values on alternate pump strokes. The pressure waveform from each stroke can then be monitored and compared with the value from the previous compressibility setting. If the new value provides lower ripple, Z can be further changed in the same direction that produced the lower ripple. If the next ripple value is worse, Z will be changed in the other direction. In this way, the pumping system seeks the optimum empirical Z for the fluid being pumped. Fluids with unknown Z values can be pumped without fear of confusion over flow, or composition variations.

Adiabatic Heating

After the compression stroke, most LC pumps use constant piston speed to deliver constant flow rates. However, even with normal liquids, the compression stroke of the pump raises the temperature of the fluid by adiabatic heating.

Just after the compression stroke, the fluid is at its highest temperature. There is no time for heat to be dissipated into the cylinder walls. The pressure must rise more than expected, prematurely opening the outlet check valve. More mass than expected, based on the constant piston speed, leaves the pump. As the fluid begins to cool, its density in the pump increases. The shrinking volume of the fluid negates some of the forward motion of the piston. This results in less mass than expected being delivered to the column. During each constant speed pump delivery stroke, adiabatic heating results in a pattern of first excessive, followed by inadequate, flow. A careful examination of the baseline, particularly with flow sensitive detectors in LC, will often reveal a pseudo-sinusoidal waveform caused by adiabatic heating.

There is no effective way to keep the fluid isothermal during a compression stroke. However, the speed of the piston can be varied to compensate for changing density. Heat generation can be accurately modelled, so an algorithm can be written to vary the piston speed during the delivery stroke, compensating for adiabatic heating.

Leak Compensation

All pump seals and check valve seats leak. Seals in good condition leak from < 1 to > 5 μl min^{-1}. Such small leaks are almost impossible to detect, but can significantly affect performance. With capillaries or micropacked columns, leaks could be a few percent of total flow. However, flow inaccuracies are generally not the biggest problem. Pump algorithms that calculate Z are based on a known fixed volume that needs to be compressed. Leaks effectively increase this volume. A pumping system that does not compensate for leaks will undercompress the fluid. The larger the leak rate, the more serious the flow ripple. Fortunately, the size of leaks can be estimated by observing the shape of the pressure waveform accompanying the compression stroke. The algorithm that seeks the optimum empirical Z automatically compensates for leaks.

Mixing Binary Fluids

In SFC using two pumps, each is likely to produce at least some pressure and flow ripple which can cause baseline noise. Since the modifier concentration in SFC is generally small, the main fluid pump and the modifier pump operate at different frequencies. When the flow rate (ml min^{-1}) approaches the stroke volume of one of the pumps, the composition can oscillate with a period: (stroke volume)/(flow rate). For example, at a flow of 1 ml min^{-1} a pump with a stroke of 100 μl will have 10 strokes per minute or one delivery stroke every 6 seconds, which is about the same as a typical peak width. If the pump noise and the peak have similar frequencies the noise cannot be suppressed.

Some pumps have variable stroke lengths to shift pump frequency away from peak frequencies. One commercial pump has a variable stroke length of 20–100 μl. At 1 ml min^{-1} flow and 20 μl stroke length, the frequency increases to 0.8 Hz, which is easier to filter from the peak frequencies.

Dual Piston Pumps

Some pumps use two alternating pistons to minimize the variation in short term fluid delivery caused by refilling. Although generally superior to single piston pumps in producing stable baselines, such pumps can sometimes create new problems. Differential leaks around each piston seal or through check valves can deliver a slightly different amount of fluid from each piston. At very low flow rates, and very low modifier concentration, the baseline can take on a square wave appearance with a period of the slower pump stroke, caused by actual variations in binary fluid composition. Shortening the stroke length, increasing the mixing volume, replacing the main seals, and check valves are all likely to improve this baseline noise.

Mixer

As in LC, mixtures of two fluids do not become instantaneously homogeneous after the mixing point. Short term composition variations can be averaged out by mixing larger volumes of the mobile phase. As with other kinds of 'filtering', combining fluid corresponding to three to five periods of the oscillation will eliminate most of the problem. A mixing column mounted downstream of the mixing point can consist of nothing more than a standard column packed with relatively large stainless steel balls. Since the intent is to obtain a homogeneous mixture, the balls should be large to spread out sharp (unwanted) changes in concentration. On the other hand, the volume of the mixer cannot be so large that it dilutes out programmed gradients.

All the volume between the mixing point and the injector must be swept during a composition gradient before the composition at the head of the column actually changes. This delay volume should be as small as possible while retaining adequate mixing. Long delays between the start of the run and the time a gradient may actually arrive at the column degrade the ability to do rapid analysis. Fortunately for SFC, flow rates are relatively high, so volumes can be swept rapidly.

When attempting to perform downstream control at low flow rates and low modifier concentration, baseline oscillation can become a problem. Under extreme conditions it may be desirable to trade gradient delay time for smoother mixing. Providing a larger mixing volume usually solves the baseline problem. As a worse case, an empty 6 mm i.d. tube 6 cm long, mounted vertically can be substituted for the mixing column.

Dynamic mixers, using a magnetic stir bar, have been developed for SFC. Although they are effective, their volumes tend to be at least as large as a passive mixer, and provide no real advantage.

The Mixing Point

With highly compressible fluids, a pressure pulse from one pump can cause back flow in the delivery line of the other pump. If a simple 'tee' is used to combine the fluids, the flow through its outlet oscillates between pure component A, a mixture of components A and B, and pure component B.

The simplest solution is to place a check valve in the modifier supply line just before the mixing point. In one embodiment, the mixing point was actually in the end fitting of a cartridge check valve. The main fluid continually swept this region. When the modifier pump was delivering, it lifted two ruby balls off two sapphire seats mounted serially and delivered the modifier directly into the main flow. This arrangement minimizes the length of time modifier is not being added to the main flow.

Purging

Pump check valves are usually designed to operate with normal liquids. Fluids with much lower viscosities tend to leak through such seals at a higher rate. Since the modifier pump outlet is pressurized by the other fluid, it can be difficult to purge gas from the pump. If gas is not removed the pump can cavitate. To eliminate this problem, the modifier pump outlet should be opened to the atmosphere, and the flow rate set to the maximum for perhaps 10 seconds. It is good practice to degas modifiers before attempting to pump them.

Flow Sensors

Because there are so many ways to generate errors in the composition of binary fluids (main piston leaks, check valve leaks, gas bubbles, mixing point problems, improper compressibility adjustment), a flow sensor can provide greater confidence in the actual value produced. They can be calibrated for specific fluids, and can be accurate to less than $\pm 1 \mu l$ min^{-1}. A sensor can also be used as a trouble shooting tool, to indicate the size of leaks.

5 Pressure Control

Second generation packed column instruments use electronic BPRs. The use of two independent control devices allows simultaneous programming of composition and pressure with a high degree of precision and accuracy. It is important that the BPR have low dead volumes if peak collection is important. The older technology, mechanical BPR used in the past sometimes had dead volumes as high as 5 ml.

BPRs should be heated to prevent the formation of solid particles of the MP as the fluid decompresses. Such particles can cause noisy baselines and erratic flow. Only modest temperature (40–80 °C) is required to prevent such problems.

Fixed Restrictors

All fixed restrictors have serious problems. Both packed and capillary columns are sometimes used with fixed restrictors, although for different reasons. With packed columns, fixed restrictors are primarily used to divert a small fraction of the total flow to GC detectors, such as the FID, NPD, or ECD. The main pressure control continues to involve a back pressure regulator.

Many types of fixed restrictors have been used. They all allow small flows (< 1–100 ml min^{-1} expanded, as little as 0.002 ml min^{-1} compressed, $P < 600$

bar). They all produce higher mass flow at higher inlet pressures. Some designs produce larger increases in mass flow than others.[15] This causes serious problems with N when such restrictors are used as the primary flow control device. Restrictors also slowly plug.

Linear

Linear restrictors consist of a piece of small, constant i.d. tubing. Because pressure drops occur approximately linearly along the length of such restrictors, the density progressively drops along the restrictor. Solutes tend to drop out of solution long before they reach the exit. Such restrictors discriminate against heavier, less volatile components. They also produce the largest increases in mass flow with increasing pressure,[15] a wholly undesirable trait. Large increases in mass flow cause a continuous loss in efficiency from the beginning to the end of pressure programs. Because of such problems, linear restrictors are unacceptable for controlling column flow, but are sometimes used to control the split flow in split injectors.

Frits

Frit restrictors consist of a porous plug of silica 1–3 cm long, deposited in the end of a piece of 50 μm i.d. tubing.[16] The frit provides multiple independent flow paths, so it is difficult to plug. However, frits can also discriminate against heavy components for reasons similar to those for linear restrictors. Finally, frits produce nearly as large an increase in flow for a given increase in pressure as the linear restrictors.

'Integral'

Integral restrictors are essentially a pinhole in the end of a piece of fused silica tubing.[17] The opening is steeply tapered from the tubing i.d. down to 1–3 μm over a length of only a few millimetres. These restrictors have the best (although still not adequate) flow characteristics.[15] However, these restrictors are also the easiest to plug. Deposition of a short very porous frit in the inlet of an integral restrictor can act as a filter to keep particles out of the pinhole.

Other

Many other versions of restrictors have been used, but they are all similar to, or intermediate between the types described above. Metal restrictors tend to cause inertness problems, particularly when hot.

6 Injection

Packed columns have similar sample capacity in both LC and SFC. In SFC, it is common to use standard LC fixed loop injection valves with either internal or external loops. On a 4.6 mm column, with **binary** mobile phases, 20 μl of most sample solvents can be injected without significant peak distortion. Direct full loop injections are the normal means of sample introduction. This means that packed column SFC has similar quantitative reproducibility to LC using fixed loop injectors.

SFC has several injection drawbacks compared with LC. Variable injection volumes in LC are often accomplished by partially filling a large fixed loop. Normally, the loop is in the flow stream. To make an injection, the loop is switched out of the flow stream, but remains filled with MP. A syringe displaces part of the MP from the loop with sample. The valve is then rotated to place the loop back in the flow stream.

Many of these devices will not work with SFC. Switching the loop loaded with MP into contact with the syringe exposes the syringe to a 'compressed gas' at up to 600 bar. Either the injection device must be capable of injecting into such a high pressure or it must be protected from the expanding gas. If a standard glass syringe is used in a needle port, the plunger or even the whole syringe could shoot across the room. The larger the loop the greater the energy stored in the fluid. To avoid such problems another valve may be required to isolate the syringe from the loop. In addition the outlet line to waste should have a larger i.d. than the sample line to try to promote venting in that direction. There has been no systematic study involving partial injections into an empty loop.

Aqueous Injections–Solvent Polarity

Packed column SFC is almost always a normal phase technique. The sample solvent should, therefore, be less polar than the MP. This introduces several potential injection problems.[18] SFC cannot deal directly with large aqueous injections. In general, some care must be exercised to ensure that the sample solvent does not interfere with the chromatography. Large injections of MeOH, even into a MP consisting of MeOH/CO_2, can result in peak distortion.

On a 4.6 mm i.d. column with a low percentage of MeOH in CO_2 as the MP, 5 μl aqueous injections can be tolerated but only at column temperatures greater than *ca.* 60 °C. Injections of 5–10 μl MeOH are acceptable at nearly any temperature. With 20 μl methanol injections, peak distortion is measurable, and 50 μl injections result in severe distortion. Decreasing sample polarity by using acetonitrile allows slightly larger injections to be made. Further decrease in solvent strength using ethyl acetate, or other solvents, allows some further minor increases in injection volume without serious distortions.

Different SPs react differently to sample solvents. Water will smear out on a silica column, distorting peaks. On an ODS column, and to a lesser extent on other more polar bonded phases, water will be largely repelled by the hydrophobic nature of the packing surface. Organics dissolved in the water can often be retained while the water passes through. Increasing the mobile phase polarity then allows elution with good peak shapes.

Very large aqueous injections can be made using a technique similar to on-line solid phase extraction (SPE). An ODS pre- or guard-column can be substituted for the loop in a six port injection valve. In the load position, water can be injected or pumped through the precolumn. After the water is introduced, the column is blown dry with CO_2 or N_2, then switched into the flow stream. Small amounts of remaining water generally cause few problems. At least 70 ml of water can be 'injected' in this way.

7 Detection

UV Detectors

In SFC, the cell must be capable of withstanding at least 400 bar with a safety margin of 1.5–2 (depending on the local regulations in different countries). Some LC–MS interfaces also create a significant back pressure (usually < 100 bar).

One novel cell design uses unusually shaped windows bevelled at a 45° angle on both front and back. Only small parts of the surfaces remain parallel to each other and perpendicular to the cell axis. The bevelled edges are mounted in matching seats machined out of vespel or similar material. The peculiar shape evenly distributes the forces, making it virtually impossible to break a lens. Performance is similar to that of ball lenses, except that the surfaces are perpendicular to the light path.

Thermal Equilibrium

Changes in temperature within the detector cell can cause refractive index (RI) problems and noisy baselines. In LC, the MP is generally at near room temperature, although there has been substantial movement toward the use of column ovens over the last 10 years. In packed column SFC, most work with modifiers is also done at relatively low temperatures (*i.e.* 30–70 °C). However, packed columns are sometimes operated as high as 200 °C with UV detection. To minimize RI problems, a heat exchanger (HX) is required.

The detector most used in the author's laboratory contains two HXs. One contains one metre of 'crinkle' tubing (tubing bent at > 90° every 5 mm) embedded in a block of aluminium, cast with heat dissipating fins. A fan blows laboratory air through the fins. A small HX with *ca.* 10 cm of embedded tubing is bolted to the flow cell to provide a final match in temperature.

There is a clear trade-off between the band broadening caused by the internal volume of the HXs, and peak-to-peak noise (N_{p-p}) caused by RI changes. The noise level varies with fluid pressure, flow rate, and the quality of the pumping system. With 'low' temperature, and 'high' densities, the N_{p-p} with SFs can be similar to the N_{p-p} obtained with normal liquids.

Since HXs are used to make $T_{cell} = T_{room}$, large variations in T_{room} defeat one of the purposes of the HX. Large swings in T_{room} can result in significant variations in the detector baseline (wander and drift). To avoid such problems, the detector should be kept out of drafts and away from heating and cooling ducts. Any device which smooths out rapid T_{room} changes will decrease drift and wander. An intriguing possibility would use fibre optics to bring light to and from the cell. Such a cell could then be mounted inside the column oven where temperature variation $\leq \pm 0.1$ °C and the same as the column.

Optimizing UV Sensitivity

If a detector offers variable slit width or bandwidth, S/N can often be improved by using relatively 'open' (large slit) systems. If the spectrum of the solutes is known it is often possible to select a bandwidth 12 or even 20 nm wide. This allows maximum light through the system, and the maximum absorbance. If the bandwidth is set too wide, however, the sample absorbs almost nothing over part of the chosen bandwidth, and the S/N ratio could actually degrade.

Lamps in UV detectors tend to put out most of their light at lower wavelengths. Setting the reference to high wavelengths, such as 450 nm, means that the amount of light in the reference beam is low, increasing noise. It is advisable to set the reference beam to a wavelength as close to the absorbance band of the solute as possible without including appreciable absorbance. As with the sample beam, a wide bandwidth improves S/N.

If a solute exhibits an absorbance maximum at 254 nm, absorbance at 300 nm is likely to be low. Setting the sample beam to 254 nm with a 12–20 nm bandwidth, and the reference beam to 340 nm with an 80 nm bandwidth, is likely to produce the highest signal to noise ratio achievable.

Most detectors also have an adjustable electronic time constant. This should be set to just pass the fastest peak without significant band broadening. Using less electronic dampening may not interfere with accurate integration but will make the chromatogram appear noisier than necessary.

Spectra

This is a topic that is complex enough to require a book of its own (see, *e.g.* refs. 19, 20). In this space only general points will be made. UV spectra are not information rich, and so by themselves are not useful for identification. However, spectra of unknown peaks can be compared with library spectra for confirmation purposes (the peak has the right retention AND the right spectrum).

Spectra taken at several points within a peak can be compared. If the spectra are identical, there is an increased chance that the peak represents a single compound. If there are differences in the spectra, it indicates co-elution of several components. Of course compounds with similar spectra, or that exactly co-elute cannot be differentiated.

There appears to be little difference in the spectra of most solutes in SFs, compared with the same solutes in normal liquids. Some bands appear sharper, more like gas phase spectra when collected from SFs, but no systematic studies have been made. At this writing it is still advisable to generate one's own library of spectra by injecting solutions of pure standards into the mobile phase of interest, rather than comparing SFC spectra with LC libraries.

It is becoming common practice in LC to present three dimensional plots of time–wavelength–absorbance. Similar capability is also available in SFC.

GC Detectors

Many of the detectors used in GC can also be used in SFC. These gas phase detectors generally cannot tolerate the full expanded flow from packed columns, which, for a 4.6 mm i.d. column typically exceeds 1000 ml min^{-1}. To use both electronic BPRs and GC detectors, the column outlet flow is split. Postcolumn splits cause fewer reproducibility problems than split injections. With split injections, both a phase transition and a split occur in a short time frame. With a postcolumn split, the fluid is homogeneous at the split point.

Interfacing GC Detectors to Packed Columns

A standard 'zero dead volume' tee with 1/16th inch fittings allow the metal tubing used with packed columns to be interfaced to fused silica restrictors. One interface that works well consists of a sheath of 1/16th inch Peek (or other elastomeric material) tubing *ca.* 2 cm long slipped over the open end of the restrictor. After passing through the sheath, the end of the restrictor should be cut off, and then pulled back until flush. The inner diameter of the sheath should be very near the outer diameter of the restrictor (Figure 2.3).

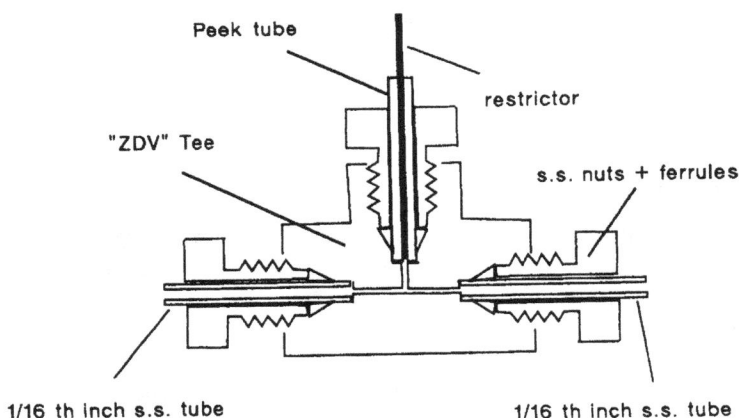

Figure 2.3 *Scheme for interfacing GC detectors to packed columns. A fraction of the total flow is diverted through a zero dead volume tee into a fixed restrictor. The sheath of Peek tubing allows the different tubing sizes and materials to be connected*

The sheath is mounted in the tee just like any other piece of 1/16th inch tubing using a metal nut and ferrule. Care should be taken not to crush the fused silica when the nut is tightened.

'GC' Detector Heated Zone Designs

Standard GC detectors can often cause serious problems in SFC without the operator recognizing the source of the problem. GC detectors have long heated zones. In SFC, high temperature causes the MP to expand and the fluid density and the solvent power of the MP can drop precipitously. Heavy, low volatility solutes can drop out of solution before the end of the restrictor. In the worst cases, such materials never elute.

Frit restrictors are the most susceptible to such problems because the fluid pressure is dissipated over the full length of the frit. These problems can be generally avoided by using a short heated zone[21], and in the more extreme cases avoiding frit restrictors. Several manufacturers have redesigned the base of some GC-like detectors to avoid pre-heating the fluid.

The Flame Ionization Detector (FID)

The FID is the most important detector for capillary SFC but is less important using packed columns. Packed columns retain solutes much more strongly than capillaries, and so more often require the use of modifiers. With most modifiers, the FID becomes useless owing to a very high, noisy background.

The real power of packed columns only comes into play when more polar modifiers are required. Since binary fluids dramatically extend the range of solutes that can be eluted, other detectors that can tolerate modifiers become more important.

The FID detector can be scaled up to allow high flow rates. In some commercial designs, at least several hundred ml min^{-1} of expanded flow, the optimum from a 2 mm i.d. column can be accommodated. To accomplish this, the jet diameter needs to be increased.

Other GC Detectors

It is not particularly helpful at this stage to produce a detailed description of all the GC detectors that have been applied to SFC. Instead, references to the literature are provided.

Among the most commonly used GC detectors in SFC, after the FID, are the electron capture[22,23] (ECD) and Nitrogen–Phosphorus[24-26] (NPD) detectors. The photoionization detector[27] (PID) has many of the characteristics of the FID. The sulfur chemiluminescence detector[28] (SCD) and the flame photometric detector (FPD)[29,30] have also been used successfully.

8 Columns

The columns used for SFC are the same as in LC. The mainstay are silica based with small, totally porous particles. The most typical use spherical 5 μm particles with 60–300 Å pores, 150–500 m^2 g^{-1} surface area, packed in stainless steel tubes 10–25 cm long, 1–6 mm i.d..

SFC is a normal phase technique, requiring polar SPs. Bare silica is widely used. The most appropriate bonded phases are of the more polar varieties, including: cyanopropyl, aminopropyl, diol, and ion exchange columns. The least widely applicable are reversed phase columns like octyl (C_8, MOS) or octadecyl (C_{18}, or ODS). These lower polarity phases tend to be most appropriate for use with hydrocarbons or compounds with long hydrocarbon tails. Polar compounds tend to tail on such phases.

Several manufacturers coat silica with a deactivating layer before applying a bonded phase. While this does produce modest improvements in peak shapes, it does not appear to be a general solution to tailing problems. With modifiers, performance appears to be little different from that of standard silica columns.

Some polymer based particles work well in SFC. However, such packing materials tend to be made of larger particles with lower inherent efficiency. Peak shapes tend to be little different on polymer or silica based columns. Polymer based packings do allow elution of more polar compounds with less polar mobile phases, but the differences are not dramatic.

One significant trend in LC has been the movement toward cartridge columns to minimize cost. The end fittings in some cartridges use Teflon seats. Although such cartridges appear to be as well packed as standard columns, high temperature can cause a problem. At high pressure and temperature above 60 °C, the Teflon can flow, causing a leak at either end of the column. In the extreme, the packing can vent into the column oven. While this is annoying and expensive, it is also a health hazard if the particles are inhaled. In SFC, temperature is much more likely to be elevated or even programmed. For temperatures of 60 °C or higher cartridge columns should not be used.

9 References

1. T.A. Berger and J.F. Deye, *Anal. Chem.*, 1990, **61**, 1181.
2. K. Anton, N.Pericles, S.M. Fields, W.H. Widmar, *Chromatographia*, 1988, **26**, 224.
3. S.T. Sie and G.W.A. Rijnders, *Anal. Chim. Acta*, 1967, **38**, 31
4. R.E. Jentoft and T.H. Gouw, *J. Chromatogr. Sci.*, 1970, **8**, 138
5. U. van Wasen and G.M. Schneider, *Chromatographia*, 1975, **8**, 274
6. M.Novotny, W. Bertsch, and A. Zlatkis, *J. Chromatogr.*, 1971, **61**, 17.
7. B. Wenclawiak, *Fresenius' Z. Anal. Chem.*, 1986, **323**, 492.
8. Y. Hirata, F. Nakata, and S. Murata, *Chromatographia*, 1987, **23**. 663.
9. B.W. Wenclawiak, *Fresenius' Z. Anal. Chem.*, 1988, **330**, 218.
10. D.W. Later, E.R. Cambell, and B.E. Richter, *J. High Resolut. Chromatogr.*, 1988, **11**, 65.
11. J. Via, L.T.Taylor, and F.K. Schweighardt, *Anal. Chem.*, 1994, **66**, 1459.
12. F.K. Schweighardt and P.M. Mathias, *J. Chromatogr. Sci.*, 1993, **31**, 207.
13. C.R. Yonker and R.D. Smith, *J. Chromatogr.*, 1986, **361**, 25.
14. C.R. Yonker, D.G. McMinn, B.W. Wright, and R.D. Smith, *J. Chromatogr.*, 1988, **396**, 19.
15. T.A. Berger, *Anal. Chem.*, 1989, **61**, 356.
16. K.E. Markides, S.M. Fields, and M.L. Lee, *J. Chromatogr. Sci.*, 1986, **24**, 254.
17. E.J. Guthrie and H.E. Schwartz, *J. Chromatogr. Sci.*, 1986, **24**, 236.
18. R.M. Smith and D.A. Briggs, *J. Chromatogr. (A)*, 1994, **670**, 161.
19. 'Diode Array Detection in HPLC', ed. L. Huber and S. George, Chromatographic Science Series, Vol. 62, Marcel Dekker, Inc., New York, 1993.
20. L. Huber, 'Applications of Diode Array Detection in HPLC', Hewlett-Packard Publication No. 12–5953–2330, Hewlett Packard, 1989.
21. T.A. Berger and C. Toney, *J. Chromatogr.*, 1989, **465**, 157.
22. S. Kennedy and R.J. Wall, *LC–GC*, 1988, **6**, 930.

23. H-C. Karen Chang and L.T. Taylor, *J. Chromatogr. Sci.*, 1990, **28**, 29.
24. P.A. David and M. Novotny, *J. Chromatogr.*,1988, **452**, 623.
25. L. Mathiasson, J.A. Jonsson, and L. Karlsson, *J. Chromatogr.*, 1989, **467**, 61.
26. J.G.J. Mol, B.N. Zegers, H. Lindeman, and U.A.Th. Brinkman, *Chromatographia*, 1991, **32**, 203.
27. P.G. Sim, C.M. Elson, and M.A. Quilloin, *J. Chromatogr.*, 1988, **445**, 239.
28. A.L. Howard and L.T. Taylor, *J. High Resolut. Chromatogr.*, 1991, **14**, 785.
29. S.V. Olesik, L.A. Pekay, and E.A. Paliwoda, *Anal. Chem.*, 1989, **61**, 58.
30. L.A. Pekay and S.V. Olesik, *Anal. Chem.*, 1989, **61**, 2616.

CHAPTER 3

Physical Chemistry of Mobile Phases Used in Packed Column SFC

1 Introduction

This chapter contains a discussion of the physical chemistry of fluids used as the mobile phase (MP) in packed column SFC. The topics include diffusivity, viscosity, solvent strength, density, phase behaviour, *etc*. This chapter is not about choosing mobile phases for chromatography, which is covered in Chapers 5–7. This material can be thought of as 'deep background' and is not necessary to develop analytical methods. It is included to provide insight leading to the chapters on systematic method development that follow.

The properties of SFs have been widely misunderstood by chromatographers. This is particularly true with binary or tertiary fluids. Phase equilibria, density, solvent strength, and adsorption effects have created considerable confusion. Fortunately, many of these phenomena are becoming better understood. In this chapter, some of the historical misconceptions about SFs are outlined, followed by discussions of present understanding.

2 Why SFC? Mobile Phase Considerations

The mobile phase (MP) characteristics of interest in packed column SFC are: high solute diffusion coefficients (D_M), and low MP viscosity, compared with the solvents used in LC, plus retention control based on physical parameters like temperature, pressure, and MP composition. With higher D_M, the optimum velocity (μ_{opt}) through the column is higher than in LC, so the same number of peaks can be separated in less time. Low viscosity means that μ can be dramatically increased to take advantage of higher D_M while at the same time lowering pressure drops (dP). High D_M and low viscosity mean that very long, efficient columns can be constructed.[1] Finally, retention control based on adjustment of physical parameters is very appealing since such control greatly improves ease of use.

Diffusion Coefficients

The speed and efficiency of a chromatographic separation is determined by physical dimensions, such as column length (L) and particle diameter (d_p), by the MP flow rate, and by the D_M of the solutes in the MP at the pressure and temperature used. A broader discussion of these topics can be found in Chapter 1. The relationships between dimensions, phases, and efficiency (N = plates = L/H = length/plate height) are formalized with either the Golay Equation,[2], describing capillary columns, or the older Van Deemter Equation[3] ($H = 0.33A + B/\mu + C\mu$, where A, B, and C are constants, μ = linear velocity), now most generally used to describe packed columns. In both equations, a relationship at the heart of the μ vs. H trade-off is the ratio D_M/μ.

The interplay between carrier flow rate and solute diffusion affects column efficiency, N. Efficiency is related to both axial and radial diffusion. In capillaries at very low μ, diffusion along the axis of the column ($B = 2D_M/\mu$) excessively broadens solute bands. In packed columns, the spaces between the particles act something like a variable cross section tube. Diffusion along the axis of the tube broadens bands similarly to the axial diffusion in capillaries. Increasing μ of the MP decreases the time the solutes have to diffuse, decreasing axial band broadening. At very high μ, axial diffusion is negligible but radial diffusion becomes a problem.

The fluid touching either the wall of a capillary or the surface of particles is stagnant even when there is a net flow of fluid through a column. As flow increases, the difference in μ from the column surface to the centre of the flow path becomes large. In capillaries, this is often represented as a parabolic flow profile in a pipe. A similar profile develops in packed columns, except that the velocity gradient develops in the spaces between particles. Molecules can diffuse **radially** (perpendicular to the direction of flow) into fluid travelling either faster or slower than its previous location. Diffusion across radial velocity gradients produces a form of band broadening ($C\mu/D_M$) called 'resistance to mass transfer'. This form of band broadening is linearly proportional to μ. The higher the flow rate, the worse the band broadening due to this mechanism.

In packed columns there are additional band broadening terms (A) which arise from the tortuous path a fluid must follow to pass through a bed of particles and diffusion into the stagnant mobile phase inside the pores of the support.

At one intermediate flow rate (the optimum linear velocity, μ_{opt}), the sum ($A + B/\mu + C\mu$) of axial diffusion, and resistance to mass transfer (radial diffusion) (and other) is at a minimum (producing H_{min}), producing the maximum column efficiency (the least total band broadening).

The D_Ms of solutes in MPs determines the relative speeds of different chromatographic techniques, as discussed in Chapter 1. A number of workers have measured [4-7] D_M values of solutes in SFs using a pseudo-chromatographic method (no retention).[8-13] Representative values in Table 3.1, show that D_Ms in SFs[14] are as much as an order of magnitude higher (more appealing) than in

Table 3.1

Aqueous binary D_M,[16] cm^2 s^{-1} (25 °C)		D_M[14] in CO_2, cm^2 s^{-1}
Citric acid (0.1 M)	0.661×10^{-5}	Naphthalene: 0.968×10^{-4}
Glycine (dil.)	1.064×10^{-5}	(25.2 °C, 171 bar, 0.90 g cm^{-3})
Glucose (0.39%)	0.673×10^{-5}	Naphthalene: 1.20×10^{-4}
α–alanine (0.32%)	0.910×10^{-5}	(54.9 °C, 239 bar, 0.80 g cm^{-3})
p-Aminobenzoic acid	0.843×10^{-5}	Naphthalene: 1.32×10^{-4}
		(47 °C, 142 bar, 0.705 g cm^{-3})
C_2H_6 in hexane,* 30 °C	6.00×10^{-5}	

* Non-aqueous liquid.

Table 3.2 *Viscosities* $\times$ 10^4 poise

Temperature °C	CO_2[14]			H_2O[16]	CH_3CN[16]
		Pressure, bar			
	100	200	400		
20	8.3	10.3	13.0	100.2	35.3
40	4.9	8.0	10.8	54.7	29.2
60	2.4	6.1	9.1	46.7	24.5

normal liquids. Solute D_M values in CO_2 appear to be nearly linear functions of density at constant T[6], and linear functions of $1/T$ at constant density.[6,7]

Even at high densities (*i.e.* 0.9 g cm^{-3}), D_Ms in CO_2 are as much as 10 times higher than D_Ms in normal liquids, like water. Therefore, μ_{opt} in packed column SFC is up to 10 times higher than in reversed phase LC. However, D_M values in SFC are as much as 10^4 times lower than in GC.

Viscosity

Supercritical CO_2[14–16] has lower viscosity than normal liquids, as shown in Table 3.2. The viscosity of CO_2 can be as little as 1/20th that of water at the same temperature. The worst case viscosity of CO_2 is 1/5th the viscosity of water at the same temperature.

Column dP is directly proportional to viscosity. The data in Table 3.2 indicate that, at the same flow rate, SFC should produce 1/5th to 1/20th the dP of reversed phase LC.

Viscosity changes[17] with temperature and pressure, as shown in Figure 3.1. The practical operating region of SFC using pure CO_2, uses reduced temperatures between 1 and 2 (reduced parameters are the actual value divided by the critical parameter, $T_r = T/T_c$, $T_r = 1–2 = 31–330$ °C, or 304–603 K for CO_2), and reduced pressures ($P_r = P/P_c$) between *ca.* 1 and 7 (71–500 atm for CO_2).

From Figure 3.1, the viscosity of CO_2 increases by a factor of 2 when P increases from 71 bar to 497 bar (at T_c). Over the entire useful range of

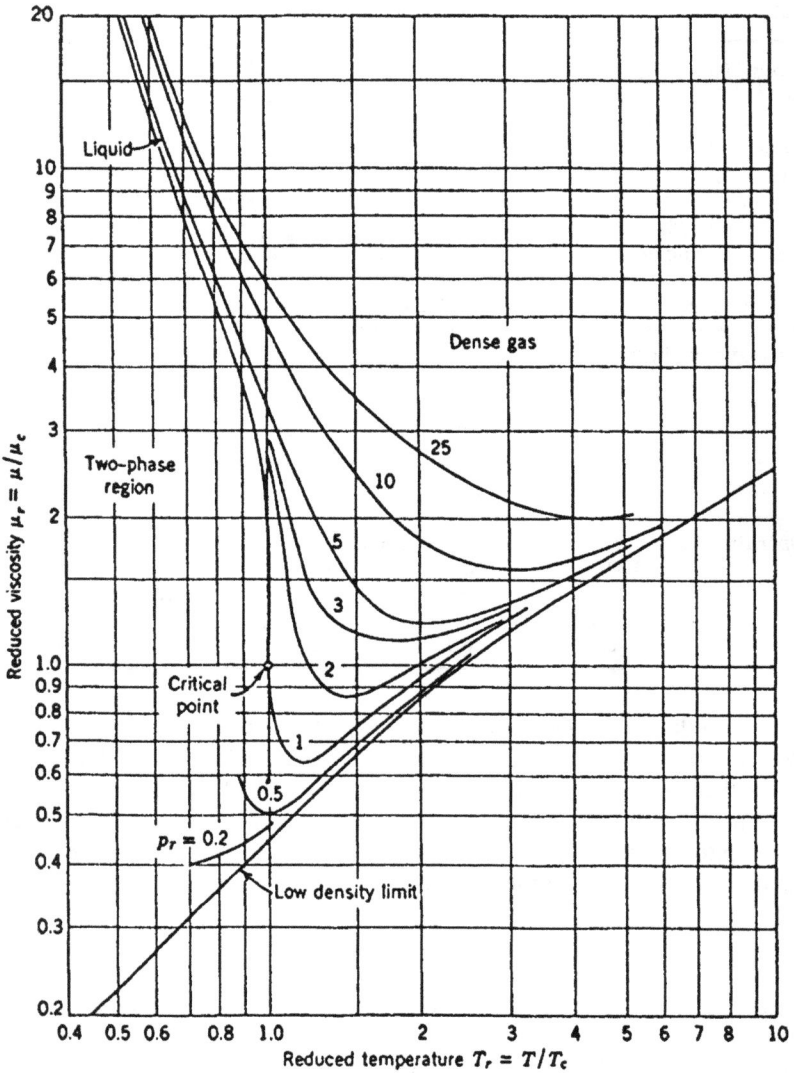

Figure 3.1 *Viscosity as a function of temperature and pressure. The parameters are presented as reduced variables, normalized to the values at the critical point, so the graph fits most gases such as CO_2. The graph indicates that the viscosities of SFs are much more like those of a gas than of a liquid. For SFC using CO_2 the useful range is $P_r = 1–8$, and $T_r = 1–2$*
(Reproduced by permission from ref. 17)

temperature and pressure, viscosity changes by less that a factor of 3. This means that dP across columns also change by no more than a factor of 3 over the entire useful range of SFC conditions.

The low viscosity of dense CO_2 has been widely verified by chromatographic experiments. Although flow rates tend to be two to three times higher in SFC, dP is typically 1/2 to 1/3rd of those in LC. Thus, the MP viscosity in SFC is of the order of 1/4th to 1/9th the viscosity typical in reversed phase LC. For example, a micropacked LC column, 1.95 m long, produced 226 000 plates in 33 minutes with $dP = 360$ bar.[18] A packed SFC column, 2.2 m long, with the same particle size, produced $> 220 000$ plates in 8.8 minutes with $dP = 160$ bar.[1] Both columns produced their theoretically maximum N, and so were operated at μ_{opt}. Thus, the SFC column produced only 44% of the dP, with $\mu = 3.75$ times higher, which indicates that the viscosity was 8.5 times lower in the SFC example than in the LC example.

3 What's in a Name? or 'There is NO Such Thing as SFC!!'

In the previous sections, it was shown that SFC exhibited higher D_M and lower viscosity than observed in LC. The following sections show that SFC is a transition technique between GC and LC. Its main advantages over the other techniques are practical, not fundamental.

Diffusion Coefficients in Near Critical Fluids

Some people still assume that D_M drops when the definition of a fluid changes from supercritical to liquid. However, even in 1983, Lauer and co-workers[7] showed that nothing dramatic happened to the D_M of PAHs when the temperature of CO_2 (and N_2O) carrier was changed from greater to less than T_c while maintaining constant density. Plots of log D_M *vs.* $1/T$ yielded straight lines with no discontinuity at T_c.

Below the T_c, but above the P_c, a fluid is defined as a liquid, although 'near critical' is somewhat descriptive. If the D_M dropped when the fluid changed from being supercritical to a liquid, then N at constant mass flow would also drop during the transition (D_M/μ would change). With constant mass flow, N does not change when the fluid is changed from sub- to super-critical (or *vice versa*). This is demonstrated[19] in Figure 3.2, using the separation of PTH derivatives of three amino acids on a cyano column, using a binary MP. Since N does not change, the ratio D_M/μ must be nearly constant.

This also means that if any position in the column produces optimum local N, then all positions in the column simultaneously produce near optimum local N. Pressure drops may cause retention and velocity gradients but do not generally affect efficiency.[1,20]

Viscosity of Near Critical Fluids

Reinspection of Figure 3.1 shows no discontinuous jump in viscosity when a fluid approaches the liquid state. The transition is smooth and continuous. In

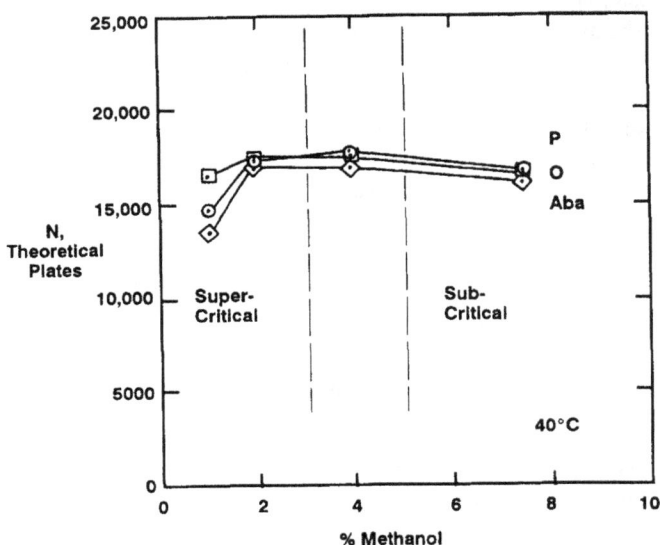

Figure 3.2 *Plot of efficiency vs. modifier concentration. At* 40 °C, *and* 200 bar *all compositions below approximately 4% MeOH/CO$_2$ are supercritical. At all compositions above ca. 4% MeOH/CO$_2$, the fluid is subcritical. Changing composition from supercritical to subcritical at constant flow, pressure, and temperature produces no change in efficiency.*
(Reproduced by permission from ref. 19).

terms of viscosity there is no apparent border between LC and SFC. In fact, there has been a recent interest in 'high temperature' LC, where the column is heated to unusually high temperature (for LC). As can be seen in Table 3.2, even modest increases in temperature result in fairly dramatic drops in the viscosity of water. Further heating results in even lower viscosity but also increases the solvent vapour pressure (P_v). To avoid boiling the solvent, workers have placed a length of narrow bore tubing on the detector outlet to provide back pressure. Such an instrumental arrangement is nearly identical to that of packed column SFC, except for the inferior means of controlling outlet pressure (p_o). It also does not allow for independent flow and pressure changes. It should further be clear from Table 3.2 that the viscosity in hot water remains substantially higher than viscosity in super-critical CO$_2$.

Chiral separation of enantiomers by packed column SFC is becoming commonplace (see Chapter 10). In many instances, the best resolution (R_s) is actually obtained using $T < T_c$ (using liquid CO$_2$). However, plots of N, or R_s vs. T show no discontinuity, and the dP remains unchanged when $T > T_c$ or $T < T_c$. Both the high μ_{opt} and the low dP usually associated with SFC are retained with near critical liquids. As the conditions are changed further away from the critical region, the fluid characteristics more nearly resemble normal liquids.

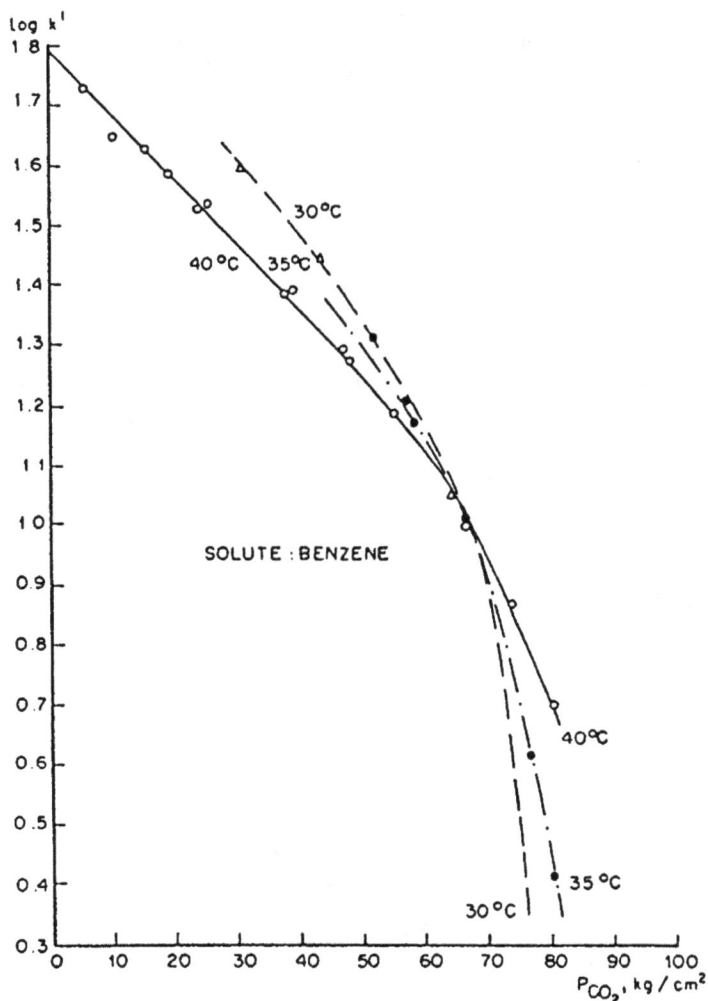

Figure 3.3 *Some GC cárrier gases cause a decrease in the retention of solutes as pressure is increased*
(Reproduced by permission from ref. 24)

Effect of Physical Parameters on Retention

The Effect of Pressure in GC

Most chromatographers would associate changes in retention due to changes in pressure exclusively with SFC. It is a general tenet in both LC and GC that pressure has little effect on k'. Surprisingly, this is not always true.

Some GC carrier gases interact with (solvate?) analytes even at modest temperature and low pressure (*i.e.* 50 °C, 1 atm). Some GC carrier gases[21-25] consistently produce lower k' values than other gases. Further, increasing the inlet pressure decreases k'. Some such results[24] are shown in Figure 3.3. The conditions were: $T > Tc$ for CO_2, but $P \ll P_c$. Thus, the carrier was a gas although, in the extreme, conditions could be described as 'near critical'. These conditions produced interactions most would (erroneously) consider to be exclusive characteristics of SFs.

Transition from SFC to LC

Lauer and co-workers[7] showed that the **retention** of various PAHs was unaffected by transitions from sub- to super-critical conditions. Plots of log k' *vs* $1/T$ yielded straight lines with no discontinuity at T_c, using either CO_2 or N_2O as carrier (transition from SFC to LC).

More recently, no noticeable deviation in retention has been observed when temperature was changed from $T > T_c$ to $T < T_c$ for numerous families of more polar solutes, using modified MPs. An example, in Figure 3.4, plots t_R *vs.* T for carbamate pesticides.[26] For the binary MP, 40 °C < T_c < 50 °C, yet there are no discontinuities or major slope changes over this temperature range.

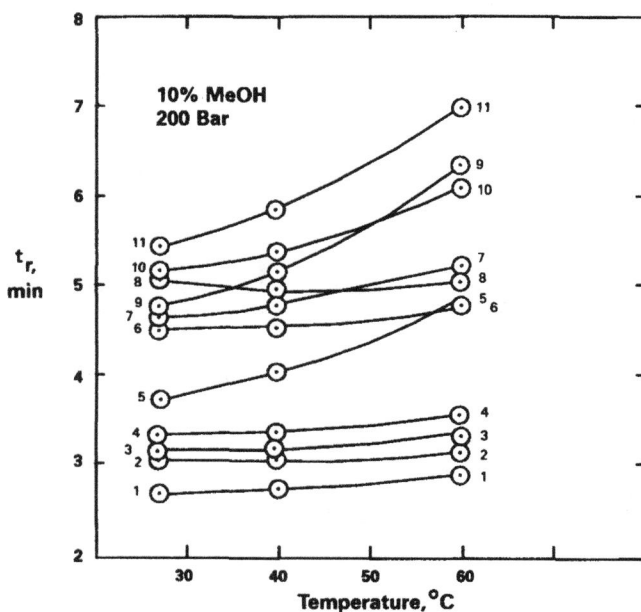

Figure 3.4   *The retention of carbamates as a function of temperature. On the left conditions are subcritical. On the right, conditions are supercritical. Note that there is no obvious transition from one defined state to the other.*
(Reproduced by permission from ref. 26)

Summary

The name SFC implies that there is some characteristic of SFs which uniquely contributes to the chromatography observed, and differentiates SFC from other forms of chromatography. However, there is no evidence that SFs possess different solvent characteristics, or physical attributes compared with subcritical gases or liquids at temperature and/or pressure near the supercritical region. There is **no physical or chemical characteristic** important to the separations that is unique to SFs. Thus, the technique is misnamed! The earlier names high pressure gas chromatography and dense gas chromatography are also clearly inadequate. The most valued characteristics of SFC belong to compressible fluids that 'solvate'. This suggests that a better name would be Compressible Solvent Chromatography (CSC?). In any event, it is probably much too late to change the name, again.

In this book, SFC includes the use of both supercritical fluids and **near critical** (subcritical) liquids and gases. The desirable characteristics all these fluids share is reasonable solvent strength with lower viscosity and higher D_Ms than are usually associated with the liquids used as LC mobile phases. Instrumentally, all the fluids are compressible, and all require elevated p_o.

4 Solvent Strength

A desirable feature of SFC is the ability to elute more polar solutes by raising the pressure. Obviously, changing the pressure is far easier than changing the composition of a fluid, such as in gradient elution LC. A quote attributed to Hannay in 1880[27] by Giddings[28] indicates that this feature has been appreciated for a very long time:

The liquid condition of fluids has very little to do with their solvent power, but only indicates molecular closeness. Should this closeness be attained by external pressure instead of internal attraction, the result is that the same or even greater solvent power is obtained...

Unfortunately, many users have been overly optimistic in interpreting this statement. Liquids condense because interactions between molecules are more intense than the thermal energy of the system. Compressed gases decompress when the pressure is removed because the intensity of intermolecular interactions is inadequate to hold the molecules together. Simply pushing molecules together makes possible more **extensive** interactions but does not force more **intense** interactions.

In the first 15 years of SFC, the solvent strength of pure fluids was grossly overestimated, in part owing to optimistic interpretation of Hannay. In the mid-1960s, Giddings[28] predicted that retention should be related to the Hildebrand solubility parameter (δ) of the solvent. With no empirical interaction parameters available, estimated values produced a δ for dense CO_2 similar to that of isopropyl alcohol.[28] If correct, pressure programming CO_2 would change solvent strength from below hydrocarbons up to the solvent strength of alcohols. This is a very wide range of solvent strength. A technique producing such dramatic changes in solvent strength in response to a simple physical parameter would severely challenge LC in terms of simplicity and cost. With the benefit of hindsight, such large changes in estimated solvent strength *vs* pressure were clearly overly optimistic. Even very dense CO_2 acts more like pentane or hexane than a more polar solvent.

Solvatochromic Dyes

Solvatochromic dyes change colour when the polarity of their solvation sphere changes[29]. Such dyes have been used extensively in both LC and SFC[29-35] to measure MP solvent strength. Each dye may measure a specific type of interaction (*e.g.* dipole–dipole or hydrogen bonding).

Each dye produces a linear solvent strength scale,[29] obtained from spectral maxima, that can be related to more common solvent strength scales like dielectric constant or the P' scale[36] from LC. Examples using the solvatochromic dye Nile Red dissolved in pure liquids are presented in Figures 3.5 and 3.6. Plotting the Nile Red scale *vs.* other solvent strength scales produces straight lines, indicating that both are measuring similar attributes.

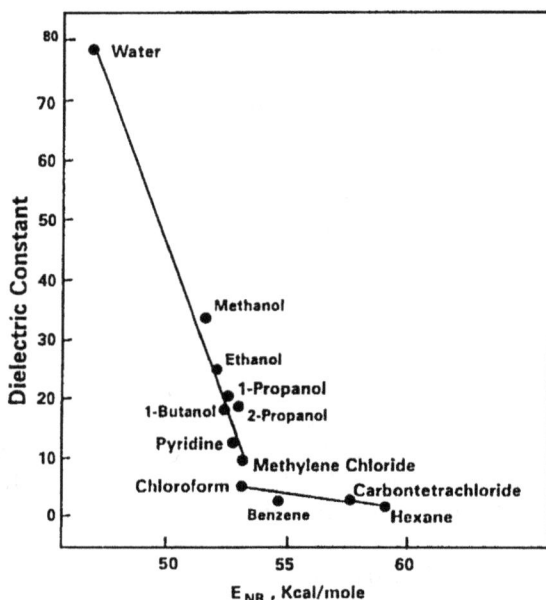

Figure 3.5   *A comparison of dielectric constants and the Nile Red energy scale.[30] The latter was obtained by dissolving the solvatochromic dye in the various solvents and observing the shift in the spectrum. Nile Red appears to produce a measurement consistent with dielectric constants*

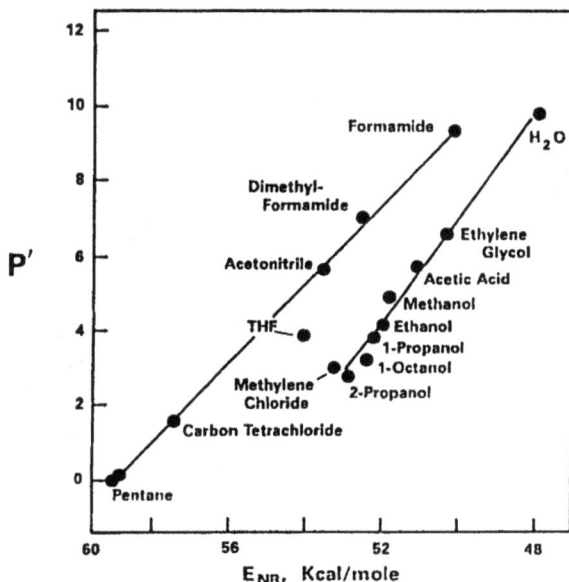

Figure 3.6 *A comparison of the P' solvent strength scale,[36] widely used in LC, with the Nile Red energy scale.[30] Note that hydrogen bonding, acidic solvents fall on a unique line*

Pure CO_2 produces an apparent solvent strength similar to hexane[30-35], using many different solvatochromic dyes. Results using Nile Red (E_{NR}) and a dye named E_{t30} are compared[31] in Figure 3.7. Solvent strength and hydrogen bonding increase from upper left to lower right in the Figure. The upper line in Figure 3.7 indicates the solvent strength of $MeOH/CO_2$ mixtures. Note that low concentrations of MeOH cause large increases in solvent strength. The E_{t30} scale on the *x*-axis is particularly sensitive to hydrogen bonding and exaggerates this aspect of MeOH relative to less hydrogen bonding modifiers. The Nile Red scale on the *y*-axis is less dependent on hydrogen bonding.

The non-linearity between modifier concentration and solvent strength can be seen more clearly in Figure 3.8. The longer vertical lines in the figure indicate the solvent strength of pure fluids: CO_2 on the left, MeOH on the right. Shorter vertical lines indicate the solvent strength of mixtures. The numbers above the shorter lines indicate the concentration of the fluid on the right end of the line in the fluid on the left end of the line (*e.g.* MeOH in CO_2). From the Figure, it is evident that the first small additions of MeOH cause large increases in solvent strength. This is apparently due to 'clustering'[37-49] of MeOH molecules. Within a cluster, modifier concentration and density are higher, creating micro-environments of locally high polarity compared with the nominal bulk composition and density.

Polar solutes interact primarily with the polar clusters of modifier whereas non-polar solutes tend to interact with the non-polar portion of the bulk MP.

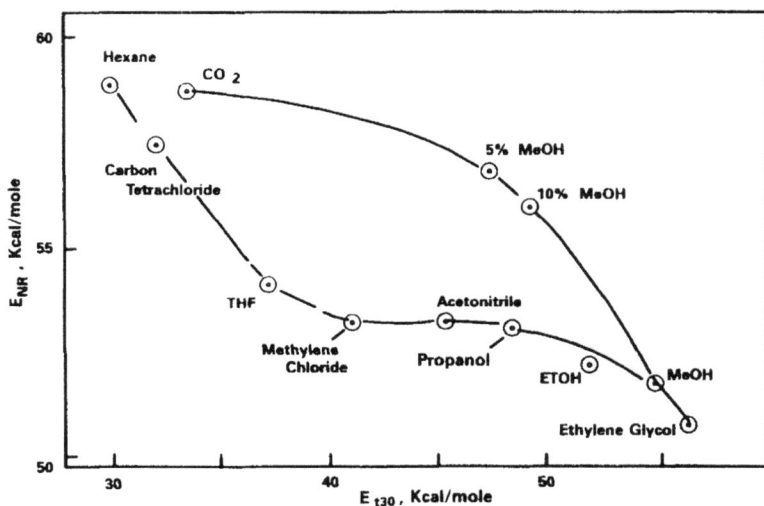

Figure 3.7   *Relative solvent strengths of various solvents on two solvatochromic energy scales, E_{NR} vs. E_{t30}. Lower line: normal liquids. Upper line: MeOH/CO$_2$* (Reproduced by permission from ref. 31)

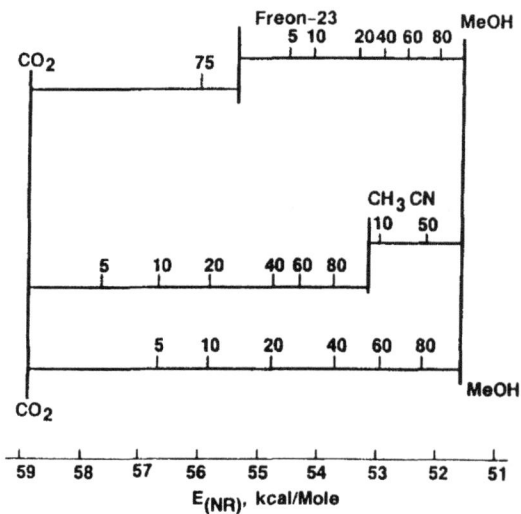

Figure 3.8   *The solvent strength of various supercritical mixtures. Each line indicates a unique binary pair. The numbers indicate the percentage of the fluid listed on the right end of each line present in the fluid listed on the left of each line. The figure is meant to indicate the non-linearity in solvent strength with percentage composition*

Johnston[50] stated that: 'a polar co-solvent (modifier) may increase markedly the solubility of a polar solute, but may not affect that of a nonpolar solute'.

Figure 3.8 was purposely drawn to resemble the elution strength scales developed by Snyder,[36] and Saunders[51] for liquid solid chromatography (LSC), also called adsorption chromatography. This 'similarity', however, may be deceptive. In LSC, the non-linearity is attributed to a surface effect (adsorption), whereas in Figure 3.8 the non-linearity is in the MP solvent strength and independent of surface effects.

On the other hand, solvatochromic dyes have indicated that the solvent strengths of the MP mixtures used in LSC are non-linear, and account for much of the non-linearity in retention. This suggests that the normal explanation of retention in LSC is incomplete or may even be incorrect. It is likely that SFC and LSC are similar, and neither is actually 'adsorption' chromatography.

5 Elution Strength

The terms 'solvent strength' and 'elution strength' are often used interchangeably but, in fact, are quite different. Solvent strength is a measure of the interactions between solutes and the MP. Elution strength indicates S–MP, S–SP, and MP–SP interactions on chromatographic retention.

As mention in Section 4, Giddings used calculated Hildebrand solubility parameters (δ). He then equated δ to elution strength (an 'elutropic' series). This set a precedent in equating solvent strength and elution strength. Further, δ of CO_2 was grossly overestimated, to be similar to that of liquid isopropyl alcohol. Modern values place δ for CO_2 at 35 °C, 200 bar at $\delta = 6.5$ cal cm^{-3}, similar to liquid isopentane. Giddings's estimated value[28] was 10.7 cal cm^{-3}.

Giddings elutropic series (actually a solubility parameter scale) is reproduced on the left of Figure 3.9. A solvent strength scale from the solvatochromic dye Nile Red is presented on the right. On the Nile Red scale (and on several other scales based on different dyes AND on modern δ scales), CO_2 is similar to pentane or hexane in solvent strength. Note that 40% MeOH in CO_2 is dramatically more polar than pure CO_2 but still less polar than liquid isopropyl alcohol.

The publication of Giddings's elutropic scale had a dramatic effect on the direction of research and opinion in SFC which, surprisingly, continues today. It implied that pressure programming could produce a very wide range of solvent strength, stretching from below hexane up to isopropyl alcohol. If true, the need for polar modifiers would be minimized, since mixtures of isopropyl alcohol in CO_2 would be no more polar than pure CO_2. Polar molecules could be eluted with pure fluids like CO_2.

One of the causes of the controversy between packed and capillary columns that raged during the early 1980s was this assumption that CO_2 was so polar. Since packed columns retained polar compounds much more intensely than capillaries, capillary chromatographers assumed that this extra retention was due to active sites, not the low polarity of the CO_2. Expectations had been set that modifiers would not dramatically change solvent strength, since CO_2 could

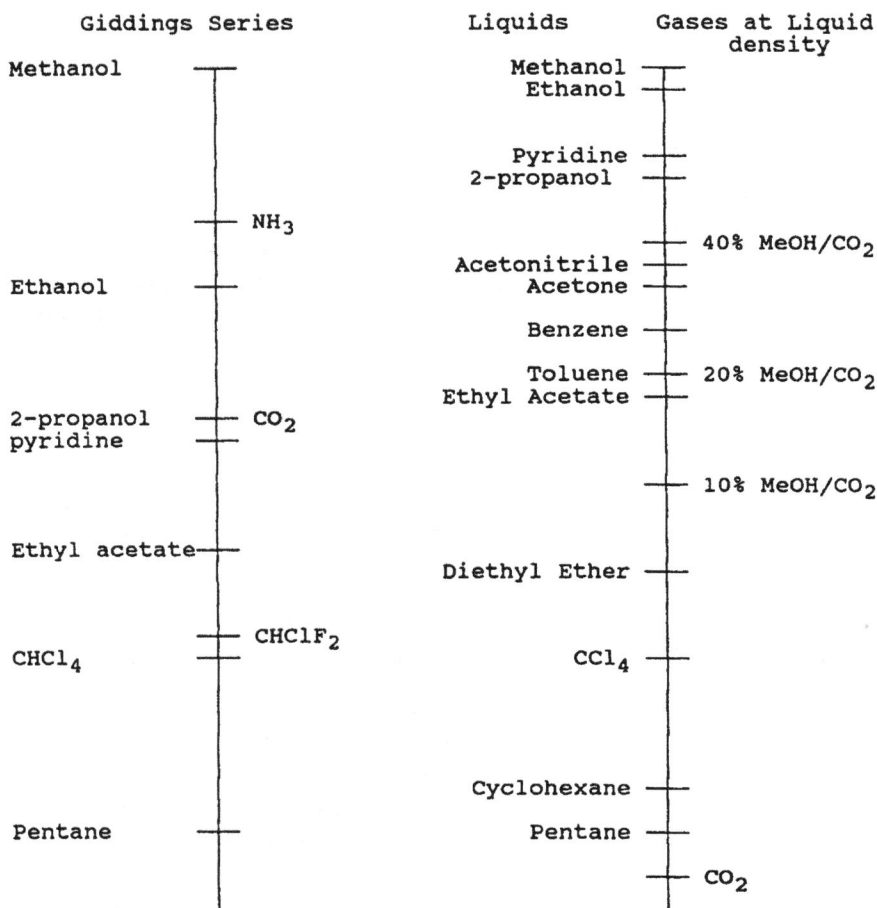

Figure 3.9 *Left: Giddings elutropic series[28] based on calculated Hildebrand solubility parameters, used with permission. Right: Nile Red energy scale[30]*

be as polar as most modifiers. When peaks of moderately polar solutes were shown to tail using pure CO_2, modifiers were often added, but only to cover active sites (deactivate 'bad' silanols) not increase solvent strength.

It was also widely believed that the addition of modifiers had a much more dramatic effect on changes in retention on packed columns than on capillaries (presumably covering a higher density of active sites). However, the only comparisons in the literature show that similar changes[52-54] were observed on both packed and capillary columns when polar modifiers were added.

The Effect of Density on Retention

Chromatographic retention can change by more than three orders of magnitude[55-57] when CO_2 density is changed from < 0.1 to > 0.9 $g\,cm^{-3}$. This

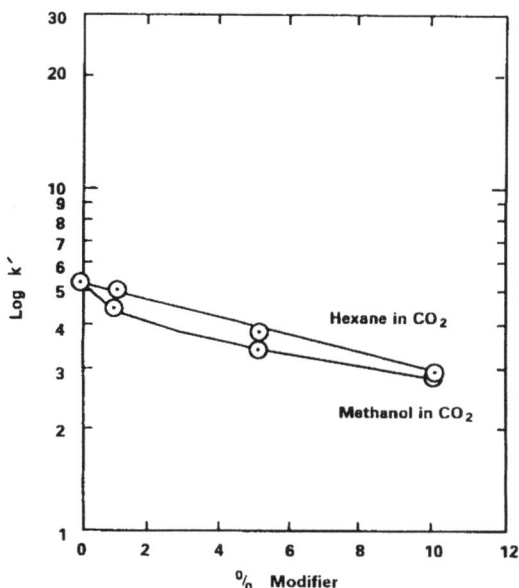

Figure 3.10 *Effect of hexane and methanol on retention of chrysene on a non-polar C_{18} column. Such results were interpreted to mean that modifiers only deactivate the support and do not dramatically change solvent strength*

appears to be a very large change in retention. Such a large change in retention was viewed by many as confirmation of the magnitude of solvent strength changes of CO_2 due to density, suggested by Giddings's elutropic series. Many saw this as proof that pressure programming in SFC spans a very wide range of solvent strength.

Further, the results in Figure 3.10 indicated that even a large concentration of a polar modifier could have little effect on elution strength. The apparent irrelevance of the presence or absence of polar methanol on retention, in Figure 3.10, was the sort of information interpreted as support for Giddings's elutropic series. Polar modifiers did not appear to enhance mobile phase solvent strength. Thus, it was assumed that CO_2 and MeOH must not be very different in solvent strength.

However, such a viewpoint lacks perspective. In normal phase LC, changing solvents from hexane to isopropyl alcohol can change retention by six to ten orders of magnitude![58] Similar large changes in retention are also feasible in reversed phase LC. Obviously, retention changes in LC are far greater than the retention changes possible using density programming in SFC. The effect of density on retention in SFC must be considered modest.

Modifier Effects Change with Column Polarity

Reversed phase C_{18} packed columns and pure CO_2 were typical choices for phases in much of the early work in SFC. This unfortunate choice of phases

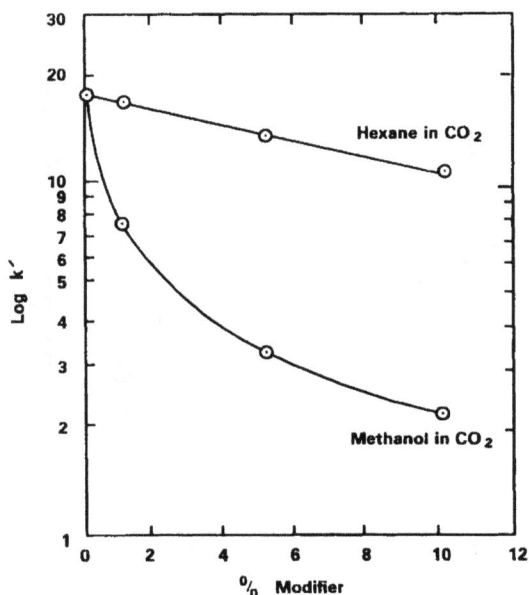

Figure 3.11   *Same as Figure 3.10, except that a polar stationary phase was used. In this case, a sulfonic acid (ion exchange) column showed dramatic differences between methanol and hexane as modifiers*

turns out to have been responsible for generating considerable confusion. On a C_{18} stationary phase, neither a hexane or methanol modifier significantly changes the retention of chrysene, as shown in Figure 3.10. This appeared to confirm Giddings's ranking of solvents. Some workers suggested that even these small changes in retention were not due to solvent strength changes but changes in density (see *e.g.* refs. 34 and 35).

However, replacing the C_{18} column with a more polar stationary phase dramatically changes apparent solvent strength of the **mobile** phase, as can be seen by comparing Figures 3.11 and 3.10. If retention is measured on polar columns, hexane and methanol modifiers perform very differently. On a sulfonic acid (or diol or amino) column, the addition of hexane to CO_2 still has only a minor effect on the retention of chrysene. Using MeOH causes major decreases in retention, as shown in Figure 3.11. Obviously, the more polar SP allows the solvent strength of the MP to be expressed in the apparent elution strength.

If the nature of the SP dramatically impacts the effect of the modifier, elution strength and solvent strength must not be equivalent.

Relative Effect of Modifiers and Density on Retention

Five years ago, modifier addition was assumed only to change density. These assertions were made **in the absence of density measurements**! Equations of state are notoriously inaccurate for calculating densities of mixtures, especially near the critical point. This prevented calculation of the effect of density with any certainty.

To counter these unsubstantiated assertions, the density of MeOH/CO_2 mixtures was measured[59] at several temperature. The retention of moderately polar molecules was then measured as a factor of percentage of MeOH in CO_2 at constant density, and at different densities but constant percentage of modifier. The results clearly show that composition has a much more profound effect on elution strength than density[59] for moderately polar molecules.

More recently, Eckert and co-workers[48] demonstrated similar findings on capillary columns. In ethane/ethanol mixtures, changes in anthracene retention were shown to be completely due to changes in **density** accompanying the addition of the ethanol. On the other hand, a similar experiment using 2-naphthol as the solute showed that retention changed more dramatically and most of the change was due to **modifier** interactions with the solute.

Following Giddings's elutropic series, hexane, which is similar to pentane, should actually dilute the solvent strength of the more polar CO_2. If retention followed Giddings calculated Hildebrand solubility parameters, then the addition of hexane should decrease solubility and increase retention.

Solubility

The addition of hexane to CO_2 increases the solubility of PAH's by up to an order of magnitude.[50] This fact, in itself, supports solvent strength scales listing CO_2 as non-polar and counters Giddings's assumptions. However, retention does not drop dramatically when even 20% hexane is added to the CO_2. If elution strength was directly related to solubility or solvent strength, the addition of hexane should decrease the retention of chrysene on C_{18}. This may be another case where an order of magnitude change is minor compared with much larger changes common in nature.

Ultra-High Pressures

Giddings also reported that very high pressure (*i.e.* 2000 atm) and ammonia as carrier gas eluted large molecules such as polymers, and polar biological molecules normally only soluble in water. Although no one has been able to reproduce this work there is still an undercurrent of interest. Several workers continue to promote the use of very high pressure which they expect to increase elution strength significantly. However, the assumptions generating the elutropic series in Figure 3.9 are the same assumptions used to propose extreme pressure, and are bogus. CO_2 is not remotely equivalent to isopropyl alcohol on any REAL solvent strength scale. The most dense CO_2 is similar to hexane in solvent strength. Much higher pressure will likely only have a minor effect on density and will not provide a solvent for polar molecules. Only larger molecular weight, but low to modest polarity solutes are likely to be more solvated by

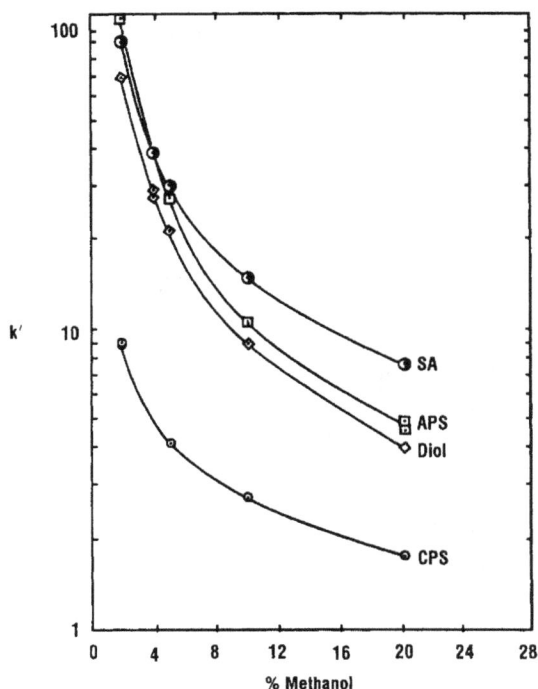

Figure 3.12   *Plot of log k' vs. percentage modifier showing the non-linear relationship for benzylamine on four different stationary phases*
(Reproduced by permission from ref. 31)

more dense fluids. Enhancements of SFC through very high pressure are likely to be modest (and dangerous).

Does Chromatography Occur at 'Infinite Dilution'?

Giddings's elutropic series was published in 1968. Over subsequent years, it became clear that CO_2 was less polar than had been thought. Despite this realization, an alternative justification for retaining the use of pure fluids and density programming evolved. It was reasoned that chromatography is carried out at essentially infinite dilution. Thus, a fluid need not be a very good solvent for a particular solute. A low polarity fluid like CO_2 might then be viable as a MP for even quite polar solutes provided the SPs were also rather non-polar and well deactivated.

It is easy to confuse **extensive** and **intensive** interactions. When the solute has weak interactions with the chosen phases, unplanned or inadvertent side interactions can overwhelm the intended interaction. A few intense interactions between 'active sites' and the solutes will always distort peak shapes or even control retention. In the extreme, a bonded phase can act as a deactivating film not participating in the separation. Uncontrolled or even undefined sites become the effective SP. Chromatography may be possible with very low solubility in the phases but the interaction between each solute molecule and the chromatographic phases are just as intense at low concentrations as at high concentrations. If the intensity of interaction depended on solute concentration, retention would change depending on concentration.

Explaining Non-linear log k' vs. Modifier Concentration Plots

Plots of log k' vs. percentage modifier are non-linear. An example of such a curve is presented in Figure 3.12. In reversed phase LC, log k' vs. percentage modifier often produces nearly straight lines.

Non-linear retention in LC has generally been assumed to indicate an adsorption mechanism, where the retention of the solute is thought to follow the adsorption isotherm of the modifier. In SFC, independent measurements by

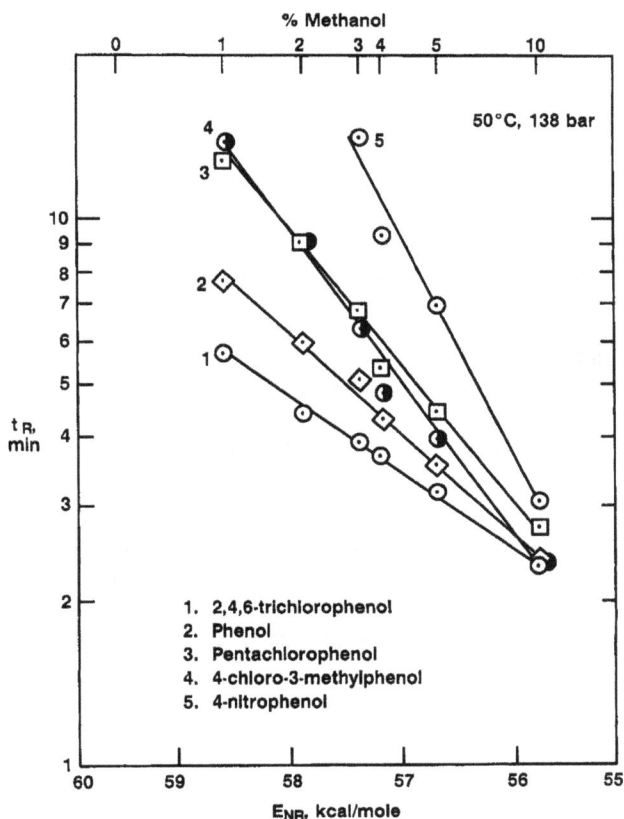

Figure 3.13 *Plot of log k' vs. E_{NR} for phenols. The linear nature of the plot suggests that mobile phase solvent strength is the primary component of changing elution strength*
(Reproduced by permission from ref. 31)

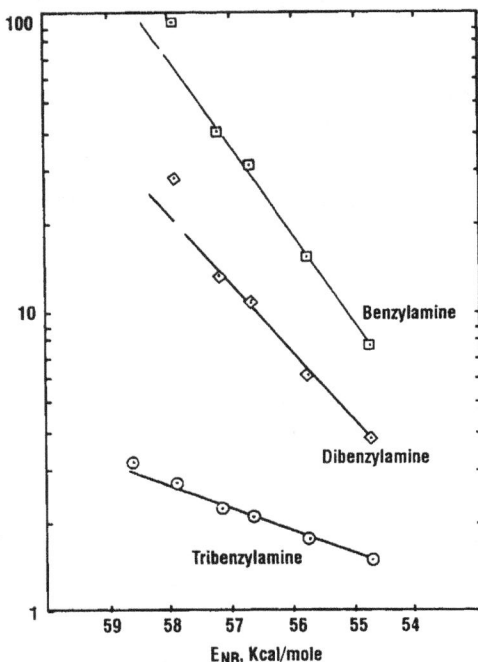

Figure 3.14   *Plot of log k' vs. E_{NR} for benzylamines, showing that Nile Red also gives a good indication of elution strength for bases*
(Reproduced by permission from ref. 31)

several groups have shown that (at constant temperature and pressure) adsorption of MeOH is essentially complete when as little as 1% MeOH is present in the MP.[60,61] Higher MeOH concentrations cause little further MeOH adsorption. Thus, adsorption of the modifier does not explain the non-linearity observed in plots of log k' *vs.* percentage MeOH at higher MeOH concentrations.

Replacing percentage MeOH with the corresponding Nile Red solvent strength scale (E_{NR}), the plots become essentially linear,[31] as shown for phenols in Figure 3.13 and for benzylamines in Figure 3.14. Numerous other families of polar solutes produce similar plots. For less polar solutes, other dyes and energy scales are probably more appropriate and the Nile Red scale may not yield linear relationships.

These results may appear to confirm the equivalence of solvent and elution strength. However, they only prove that, for appropriately matched phases and solutes, the two are proportional.

Limited Density Range in Binary Fluids

It is attractive to change retention by changing a physical parameter like temperature or pressure, or both. However, Figure 3.15 shows that low

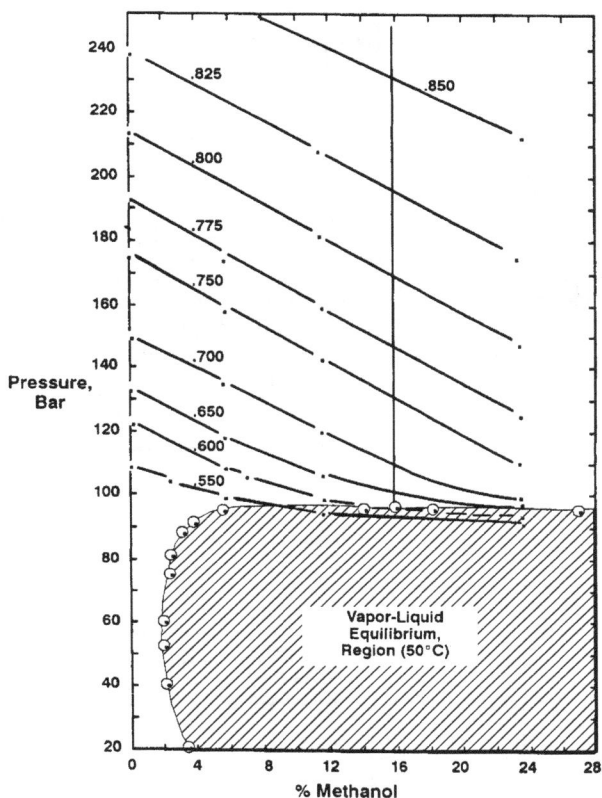

Figure 3.15   *The liquid–vapour equilibrium line for MeOH/CO$_2$ mixtures at 50 °C. The solid vertical line indicates the dividing line between supercritical and subcritical conditions but has no physical reality. The diagonal lines indicate conditions producing constant density. The numbers just above each diagonal line indicates its density, in g cm^{-3}*
(Reproduced by permission from ref. 19)

densities can seldom be achieved using modified fluids at modest temperature.

The nearly diagonal lines[19] at the top of Figure 3.15 are experimentally measured, constant density lines. Densities less than *ca.* 0.5 g cm^{-3} can only be achieved below *ca.* 2% MeOH. Above 2% MeOH, fluids with densities less than 0.5 g cm^{-3} are unstable. A **single** low density (< 0.5 g cm^{-3}) phase cannot exist (such conditions lay within the two phase region). Consequently, density programming is less useful with modified fluids than it is with pure fluids since a much narrower range of densities is available (0.5–1.0 g cm^{-3} for MeOH/CO$_2$ *vs. ca.* 0.05–1.0 g cm^{-3} for pure CO$_2$).

At higher temperature (*i.e.* > 100 °C), a wider range of densities is available. Higher pressure is required, however, to achieve high densities.

5000 (34 480)
4000 (27 580)
3000 (20 680)
2000 (13 790)

1000 (6900)
800 (5520)
600 (4140)
500 (3450)
400 (2760)
300 (2070)
200 (1380)

100 (690)
80 (552)
60 (414)
50 (345)
40 (276)
30 (207)

20 (138)

10 (69)

PRESSURE IN LBS. PER SQ. IN. ABSOLUTE (kPa ABS)

LIQUID REGION

CRITICAL POINT

SOLID REGION

VAPOR REGION

TRIPLE POINT

14.7 PSIA

°F. -160 -140 -120 -100 -80 -60 -40 -20 0 20 40 60 80 100 120 140
°C. -106.7 -95.6 -84.4 -73.3 -62.2 -51.1 -40 -28.9 -17.7 -6.7 4.4 15.6 26.7 37.8 48.9 60

TEMPERATURE

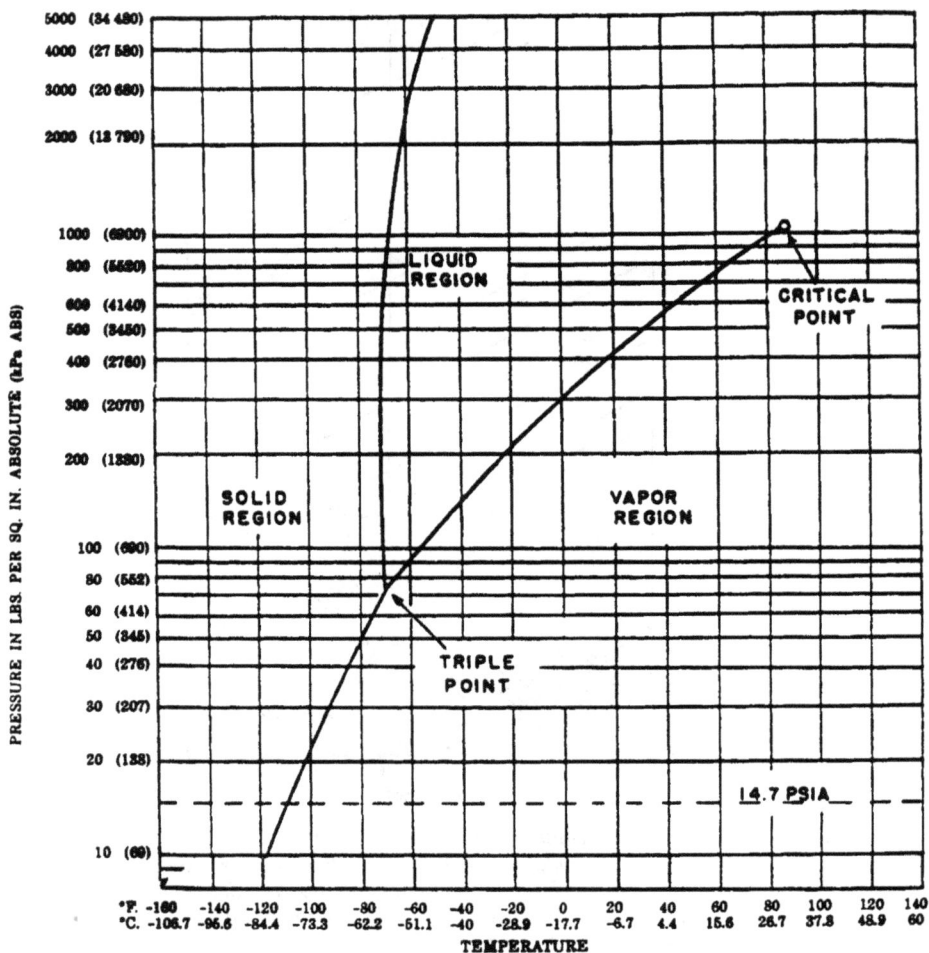

Figure 3.16 *Phase diagram for CO$_2$. The solid lines indicate conditions were two phases are in equilibrium. Away from the lines only one phase always exists* (Reproduced by permission from ref. 62)

6 Phase Behaviour

Near critical fluids can sometimes break down into a gas phase and a liquid phase. Since multiple MPs degrade chromatography, conditions producing two phases need to be understood and avoided. Many workers have been overcautious about phase separations, especially with binary fluids. Some have even suggested that binary fluids should be avoided owing to uncertainty in phase behaviour. However, extensive phase behaviour information is available in the literature. It is trivial to avoid two phase regions, and phase separations are much less common than has been widely believed.

What's Wrong with Phase Diagrams?

One kind of phase diagram consists of plots of temperature *vs.* pressure in which boundaries between phases are indicated by solid lines. A phase diagram[62] for CO_2 is shown in Figure 3.16. The vapour–liquid equilibrium (VLE) line (gas and liquid in equilibrium) extends from the 'triple point' of CO_2, where three phases (gas, liquid, and solid) co-exist, to its 'critical point', above which only a single phase can exist.

Unfortunately, most of the phase diagrams presented as part of the SFC and SFE literature have been drawn with **extra** solid or dashed lines, intended only to outline the region of the plot **defined** as supercritical. Many books about SFs even have such diagrams (with the extra lines) on their cover. These lines have been widely misinterpreted. Until recently, it was almost universally believed in chromatography circles that a transition from supercritical to subcritical conditions was **always** accompanied by a phase transition or a major change in viscosity, density, and/or D_M. These extra lines do NOT represent phase boundaries, only definition boundaries.

'Supercritical' is a **defined** state. It possesses no physically unique characteristics. There are only three states of matter: solids, liquids, and gases. There is no phase transition when a gas or a liquid changes into a SF. Crossing these extra lines **NEVER** causes a phase transition or even a change in viscosity, density, or D_M. Such lines should not be drawn and should be ignored.

Using high pressure view chambers, many workers have observed two phases in equilibrium or the separation of one phase into two. If two phases are in equilibrium, NEITHER can be supercritical (by definition). The formation of two phases always involves crossing the **real** solid line separating gases and liquids in phase diagrams. To form two phases directly out of a **pure** SF, both temperature and pressure must be **simultaneously** changed to cross exactly through the critical point. Changing only temperature or only pressure of a SF can not cause a phase transition if the fluid was really supercritical. The definition of the fluid can change, but no second phase forms.

If a pure fluid was supercritical but the pressure is changed to $< P_c$, the fluid definition changes to a gas without a phase transition. If the fluid is supercritical, but the temperature is changed to $< T_c$, the definition changes from SF to liquid. There is no phase transition.

Modified Fluid Phase Behaviour

The erroneous idea that transitions from supercritical to subcritical conditions produce two phases is even more entrenched when binary mobile phases are considered. The addition of a modifier changes phase behaviour, but in a predictable manner. The VLE behaviour of many mixtures is available in the chemical engineering literature (*e.g.* refs. 63–65 for MeOH/CO_2) but is usually overlooked by chromatographers.

As the concentration of MeOH in CO_2 increases, both T_c and P_c of the mixture increase, represented graphically[66] in Figure 3.17. At 48% MeOH in

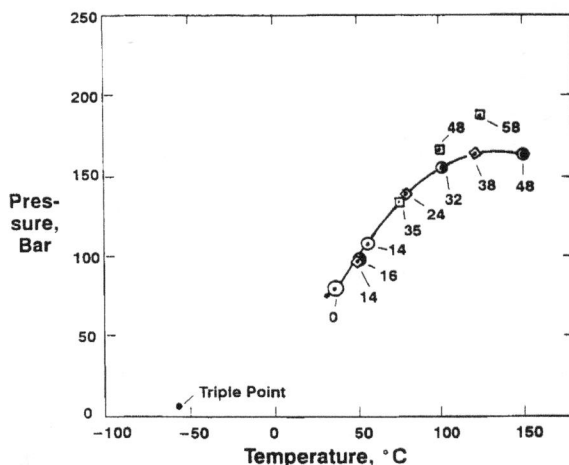

Figure 3.17 *The critical point of binary mixtures of MeOH in CO_2 increases with increasing modifier concentration up to ca. 48% MeOH*
(Reproduced by permission from ref. 66)

CO_2, $T_c \approx 150\,°C$ and $P_c \approx 150$ bar. Some have interpreted this to mean that the operating temperature and pressure must be $> 150\,°C$ and 150 bar **at all times** so that two phases cannot possibly form. Such an interpretation is ludicrously conservative and precludes the use of the most useful conditions for packed column SFC, including those used in most of the published literature.

Phase Behaviour of Methanol/Carbon Dioxide Mixtures

In Figure 3.18, the VLE behaviour of MeOH/CO_2 mixtures at different temperatures is represented[66] in a plot of percentage MeOH *vs.* pressure. Each line in the figure encloses a two-phase region which is characteristic of a specific temperature. The area outside each line contains a single phase at that temperature.

In the space under or inside a line in Figure 3.18 a single phase cannot be made. Instead, two phases always exist in equilibrium. If two MPs exist anywhere inside a chromatographic column, peak shapes are generally poor. This is demonstrated[66] in Figures 3.19 and 3.20. In Figure 3.19, the VLE line of MeOH/CO_2 at 36.85 °C is shown with a shaded two-phase region. Chromatograms collected using the conditions along Line A in the figure are presented in Figure 3.20. When p_0 was set to values inside the two-phase region of Figure 3.19, chromatograms in Figure 3.20 showed that the peaks broke up and the baseline became noisy.

Outside the two phase region, even though a single phase exists, not all conditions are supercritical. This can be explained more clearly using another Figure with the same form as Figure 3.19.

The phase behaviour[19] of MeOH/CO_2 mixtures at 50 °C is shown in

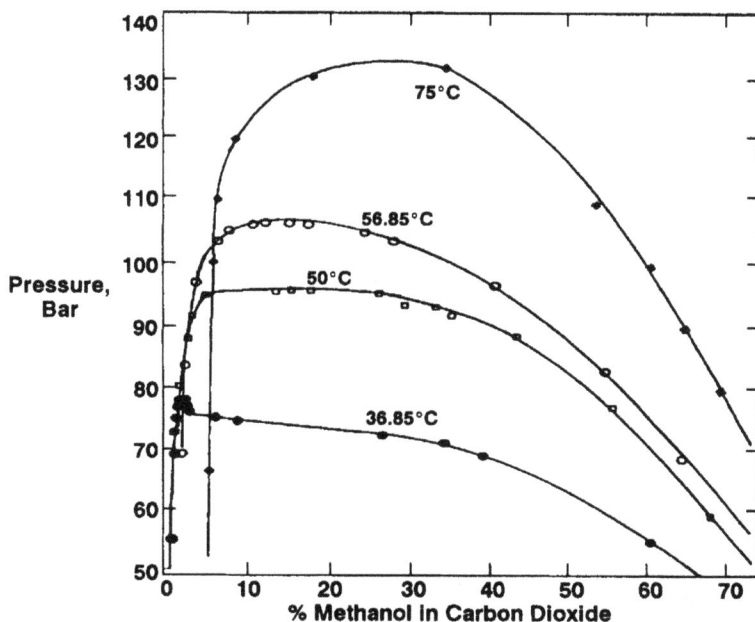

Figure 3.18 *Liquid vapour equilibrium lines for MeOH/CO₂ at various temperatures. At each temperaturea, a unique line encloses a region where two phases always exist*
(Reproduced by permission from ref. 66)

Figure 3.15. The maximum on the curve actually **defines** the critical point of the composition producing the maximum. It is difficult to determine the precise maximum but $T_c = 50\,°C$ for *ca.* 16% MeOH in CO_2. Extending a horizontal line from the maximum on the curve to the *y*-axis indicates that for 16% MeOH in CO_2, at 50 °C, $P_c \approx 97$ bar.

All compositions having more than *ca.* 16% MeOH in CO_2, at 50 °C and outside the two-phase region (including $P < 97$ bar), are **subcritical**. Such fluids are defined as liquids. Most (but not all) compositions having $< 16\%$ MeOH in CO_2 at 50 °C and outside the two-phase region are supercritical.

On the left hand side of the Figure there is a region of low modifier concentration where pressure can be decreased to 1 bar without entering a two phase region. In part of this region, some fluid compositions are **defined** as a gas. At higher pressure, the fluid is supercritical. There is no phase boundary between gas and SF.

A two-dimensional plot cannot convey enough information to describe completely the boundaries between sub- and super-critical conditions for all compositions. The good news for chromatographers is that it is irrelevant whether the fluid is a liquid, a gas, or a supercritical fluid, as long as a single fluid exists (ANYWHERE outside the two phase regions in the Figures).

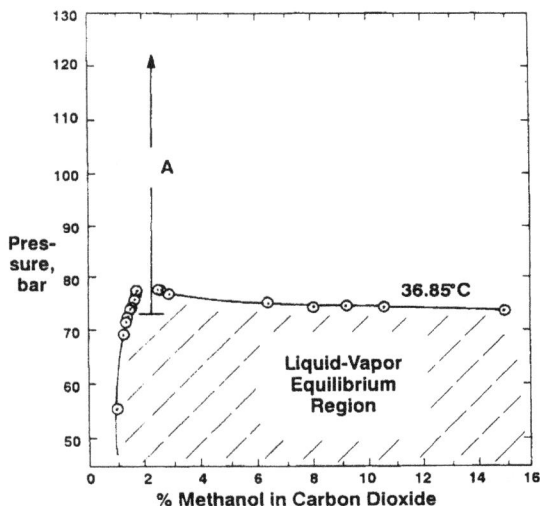

Figure 3.19 *The liquid–vapour equilibrium line of MeOH/CO$_2$ mixtures at 36.85 °C. The vertical line marked 'A' indicates the range of conditions used to collect the chromatograms in Figure 3.20*
(Reproduced by permission from ref. 66)

At a fixed temperature and pressure, the composition can be programmed from a state that is clearly supercritical to another state which is clearly subcritical without a phase change. Not only is it possible, it is typical. This can also be seen from Figure 3.15. Starting at 0% modifier (the left margin) and progressively increasing modifier concentration (proceeding horizontally across the Figure) corresponds to programming composition while holding temperature and pressure constant. If $P > 97$ bar (see Figure 3.15), the two-phase region is never encountered and a single phase exists throughout the experiment. Initially the phase is supercritical. Around 16% modifier, the same phase is defined as a liquid.

The diagonal lines in Figure 3.15 indicate conditions producing constant density. Note that the fluid density tends to increase as MeOH concentration increases (at constant temperature and pressure). The bold vertical line in the Figure represents the dividing line between supercritical and subcritical compositions (the line has NO physical reality and should actually NOT be drawn on such curves). To the left of the bold line (and somewhat **above** the shaded area), all compositions are supercritical (fluids to the left of the shaded area are gases). The actual conditions resulting in changing the fluid definition from SF to gas are NOT indicated in the figure. Just above the two phase region, some of the compositions are already gases. If, at a specific low modifier concentration, the pressure is dropped, the definition of the fluid changes from supercritical to gas **before** the pressure reaches the line defining the two-phase region. If the solid line defining the two-phase region is crossed, the fluid changes from a (subcritical) gas to a gas and a liquid in equilibrium.

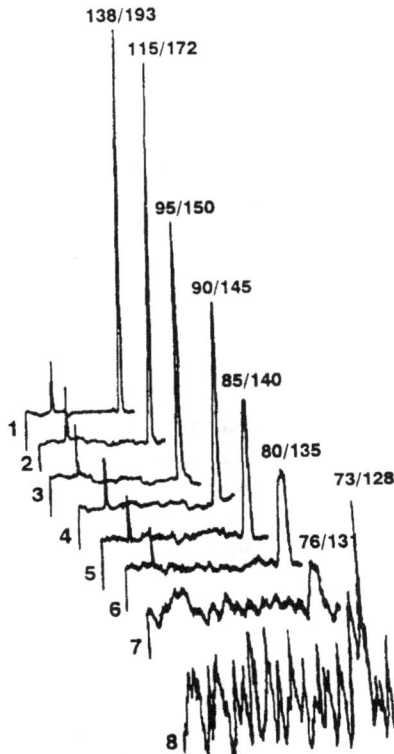

Figure 3.20 *Chromatograms collected using conditions along line 'A' in Figure 3.19, indicating phase separation at low outlet pressures (inside the shaded area in Figure 3.19). The numbers next to the chromatograms are inlet/outlet pressures in bar*
(Reproduced by permission from ref. 66

To the right of the bold line (and **outside** the shaded area), all the compositions are subcritical (liquids). If pressure is decreased until the boundary of the two-phase region is reached, the fluid changes from a (subcritical) liquid to a liquid in equilibrium with a gas.

Inside the shaded region, two subcritical phases always exist. The composition of each phase can be found by drawing a horizontal line (corresponding to the pressure of interest) that intersects both borders of the shaded region. Dropping vertical lines from the intersections to the *x*-axis indicates the compositions of the two phases (it may be necessary to use Figure 3.18 to see the intersection at higher compositions since both Figure 3.19 and 3.15 are truncated).

7 Conclusions

The differences between SFC and LC or GC are more practical than theoretical. The primary differences are higher diffusivity and lower viscosity than liquids, allowing faster analysis and/or higher resolution than LC. The differences between the techniques have been exaggerated.

Binary fluids and packed columns retain most of the advantages of capillary columns with pure fluids. In addition, modified fluids allow a much wider range of solvent strength than pure fluids. Potential problems with phase separation of binary fluids have been grossly overemphasized.

Mobile phase phenomena in supercritical fluid chromatography have been widely misinterpreted. Much of this confusion stems from the assumption that retention is simply a measure of mobile phase solvent strength. Another way of saying this is that elution strength and solvent strength are thought of as being identical. They are not. Mobile phase solvent strength is only one aspect controlling elution. It can be thought of as solute–solvent or S–MP interactions.

The presumed equivalence of solvent strength and elution strength implicitly assumes that S–SP interactions occur between the solute and the purchased stationary phase, and such interactions are constant. However, there are extensive interactions between the MP and the SP which change with physical conditions such as temperature and pressure.

8 References

1. T.A. Berger and W.H. Wilson, *Anal. Chem.*, 1993, **65**, 1451.
2. M.J.E. Golay, in 'Gas Chromatography–Amsterdam 1958', ed. D.H. Desty, Academic Press, New York, 1958.
3. J.J. van Deemter, F.J. Zuiderweg, and A. Klinkenberg, *Chem. Eng. Sci.*, 1956, **5**, 271.
4. Z. Balenovic, M.N. Myers, and J.C. Giddings, *J. Chem. Phys.*, 1960, **52**, 915.
5. I. Swaid and G.M. Schneider, *Ber. Bunsenges. Phys. Chem.*, 1979, **83**, 969.
6. R. Feist and G.M. Schneider, *Sep. Sci. Tech.*, 1982, **17** (1), 261.
7. H.H. Lauer, D. McManigill, and R.D. Bored, *Anal. Chem.*, 1983, **55**, 1370.
8. J.C. Giddings and S.L. Seager, *J. Phys. Chem.*, 1960, **33**, 1579.
9. J. Bohemen and J.H. Purnell, *J. Chem. Soc.*, 1961, 2630.
10. G. Taylor, *Proc. R. Soc. London, A*, 1953, **219**, 186.
11. G. Taylor, *Proc. R. Soc. London, A*, 1954, **223**, 446.
12. G. Taylor, *Proc. R. Soc. London, A*, 1954, **225**, 473.
13. R. Aris, *Proc. R. Soc. London, A*, 1956, **235**, 67.
14. 'Handbook of Chemistry and Physics', 69th Edn., CRC Press, Boca Raton, Fla.
15. 'Gas Encyclopedia', Elsevier, Amsterdam 1976.
16. J.H. Perry 'Chemical Engineers Handbook', 4th Edn. McGraw Hill, New York, 1963.
17. O.A. Uyehara and K.M. Watson, *Nat. Petroleum News, Tech. Section 36*, Oct. 4, 1944, 764.
18. K.E. Karlsson and M. Novotny, *Anal. Chem.*, 1988, **60**, 1662.
19. T.A. Berger, *J. High Resolut. Chromatogr.*, 1991, **14**, 312.
20. T.A. Berger and L.M. Blumberg, *Chromatographia*, 1994, **38**, 5.
21. A. Goldrup, G.R. Luckhurst, and W.T. Swanton, *Nature (London)*, 1962, 333.
22. A. Karmen, I. McCaffrey, and R.L. Bowman, *Nature (London)*, 1962, 575.
23. D.H. Desty, A. Goldup, G.R. Luckhurst, and W.T. Swanton, in 'Chromatography-1962', ed. M. van Swaay, Butterworths, London, 1962, p. 67.
24. S.T. Sie, W. Van Beersum, and G.W.A. Rijnders, *Sep. Sci.*, 1966, **1**, 459.
25. S. Wicar and J. Novak, *J. Chromatogr.*, 1974, **95**, 13.
26. T.A. Berger, W.H. Wilson, and J.F. Deye, *J. Chromatogr. Sci.*, 1994, **32**, 179.
27. J.B. Hannay, *Proc. R. Soc. London*, 1880, **30**, 484.
28. J.C. Giddings, M.N.Myers, L.M. McLaren, and R.A.Keller, *Science*, 1968, **162**, 67.
29. J.G. Dorsey, *Chromatography*, 1987, **2** (5), 37.
30. J.F. Deye, T.A. Berger, and A.G. Anderson, *Anal. Chem.*, 1990, **62**, 615.
31. T.A. Berger and J.F. Deye, in 'Supercritical Fluid Technology', ed. F.V.Bright and M.E.P. McNally, ACS Symposium Series 488, American Chemical Society, Washington, DC, 1992, Chapter 11.

32. J.A. Hyatt, *J. Org. Chem.*, 1984, **49**, 5097.
33. M.E. Sigman, S.M. Lindley, and J.E. Leffler, *J. Am. Chem. Soc.*, 1985, **107**, 1471.
34. C.R. Yonker, S.L. Frye, D.R. Kalkwarf, and R.D. Smith, *J. Phys. Chem.*, 1986, **90**, 3022.
35. S.L. Frye, C.R. Yonker, D.R. Kalkwarf, and R.D. Smith, in 'Supercritical Fluids: Chemical and Engineering Applications', ed. T.G. Squires and M.E. Paulaitis, American Chemical Society, Washington, DC, 1987, Chapter 3.
36. L.R. Snyder, *J. Chromatogr.*, 1974, **92**, 223.
37. J. Figueras, *J. Am. Chem. Soc.*, 1971, **93**, 3255.
38. S. Kim and K.P. Johnston, *AIChE J.*, 1987, **33**, 1603.
39. K.P. Johnston, S. Kim, and J.M. Wong, *Fluid Phase Equilib.*, 1987,**38**, 39.
40. U.K. Deiters, *Fluid Phase Equilib.*, 1982, **8**, 123.
41. U.K. Deiters, *Fluid Phase Equilib.*, 1983, **13**, 109.
42. W.B. Whiting and J.M. Prausnitz, *Fluid Phase Equilib.*, 1982, **9**, 119.
43. P.M. Mathias and T.W. Copeman, *Fluid Phase Equilib.*, 1983, **13**, 91.
44. Y. Hu, E.G. Azevedo, and J.M. Prausnitz, *Fluid Phase Equilib.*, 1983, **13**, 351.
45. J.M. Walsh, G.D. Ikonomou, and M.D. Donohue, *Fluid Phase Equilib.*,1987, **33**, 295.
46. K.-S Nitsche and P. Suppan, *Chimia*, 1982, **36**, 346.
47. J.M. Dobbs, J.M. Wong, R.J. Lahiere, and K.P. Johnston, *Ind. Eng. Chem. Res.*, 1987, **26**, 56–65.
48. M.P. Ekart, K.L. Bennett, S.M. Ekart, G.S. Gurdial,.
    C.L. Liotta and C.A. Eckert, *AIChE J.*, 1993, **39**, 235.
49. C.A. Eckert, B.L. Knutson, *Fluid Phase Equilib.*, 1993, **83**, 93.
50. J.M. Dobbs, J.M. Wong, and K.P. Johnston, *J. Chem. Eng. Data*, 1986, **31**, 303.
51. D.L. Saunders, *Anal. Chem.*, 1974, **46**, 470.
52. A.L. Blilie and T. Gribrokk, *Anal. Chem.*, 1990, **62**, 1181.
53. J.M. Levy, Ph.D. Thesis, Case Western Reserve University, 1986.
54. C.R. Yonker, R.D. Smith, *J. Chromatogr.*, 1986, **361**, 25.
55. U. van Wasen and G.M. Schneider, *Chromatographia*, 1975, **8**, 274.
56. C.R. Yonker, D.G. McMinn, B.W. Wright, and R.D. Smith, *J. Chromatogr.*, 1987, **396**, 19.
57. T.A. Berger, *Chromatographia*, 1993, **37**, 645.
58. L.R. Snyder, 'Principles of Adsorption Chromatography', Marcel Dekker, New York, 1968.
59. T.A. Berger and J.F. Deye, *Anal. Chem.*, 1990, **62**, 1181.
60. J.R. Strubinger, H. Song, and J.F. Parcher, *Anal. Chem.*, 1991, **63**, 104.
61. C.H. Lochmuller and L.P. Mink, *J. Chromatogr.*, 1989, **471**, 357.
62. 'Handbook of Compressed Gases', 3rd Edn., Compressed Gas Association, Inc., Van Nostrand-Reinhold, New York, 1990, p. 288.
63. D.B. Robinson, D.-Y. Peng, and S.Y.-K. Chung, *Fluid Phase Equilib.*, 1985, **24**, 25.
64. J.H. Hong and R. Kobayashi, *Fluid Phase Equilib.*, 1988, **41**, 269.
65. E. Brunner, W. Hultenschmidt, and G. Schlichtharie, *J. Chem. Thermodyn.*, 1987, **19**, 273.
66. T.A. Berger and J.F. Deye, *Chromatographia*, 1991, **31**, 529–534.

CHAPTER 4

Physical Chemistry of the Stationary Phase

1 Introduction

Chapter 3 was presented to describe the physical chemistry of the mobile phases (MPs) used in SFC. The intent was not to catalogue which fluids were used to separate specific solutes. Detailed chapters on specific application areas are provided at the end of the book. Instead, the material was presented to provide the reader with a context with which to understand what attributes of the fluids are important. Similar concepts will be discussed in this chapter but in terms of the packing material and stationary phase (SP). This does NOT include specific discussions of the chemical interactions of a specific SP. A discussion of phase selection can be found in Chapter 6.

As for Chapter 3, a thorough understanding of the material in this chapter is not necessary for practical method development. However, some of the steps suggested in Chapters 6 and 7 might appear arbitrary or even counter-intuitive without at least a brief review of this material.

The material presented in this chapter is likely to surprise many readers and is probably the most controversial aspect of this book. This chapter contains an extended discussion of adsorption of MP components onto the SP. Much of the behaviour of the stationary phase is interpreted in terms of the effect the adsorbed MP components have on peak shape and retention. In addition there is a long section on additives in the MP. That material is presented here instead of in Chapter 3 because most of the action of the additives involves interaction with the SP.

2 The Nature of Packing Materials

The most commonly used packing materials in SFC are the same as used in HPLC and packaged in the same ways. The particles are made of totally porous, high surface area, hydrated silica.[1] The more modern packings are spheres between 1 and 10 μm in diameter. Most analytical packings have pore diameters between 60 and 300 Å. In LC, the mobile phase forms stagnant pools inside these pores. In SFC, a somewhat higher level of convection appears to occur within these pores.

Different silicas contain varying trace amounts of impurities such as metals, which can have a major impact on chromatographic selectivity and peak shapes. Metal ions could form sites more acidic than pure silica. Some manufacturers go to great lengths to remove all traces of metals, while others attempt to produce uniform batches of material with known levels of metals present. Packing materials and columns have dramatically improved over the last 10 years.

Small porous packings have interstitial spacings between the particles much smaller than the particle diameter. The fluid subsequently must follow a tortuous path through flow channels only a few micrometres wide. Despite the low viscosity of supercritical fluids, such flow creates significant pressure drops. For example, 5 μm packings will cause a pressure drop of 10–30 bar over the length of a 20–25 cm long column operated at optimum flow rate.

Alumina has also been tried as a support material in SFC but results have been disappointing. It is difficult to bond organic SPs to alumina, and it tends to be made up of small platelets yielding poor pore shapes. Zirconia is still too new for its role to be assessed.

Polymer based particles show some potential for use in SFC. However, they tend to be less efficient than silica based columns and do not offer significant differences in activity compared to silica.

Surface Area

Commercial porous silica packings typically have a surface area of 100–500 $m^2 g^{-1}$, although packings with much smaller surface area are available. Particles with very large pores, such as 4000 Å, can have surface areas less than 10 $m^2 g^{-1}$, approaching values expected for pellicular packings. Most analytical packings have a void volume or pore volume in the vicinity of 1 $cm^3 g^{-1}$. Contrary to one report,[2] retention in SFC is generally directly proportional to surface area,[3] as shown in Figure 4.1. A typical 4.6 × 200 mm packed column contains *ca.* three grams of packing although the density of the material, and the subsequent weight of packing in a column varies widely. The ratio of mobile phase volume to stationary phase surface area can be more than 10^6 times larger on a capillary than on a packed column. This means that solute molecules have up to 10^6 times more opportunity to interact with the surface on a packed column compared with a capillary column.

Phase Ratio

All solutes exhibit a distribution coefficient (K_D) which describes how that solute will partition between two chromatographic phases. It is the ratio ($K_D = [C_{SP}]/[C_{MP}]$) of the **concentrations** of the solute in each phase under the specific conditions of interest.

Chromatographic retention has both extensive and intensive aspects. The distribution coefficient (K_D) can also be thought of as the product of a phase ratio (β), describing the relative volumes of the two phases (extensive), and a

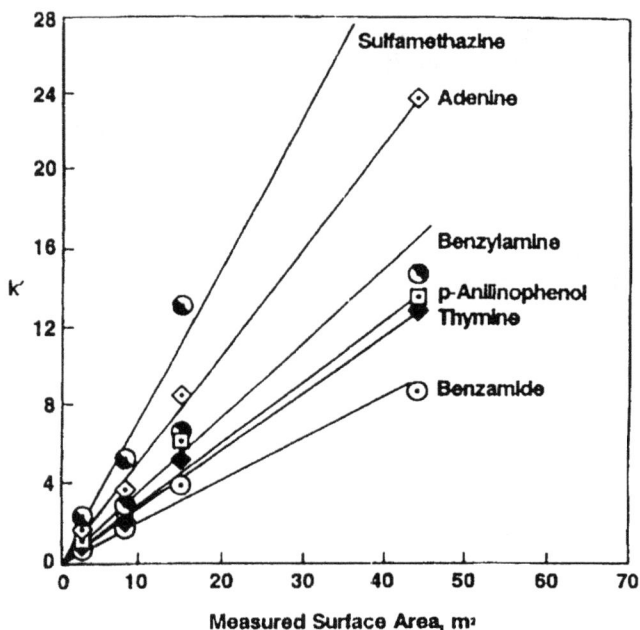

Figure 4.1 *Retention is a linear function of column surface area in packed column SFC. Both acidic and basic solutes gave similar results. The columns were all packed with 7 μm Nucleosil Diol packings with different pore sizes. Packings exhibited 100, 300, 500, and 4000 Å pores*
(Reproduced by permission from ref. 3)

partition ratio (k or k') which describes the ratio of the mass of the solute in each phase (intensive):

$$K_D = (k')(\beta) = (m_s/m_m)(V_m/V_s)$$

In general, the chromatographic literature does not discuss packed column retention in terms of phase ratio. However, a case can be made for using the concept to try to compare capillaries and packed columns. The phase ratio ($\beta = V_m/V_s$) is the volume of the MP (V_m) divided by the volume of the SP (V_s).

The partition ratio ($k' = m_s/m_m$) is the mass of solute in the SP divided by the mass of solute in the MP. k' is, consequently, a direct measure of the relative intensity of interactions between the solute and each phase. It is usually calculated in chromatography using the retention time of the solute (t_R) and the transit time of an unretained solute (t_0) [$k' = (t_R - t_0)/t_0$].

K_D is a constant. If β is changed, k' must change to compensate.

The relative surface areas of packed and capillary columns suggest that retention should be up to 10^6 times greater on packed columns. However, the application of chemically bonded organic SPs to the surfaces of the supports tends to mitigate the differences in surface areas between capillaries and column packings.

On packings, bonded SPs tend to be limited to monomolecular coatings less than a nanometre thick. Capillaries are generally coated with thick films of a liquid SP. Film thicknesses (d_f) of a few tenths of a micrometre are typical.

Even with bonded SPs, packed columns have βs 5–100 times smaller than capillaries. Packed column k' would, therefore, be 5–100 times higher, assuming that both column types used chemically identical (but different thickness) phases [same K_D, so $(\beta)(k') = $ constant]. Such differences in retention are a major reason that packed and capillary columns are difficult to compare, and in fact have different uses.

With a large β, the interaction between the solute (S) and SP must be much more intense than the interaction between the solute and the MP. If the S–SP interaction were not very intense, weak S interaction with a small volume of SP would be overwhelmed by the S interaction with a much larger volume of the MP. With weak interactions and a large β, the solute would be nearly unretained. Capillaries require a large difference in polarity between the phases because they are made with a large phase ratio.

Packed columns, with a small β, need less intense S–SP interactions to achieve retention similar to capillaries using the same MP. To avoid excessive retention on a packed column, the MP and SP need to be more like each other than the phases used with capillaries. If pure CO_2 is used as the MP, a packed column requires a very non-polar stationary phase (since CO_2 is very non-polar). When polar columns are used, a more polar MP is required to avoid excess retention.

Both CO_2 and modifiers adsorb[4-6] extensively onto the support, depending on MP and SP identity, pressure, and temperature. The effective SP in SFC can be composed of the phase bonded to the support material (or the bare support) **plus** an adsorbed film of MP components. Adsorption tends to decrease β but also makes the two phases more similar chemically. These phenomena are discussed extensively, below.

Active Sites

Excess Retention

Using pure CO_2 as the MP in SFC, tailing and even the lack of elution has been a common problem. Greater retention on packed columns compared with capillary columns has been blamed on 'active sites', assumed to be present in greater numbers and at much greater density on packed columns.[7-14] However, the identity and quantity of such 'active sites' is simply speculation. There is no consensus on what the term 'active sites' means, or on the population density of 'active sites'. It is clear that differences in β explain the differences in retention.

Change in retention due to the addition of a modifier to the MP has also been widely assumed to be greater on packed columns than on capillaries. If true, this would support the concept of 'active sites' increasing retention on the packed column. The modifier would primarily cover up or deactivate the active sites. However, the literature clearly shows that modifiers have at least as large an impact on the retention of solutes on capillaries as on packed columns. The retention of polycyclic aromatic hydrocarbons (PAHs) decreased by 15–32% on packed columns,[8,15] but 26–28% on capillaries[9,14] when 2% isopropanol or MeOH was added to CO_2.

The differences in absolute retention between packed and capillary columns can be completely explained by differences in surface area and β[16]. Nevertheless, the literature continues to speculate prominently on the role of 'active sites', despite the lack of direct cause and effect evidence.

Tailing

Secondary retention mechanisms cause tailing of some sample components but not others. Metal ions or a sub-population of oriented silanols are often blamed for such tailing. Silica is susceptible to attack by bases. Basic solutes often tail in both SFC and LC. Some chromatographers lump all tailing mechanisms together and pronounce them due to 'active sites'.

Several manufacturers have attempted to deactivate silica by coating its surface with a very thin layer of an organic polymer and then attaching a standard SP to this coating. In the United States, one of the better known materials of this type is Deltabond, produced by Keystone Scientific.

Polymer coated silica columns (like Deltabond) behave nearly the same as standard columns although peak shapes tend to be somewhat better. Nevertheless, such columns do not allow more polar solutes to be eluted with pure fluids, like CO_2. When modifiers are used, polymer coated and 'bare' silicas act similarly.

Several manufacturers have produced polymer based particles that contain no silica. However, these materials also produce tailed peaks, indicating that tailing is not a silica phenomenon. Base deactivated columns produce the same results as standard silica columns in SFC although selectivity can be slightly different. Several tailing mechanisms appear to be related more to MP or SP phenomena rather than 'active sites'.

Selectivity

Hydrocarbon group separations are sensitive to the nature of the support material. Such separations are performed on bare silica with pure CO_2 as the MP. Resolution between groups is primarily a function of chemical differences, not efficiency, so selectivity ($\alpha = k_2/k_1$) is the primary means of separation. Resolution of test mixes on different bare silicas can vary by more than a factor of 4.5. Some produce extremely good α, while others produce almost no group α. Unfortunately, none of the information supplied by the various manufactures allows the user *a priori* to determine which silica will produce an optimum separation. In this case, 'active sites' are probably metal ions occluded in the silica. These metal ions probably provide an alternative retention mechanism

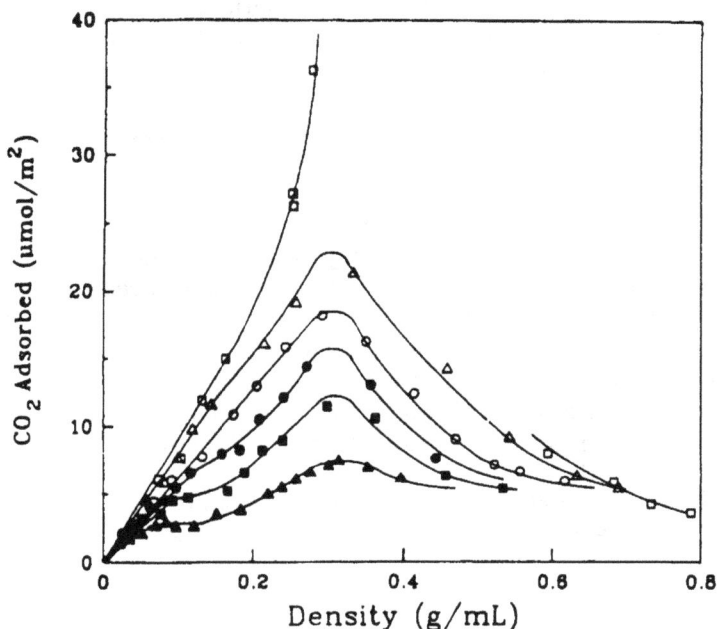

Figure 4.2 *Adsorption (actually surface excess) of CO_2 on an ODS column. Adsorption decreases as the temperature is increased. The maximum occurs near the reduced density.[4] From bottom to top: 100, 70, 60, 50, 40, and 30 °C*

for some of the solutes. Such mixed mechanisms degrade the α of the silica toward the groups.

3 Adsorption of Mobile Phase Components

When MP components adsorb onto packings they form a dense, liquid like film on the surface. This film is usually more dense than the MP but consists of the same chemical species. This adsorption should be considered in discussions of β and phase polarity.

Adsorbed Carbon Dioxide Acts as a Stationary Phase

Using radio tracer pulse chromatography,[4] the adsorption of pure CO_2 onto several silica based packed column SPs has been measured as a function of temperature and pressure. A representative plot of surface excess *vs.* density for CO_2 on an ODS column is presented in Figure 4.2.

Maximum adsorption of CO_2 occurs at a density of 0.3 g cm^{-3}, between 40 and 100 °C on both bare silica and ODS (C$_{18}$). Adsorption decreases with increasing temperature. At 40 °C, the maximum adsorption on ODS was 22 μmol m^{-2}, and on silica 30 μmol m^{-2}. If the density of the adsorbed film is 1.0 g cm^{-3}, it would be up to 1.3 nm thick (multiple monolayers). The surface of silica contains *ca.* 8 μmol m^{-2}. If the surface were occupied by one CO_2 molecule per silica molecule the film would be up to three monolayers thick at 40 °C. At lower temperature, the film sometimes exceeds four monolayers. These numbers represent surface excess, so even more material is actually present on the surface. The full thickness would, therefore, be even greater.

Increasing the mobile phase density decreases surface excess. At a density of 0.6 g cm^{-3}, surface excess on the silica column decreased from 30 to 15 μmol m^{-2}. Since the mobile phase density is doubled, the number of MP molecules per unit of volume is doubled. This makes it unclear whether the number of molecules adsorbed on the SP actually has changed since the adsorption is described in terms of 'surface excess'.

The effect of temperature is less ambiguous. Adsorption is at a maximum at low temperatures $(T > T_c)$ and decreases as temperature is raised. The maximum adsorption on ODS dropped from 22 μmol m^{-2} at 40 °C, to 8 μmol m^{-2} at 100 °C.

In capillary SFC, simultaneous use of low temperature and low densities is generally avoided. Peak shapes on packed columns tend to be better when temperature is low but density is high. These practical limits in operating regions are probably related to adsorption phenomena, as will be shown in a later section.

Phase Ratio of an Adsorbed Film

The adsorbed film has a volume that can be used to calculate a β independent of the volume of the bonded SP. The volume of the adsorbed film (V_{ads}) of CO_2 and the packing void volume (V_{void}) can be estimated, allowing the calculation of a phase ratio ($\beta_{ads} = V_{void}/V_{ads}$).

First, the weight of the adsorbed film is determined. If the density in the adsorbed layer is assumed to be 1 g cm^{-3}, the volume of the adsorbed film can be determined (*e.g.* the maximum on ODS is 0.968 μl m^{-2}). In one reference[5] the void volume of a packing was measured as a function of the amount adsorbed. The void volume on ODS showed a decrease of between 1 and 3.5 μl for a change in the amount adsorbed from 1 or 2 to 20 μmol m^{-2}. Thus, the decrease in V_{void} is in reasonable agreement with the calculated increase in V_{ads}, and indicates that V_{ads} is a significant fraction of V_{void}.

Volumes Adsorbed

Silicas with 100 Å pores have nominal surface areas in the vicinity of 350 m^2 g^{-1}. While the adsorbed layer contained 968 μg m^{-2} of CO_2, one column (2 × 140 mm) had an *in-situ* measured surface area of 28.9 m^2 and V_{void} of *ca.* 0.3 cm^3. If the adsorbed film has a density of 1 g cm^{-3}, then the total V_{ads} of CO_2 would be 28 μl. At 22 μmol m^{-2} adsorbed on ODS, the measured V_{void} was 10 μl m^{-2}, or a total of 289 μl. The ratio of the V_{void}/V_{ads} was, therefore, *ca.* 10.3 (289/28).

SFC With Pure Carbon Dioxide is Normal Phase

The adsorbed CO_2 is essentially a condensed fluid with a likely density near 1.0 g cm^{-3}, even when the MP density is *ca.* 0.3 g cm^{-3}. Since solvent strength is proportional to density, the adsorbed film should have higher solvent strength than the less dense MP. Thus, the adsorbed film should modify both the volume and 'polarity' (or solvent strength) of the SP.

If the SP is covered with a denser, more polar film of adsorbed MP, SFC must almost always be a normal phase technique. To perform reversed phase packed column SFC, one would need to prevent adsorption of MP components or create a situation where there is a **preferential** adsorption of **less polar** (!) MP components from a more polar main fluid. Neither seems likely using the common MPs used presently. Alternatively, a very non-polar SP, with a substantially more polar MP, would provide a reversed phase type of mechanism. However, it is unlikely that a fluid as polar as water, but with substantially less severe critical parameters will be found (intense intermolecular interactions without the fluid condensing!). In the absence of such a fluid, there is little reason to try to develop reversed phase SFC.

Swelling

The term 'swelling' is sometimes used to indicate that the MP interacts with the SP.[17-19] This term is somewhat unfortunate since it can imply a physical effect like a sponge soaking up a spilled liquid. Such swelling is often viewed as a simple increase in the volume of the bonded SP.

In SFC, the adsorbed film can be substantially more dense than the MP. Since solvent strength is related to density, the adsorbed film and the MP have different solvent strengths. Thus the adsorption of the MP can dramatically change both the volume and the polarity (or chemical nature) of the SP. The adsorbed CO_2 film is at least as thick as the bonded phase film thickness, at least doubling the volume of the phase. Further, the polarity of this multilayer film is real and potentially quite different from that of the bonded phase.

Chromatographic Evidence of Adsorbed Carbon Dioxide Acting as Part of the Stationary Phase

The Van Deemter equation ($H = A + B/\mu + C_m\mu + C_s\mu$) relates speed to efficiency ($N = L/H$). At low linear velocity (μ), efficiency is dominated by longitudinal diffusion (B/μ) along the axis of the column. At high flow rates, efficiency is normally dominated by resistance to mass transport ($C_m\mu$) in the MP. The SP is generally made thin enough so that it has no impact on the speed–resolution trade-off ($C_m\mu \gg C_s\mu$). When the SP is thicker, solutes spend additional time diffusing in the film ($C_m\mu \approx C_s\mu$). If MP μ is not decreased, the volume of mobile phase containing the solute increases, and efficiency is lost. With thick films, $C_s\mu$ becomes important or even dominant. Theoretical

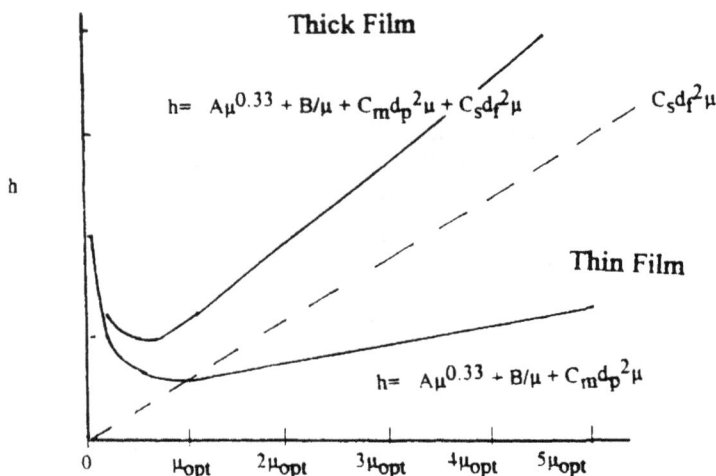

Figure 4.3   *Idealized van Deemter plots intended to indicate the effect of a thick film of adsorbed CO_2 acting as part of the SP. The lower curve indicates typical relationships between plate height (h) and linear velocity (μ). A is a packed column term, B involves longitudinal diffussion, C_m involves radial diffusion (resistance to mass transfer). The upper curve indicates the change in h resulting from the use of a much thicker SP. C_s is a constant containing the diffusion coefficient of the solute in a thick SP, d_f is the film thickness of the SP*

differences between thin films and thick films are graphically summarized[20] in Figure 4.3.

Carbon dioxide strongly adsorbs onto column packings. At $T \approx T_c$ and $P \approx P_c$, adsorption is at a maximum. Van Deemter curves were collected using a silica column held at 40 °C with various low outlet pressures, as shown in Figure 4.4. At $P > 100$ bar, the curves were flat,[20] indicating normal performance. At all higher pressure, the column produced typical high efficiencies over a wide range of flow rates.

At $P < 100$ bar, the slopes of the curves (above the optimum velocity) increased dramatically, and efficiency degraded. This is consistent with the formation of a thick stationary phase (larger $C_s d_f^2 \mu$ term), as suggested in Figure 4.3. Peak shapes remain symmetrical and Gaussian. The loss in efficiency is difficult to explain, other than by accepting that the adsorbed film of CO_2 acts as part of a very thick SP.

4 Adsorption From Binary Mixtures

Modifiers strongly adsorb. The adsorption of modifiers does not displace adsorbed CO_2[5] and the total amount of adsorbed material increases. When MeOH is added to the MP, the maximum amount of CO_2 adsorbed **increases** marginally.

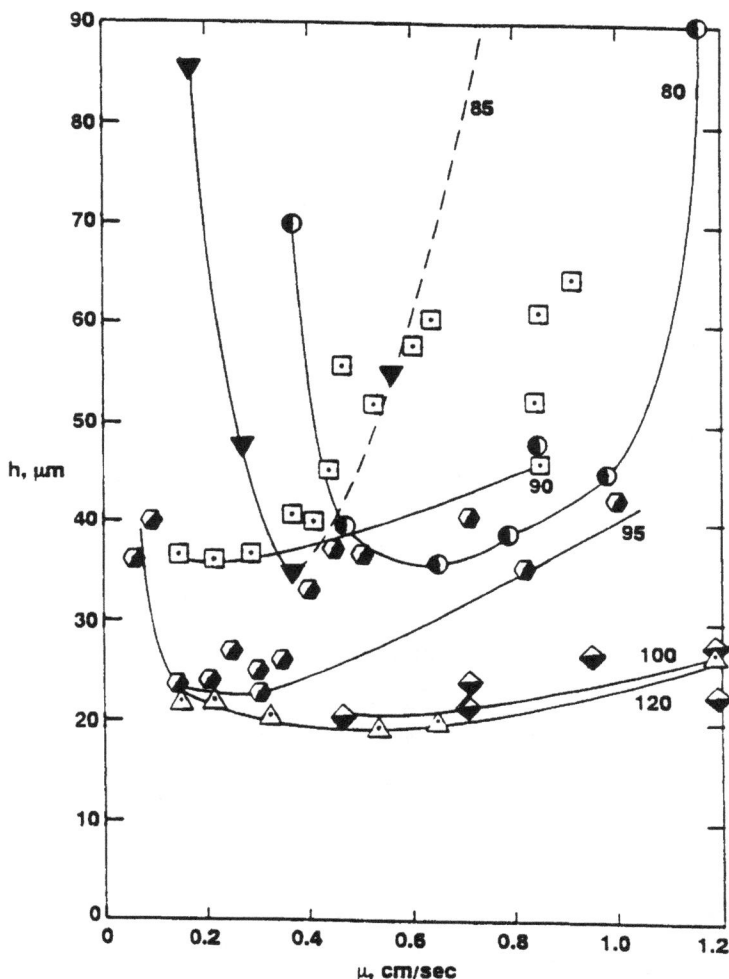

Figure 4.4 *Experimental van Deemter plots of silica columns with pure CO_2 at low temperature and pressure. At low pressure, adsorption is extensive and the curves take on the shape indicating a thick stationary phase. At slightly higher prressure (> 100 bar), adsorption decreases and the columns behave as though there was only a thin SP. Temperature = 40 °C. Outlet pressure indicated beside each curve*
(Reproduced by permission from ref. 20)

The MP density at which maximum adsorption of CO_2 occurs can change when MeOH is added to the MP, as can be seen by comparing Figures 4.5 and 4.2. With pure CO_2, the maximum adsorption occurs on ODS and silica at MP densities of 0.3 and 0.5 g cm^{-3}, respectively. With 2% MeOH added, the maximum adsorption of CO_2 on silica remains at 0.5 g cm^{-3} but increases on ODS to *ca.* 0.5 g cm^{-3}. In addition, the maximum is broader (approximately constant adsorption from a density of 0.3 to near 0.6 g cm^{-3}).

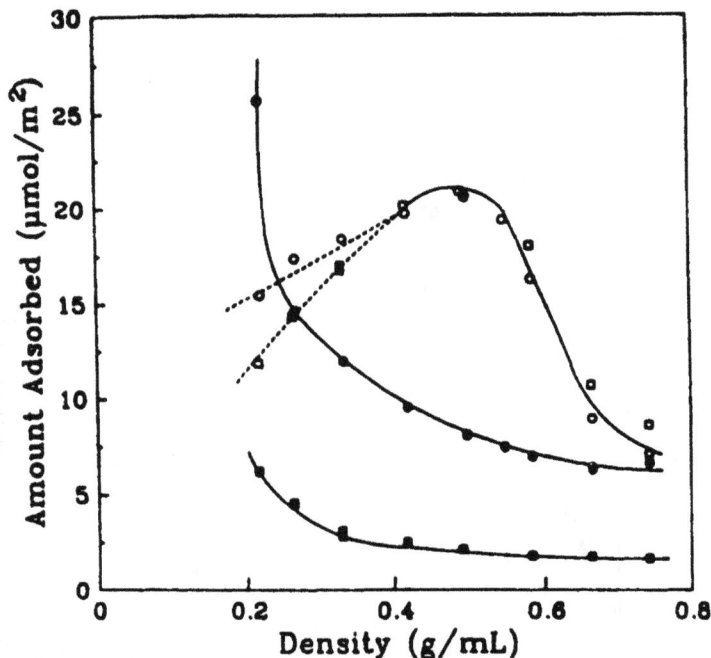

Figure 4.5   *The adsorption of CO_2 and MeOH on ODS and silica. Open circles = CO_2 on silica, open squares = CO_2 on ODS, shaded circle MeOH on silica, shaded square = MeOH on ODS*
(Reproduced by permission from ref. 5

Methanol adsorbs extensively from binary MeOH/CO_2 mixtures. The most surprising aspect of this adsorption is the high surface coverage on the packing achieved from low concentrations in the MP. Monolayer adsorption of MeOH occurs from as little as 1% MeOH in CO_2, as shown[6] in Figure 4.6.

A steep rise in MeOH adsorption occurs[5] as the MP density decreases (see Figure 4.5). Such adsorption can be extensive, even exceeding the quantity of CO_2 adsorbed. At low densities, the two phase region is approached (see phase diagrams in Chapter 3), and the fluid becomes difficult to use chromatographically.

Adsorption (or at least surface excess) decreases when the density increases. Silica adsorbs 10 μmol m^{-2} of MeOH at a density of 0.4 g cm^{-3}, decreasing to 7 μmol m^{-2} at 0.8 g cm^{-3}.

Adsorption on less polar SPs is less extensive and also less a function of density. For example, the adsorption of MeOH on an ODS column amounts to only 2–2.5 μmol m^{-2} at densities between 0.4 and 0.8 g cm^{-3} (50 °C, 2% MeOH).

At the point of maximum **total** adsorption (the sum of both MeOH and CO_2) (at a density of 0.5 g cm^{-3}), MeOH is 9% of the adsorbed film on ODS and 28% of the adsorbed film on silica.[5] Total adsorption (CO_2 plus MeOH) reached 23 μmol m^{-2} on ODS and 29 μmol m^{-2} on silica. Most workers accept

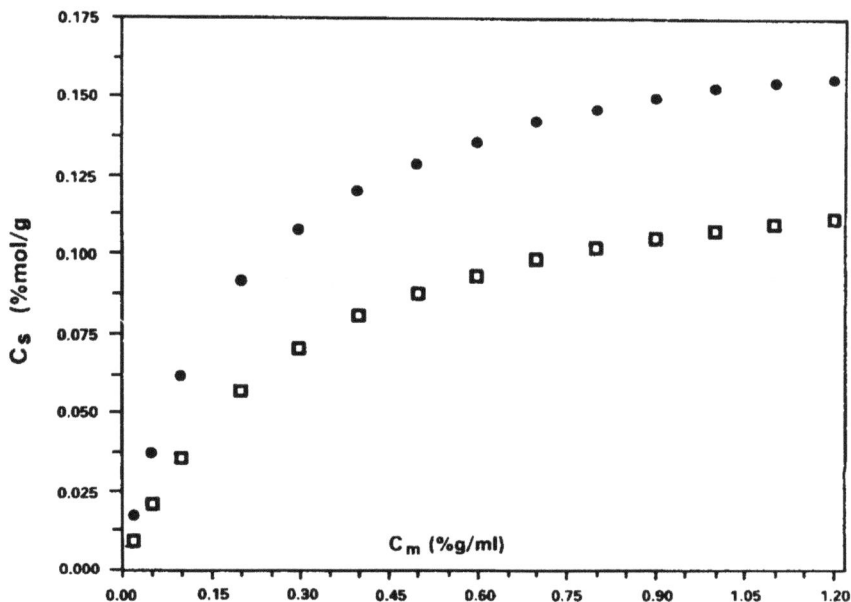

Figure 4.6 *The upper curve represents adsorption of methanol from MeOH/CO$_2$ mixtures. Note that adsorption is leveling off at 1.2% suggesting that adsorption is complete*
(Reproduced by permission from ref. 6)

that a silica surface contains 8 μmol m^{-2} of sites. So the adsorbed film is at least three monolayers thick.

As MP density increased, the total quantity adsorbed decreased, but the concentration of the polar modifier in the adsorbed film increased. On ODS, MeOH increased to 22% of the total (9 μmol m^{-2}) and nearly 50% MeOH (15 μmol m^{-2} adsorbed) on silica at a MP density of 0.8 g cm^{-3}. These results were again obtained using 2% MeOH in CO$_2$. The MeOH concentration in the adsorbed film was, therefore, up to 25 times higher than in the MP. Clearly, the adsorbed MeOH must make the environment on the surface of the SP much more polar than in the bulk fluid. The adsorbed layer, thus, changes the volume AND the polarity of the SP.

Since the adsorbed layer is more polar than the MP (higher MeOH concentration AND higher density), packed column SFC using modified mobile phases will almost always be a normal phase technique. In order to perform reversed phase separations, a low polarity stationary phase would be required on which no polar mobile phase components could adsorb.

In the previous chapter it was shown that the addition of polar modifiers dramatically increased the solvent strength or polarity of supercritical fluids used as MPs. Above, it was shown that modifiers can dramatically increased the polarity and volume of the SP. Chromatographic retention reflects the sum of these MP and SP interactions.

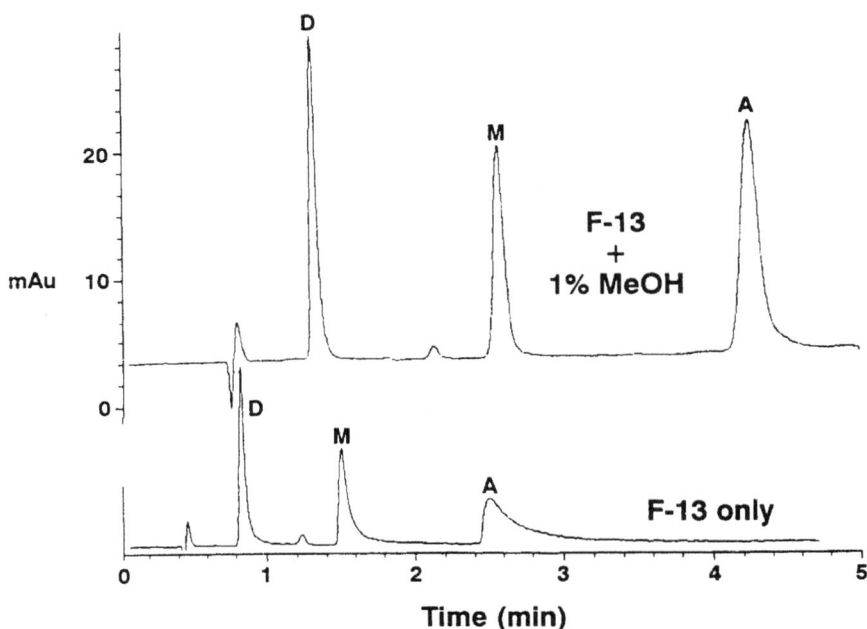

Figure 4.7   *Aniline retention doubled when MeOH was added to the Freon-13 MP*
(Reproduced by permission from ref.31)

Adsorbed Binary Fluids Sometimes Increase Retention

The retention of anilines almost doubled, on a Deltabond CN column, when
1% MeOH was added to Freon-13, as shown in Figure 4.7. This increase in
retention, despite the increased MP solvent strength, is difficult to explain. The
most likely explanation is an increase in the polarity or volume of the SP, due to
adsorption, exceeding the increase in polarity of the MP.

5 Polar Additives

In LC, the traditional explanation for tailing or lack of elution involves the
presence of 'active sites' usually meaning some sub-set of silanols or metal ions[1]
on the traditional silica packing material. Additives are very polar substances
added to the MP, at low concentrations, some of which are expected to improve
peak shapes by covering up, adsorbing on, or reacting with active sites.

An alternative approach to dealing with active sites involves end capping, or
polymer coating schemes meant to permanently cover active sites, eliminating
the need for an additive in the MP. An even more drastic step is to replace the
silica with a polymer based support which is expected not to have active sites.

Virtually all peak distortions in SFC are attributed to the existence of active
sites on the SP. Although it is convenient to have a universal explanation of
poor peak shapes **it does not provide any insight into how to solve the problem.**

We believe the poor success rate of many of the techniques intended to avoid peak tailing is due to the fact that the wrong problem is being attacked. Active sites are only one of many contributors to tailing.

In SFC, end capping, deactivation, and polymer based supports have not been successful in eliminating peak tailing. Ion pairing approaches have also been largely unsuccessful. A few approaches using additives have been quite successful. Effective additives are generally too polar to be miscible with supercritical fluids.[21] Instead, they are added to a modifier, and then the modifier plus additive is pumped as a single fluid.

Polar solutes, like aliphatic amines[22,23] and polyfunctional carboxylic acids,[24,25] will not elute or else elute with severely distorted peaks without an appropriate additive in the mobile phase. In many such situations, the addition of a strong acid or base to the MP dramatically improves peak shapes. A typical example is provided in Figure 4.8 showing an extremely broad, flat topped peak for benzylamine, which becomes a sharp high efficiency peak with the addition of a basic additive.

Additives are not necessarily effective in improving peak shapes. Multifunctional hydroxysteroids (*e.g.*, hydrocortisone, or estriol) are much less polar than polyfunctional carboxylic acids.[24,25] Nevertheless, peaks[27] were distorted or they did not elute using pure CO_2 and non-polar columns such as ODS (C_{18}), MOS (C_8), or phenyl. The addition of MeOH to the MP improved some peaks

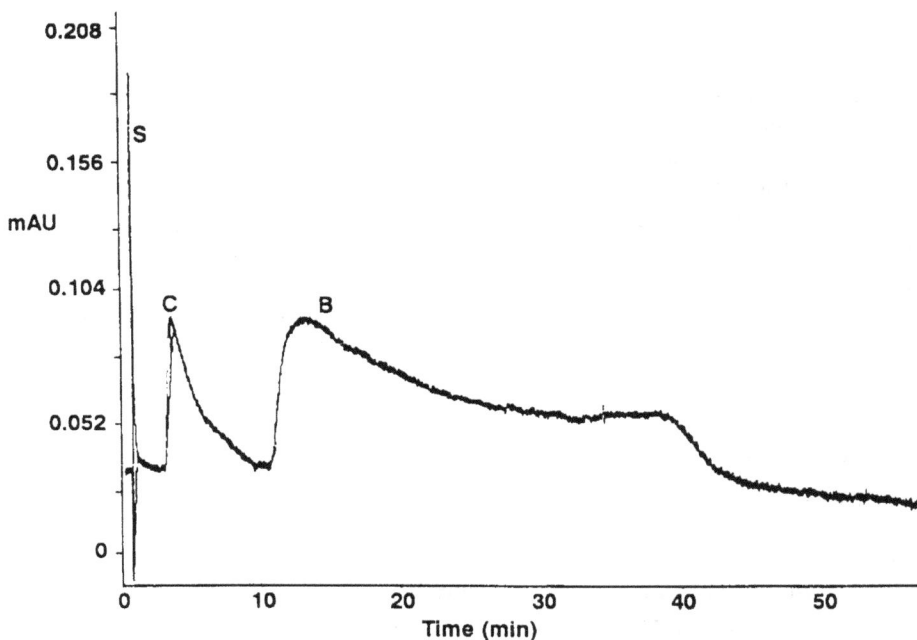

Figure 4.8 *An example of extremely poor peak shape without additive. Benzylamine from Deltabond C_8 using pure CO_2, 40 °C, 182 bar outlet. See also Figures 6.2 and 6.4*
(Reproduced by permission from ref. 22)

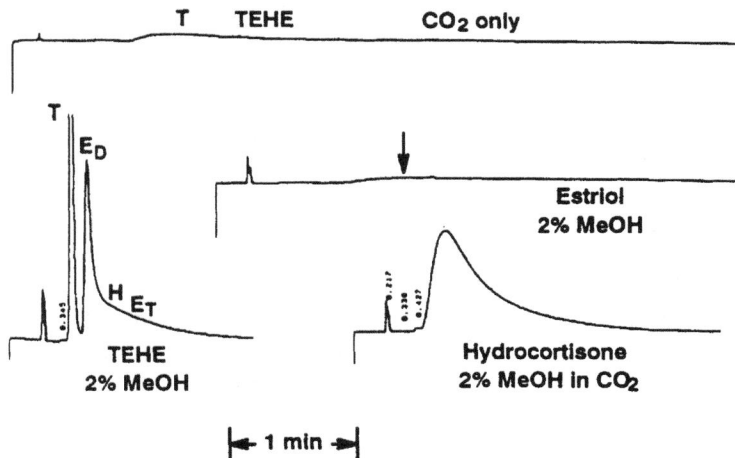

Figure 4.9 *The elution of hydroxysteroids from low polarity stationary phases. Some peaks continue to tail even when enough MeOH is added to make retention inadequate on a C_{18} column*
(Reproduced by permission from ref. 27)

but others remain badly distorted,[27] as shown in Figure 4.9. This behaviour superficially resembles the case in Figure 4.8. However, polar additives did NOT further improve peak shapes on these columns, as indicated in Figure 4.10.

Increasing the **stationary phase** polarity resulted in a very different result. The same compounds readily eluted with excellent peak shapes from more polar phases, such as silica, cyanopropyl, and diol, using MeOH modified CO_2 WITHOUT additives.

The failure of the additives to improve peak shapes on the less polar columns is difficult to explain in terms of active sites. If tailing on the less polar columns is caused by the solutes participating in two retention mechanisms (solute–bonded phase and solute–active site), then the solutes interact with these active sites but the more polar additives do not. The range of behaviours suggests multiple tailing mechanisms.

The Roles of Additives

The use of additives dramatically extends the range of solute polarity amenable to CO_2 based SFC. Additives appear to have multiple functions. Each of the functions of additives is discussed in a separate section, below. The intent is to indicate the much more complex nature of additive action and, hopefully provide insight into how to improve peak shapes.

Additives appear to perform at least four different functions.[21] These include: coverage of active sites, changing the polarity of the SP, suppression of ionization or ion pair formation by solutes, and changing the polarity of the MP.

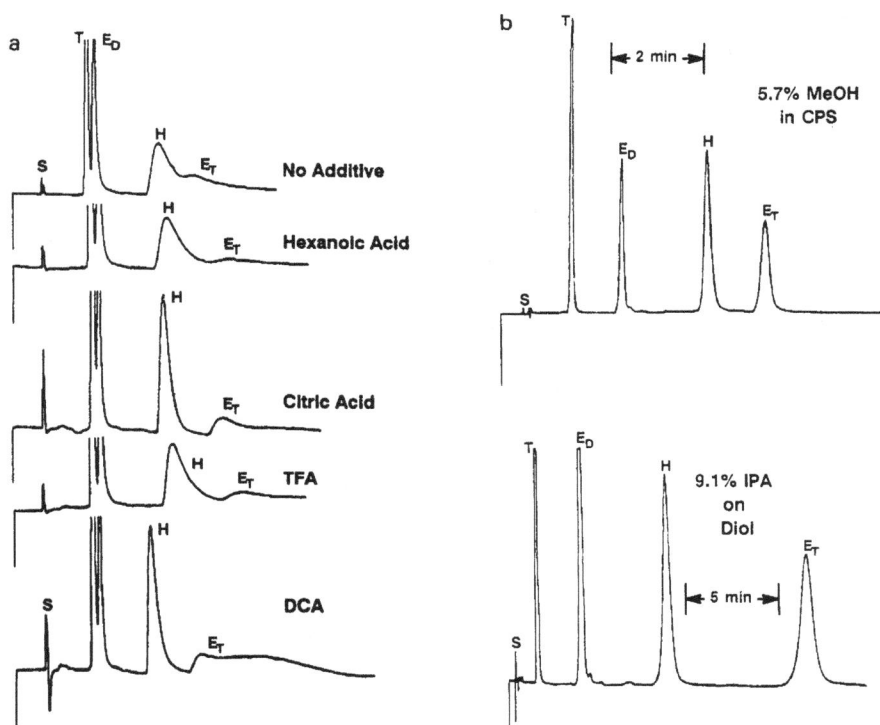

Figure 4.10 (a) *Additives fail to improve peak shapes of some of the solutes on a phenyl column.* (b) *On more polar columns (e.g. cyanopropyl, diol), MeOH/CO_2 mixtures produce beautifully symmetric, high efficiency peaks without additives. The failure of additives to improve peak shapes on the less polar phenyl suggests that tailing is unrelated to 'active sites'* (Reproduced by permission from ref. 27)

The square topped peak for benzylamine in Figure 4.8 implies a solubility problem, although other explanations are possible. The very long, low, mesa like shape suggests that either the local capacity of the SP (solubility in the SP) is extremely low or the 'capacity' of the MP to transport the solute is inadequate (low solubility in the MP). Alternatively, the solute could have several forms: an insoluble (perhaps ionic?) form and a (undissociated?) soluble form. The insoluble form would drop out of solution at the head of the column, but would be in equilibrium with the soluble form which would be transported down the column. The addition of the additive could: increase the solvent strength of the MP, increasing solubility; raise the polarity of the SP, increasing its capacity; or suppress the ionization (or dissociation or ion pairing) of the solute, allowing a single form to be transported by the MP. There has been surprisingly little work done to understand the mechanisms involved. Much of the evidence that exists on the functions of additives is presented in the following sections.

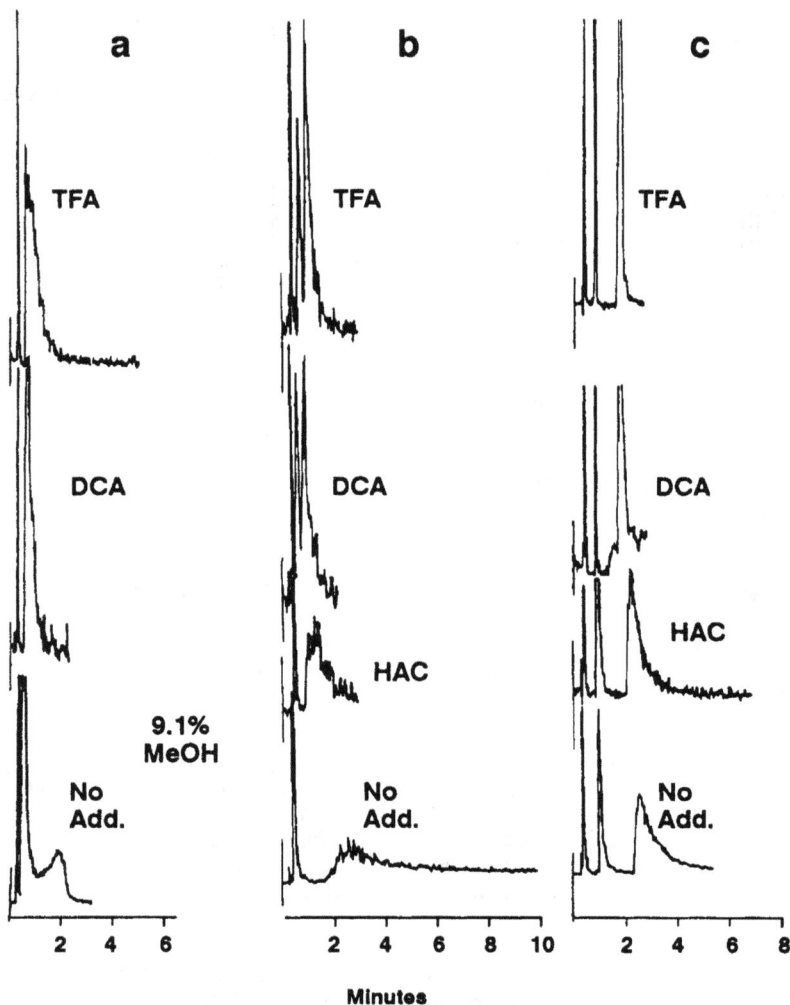

Figure 4.11 *Peak shapes of benzoic, phthalic and trimellitic acids on MOS (1.2% MeOH in CO_2), phenyl (0.8%) and cyano (1.6%) SPs using various additives in a MeOH/CO_2. Increasing additive 'polarity' improves peak shapes. Key: HAC = acetic acid, DCA = dichloroacetic acid, TFA = trifluoroacetic acid. Other acids used with similar effects but not shown, citric acid, chloroacetic acid, trichloroacetic acid. Columns were 2×100 mm, 5 μm d_p, 40 °C, 130 bar, 3 ml min^{-1} (ca. $6 \times \mu_{opt}$)*
(Reproduced by permission from ref. 21)

Active Sites–Additive Polarity

A mixture of benzoic, phthalic, and trimellitic acids produced severely tailed peaks and no elution of some peaks on MOS, phenyl, and CN columns using binary MeOH/CO_2 mixtures.[21] The addition of progressively more acidic additives produced a trend toward better peak shapes and less retention, as shown for three SPs in Figure 4.11. Acetic acid (HAC) and chloroacetic acids (CA) only marginally improved peak shapes on any of the columns even though they were present in the MP at much higher concentrations than the solutes. Dichloroacetic acid (DCA) (warning: extremely toxic) noticeably improved peak shapes, while the even more acidic trichloroacetic (TCA) and trifluoroacetic (TFA) acids produced symmetrical solute peaks on all but the least polar columns (MOS and phenyl).

Results on diol and sulfonic acid columns were also studied[21] and continued the general trend of Figure 4.11. With no additive, the solutes did not elute or severely tailed. Increasing SP polarity produced better peak shapes but only with the presence of the most polar additives. Low polarity additives did not produce good peak shapes.

One should probably ask: which should be more polar, a silanol or a sulfonic acid? If compounds tail on a diol or sulfonic acid column when no additive is present, it is difficult to attribute such tailing to the simultaneous presence of a less populous and less polar secondary phase interaction (why should solutes be more retained on silanols than on the more polar bonded phases?).

Changing the Stationary Phase Polarity

Some solutes elute with binary MPs with symmetrical peaks and reproducible retention times. The subsequent addition of an additive sometimes significantly changes retention time, without changing peak shape. Symmetrical peaks imply that a single retention mechanism is operating. Since the peaks remain symmetrical throughout the process it is difficult to think of the process in terms of competing mechanisms.

Another piece of information provides additional insight. Additives can be very strongly held by the stationary phase. One measurement technique that can be used to determine the amount adsorbed is very much like a titration with an endpoint. The sudden introduction of an additive does not result in a baseline offset just after the column transit time. Instead, a substantial time can pass with no additive apparent in the column effluent. Eventually the background undergoes a rapid, step change to a new absorbance level. This suggests that additive progressively fills all available sites as it passes from the column inlet to the column outlet. If all available sites are filled before any additive proceeds down the column, then a mixed phase does not form. Instead, the column acts as though there were two columns in series. As time passes, the second column, without any additive adsorbed, gets shorter and the first column, with complete additive coverage, becomes longer. Additives can cause an increase, a decrease, or no change in solute retention.

Surface Coverage by Additives

The quantity of additives that adsorb onto various SPs has been measured.[21] Low to moderately polar SPs adsorbed only small amounts of additive. A MOS

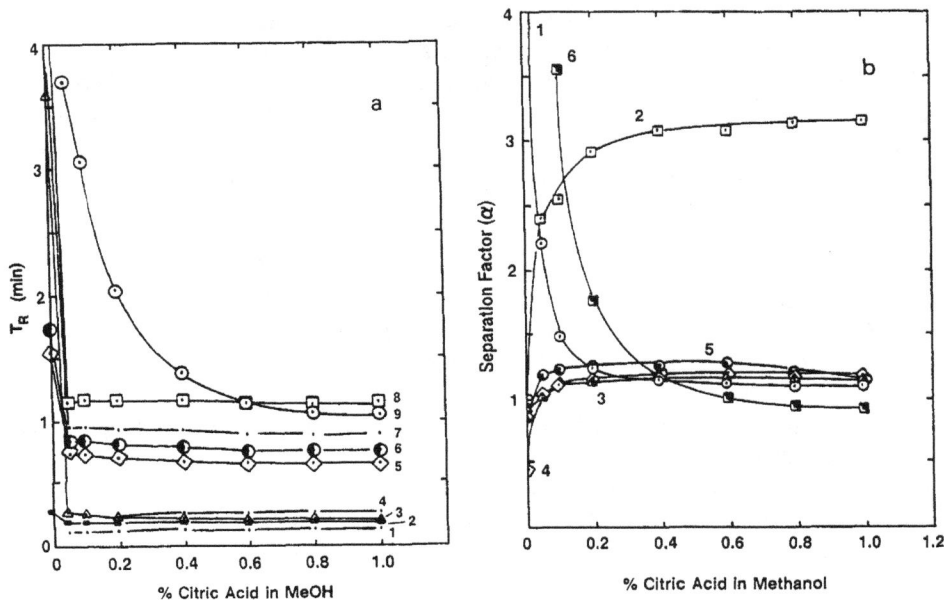

Figure 4.12 *Retention time (t_R) and selectivity (α) as a function of additive concentration. MeOH concentration, μ, temperature, and pressure all remained fixed throughout. A few solutes eluted rapidly with good peak shapes without an additive. The retention times and peak shapes of most solutes dramatically changed when the first small amount of an additive was placed in the MP. Higher concentrations had no additional effect on any but two solutes. The retention times and peak shapes of those two solutes continued to change with increasing concentrations of additive. Only those two solutes had lower aqueous pK_a values than the additive*
(Reproduced by permission from ref. 21)

(C_8) and a cyanopropyl column each adsorbed only enough additive to produce a surface coverage on the SP of approximately 0.4–0.6% of a monolayer (similar to some estimates of the population of metal ions on some silicas). Under the same conditions, much more polar sulfonic acid and diol columns adsorbed much larger amounts of additive, creating surface coverages of up to 21%. Either the surface coverage of silanols or other 'active sites' is very low on the MOS AND the cyanopropyl columns and very high on the others, or adsorption is more complicated than simple coverage of active sites.

Additive Concentration–Ionization Suppression?

With very polar additives that are clearly more acidic than acidic solutes, the first small addition of additive improves peak shapes and sometimes shifts retention. The addition of higher concentrations of the same additive tends to have little further effect. However, if the additive is of similar acidity to the solutes, the concentration of the additive appears to be important in determining peak shape and retention of the solutes.

Benzoic acid and six other hydroxybenzoic and polycarboxylic acids were separated[21] on a diol column using $MeOH/CO_2$ with various low concentrations of citric acid as an additive. Without any additive, three of the solutes eluted with good peak shapes, some peaks were relatively symmetrical but long retained, while a few others did not elute, as indicated in Figure 4.12.

The diol functional groups from the bonded phase are likely to be at least as polar as silanols, are more accessible, and are obviously present in larger numbers. The presence of active sites might be construed as explaining poor peak shapes or no elution. However, why do similar solutes elute with good peak shapes in spite of the presence of the same active sites?

The addition of a small amount of citric acid to the mobile phase dramatically changed the retention of all the solutes. For most of the solutes, further increases in the additive concentration had minimal effect on retention or peak shapes. Such behaviour could be interpreted as the additive saturating active sites, preventing further interaction between the active sites and the solutes. It can also be interpreted as a change in the SP polarity.

Several solutes continued to exhibit poor peak shapes and long retention after the first small concentration of additive was used, as indicated in the plot of selectivity ($a = k'_2/k'_1$) vs percentage additive in Figure 4.12b. The 1,2-disubstituted compounds were the only solutes present with an aqueous pK_a lower (more acidic) than the additive (citric acid) and were the only solutes sensitive to the concentration of the additive. Those solutes behaved as though they were in equilibrium with the additive. The other solutes were all weaker acids and responded to (were protonated by?) the first small addition of the additive.

Other measurements[21] indicated that the surface coverage of the SP by additive is independent of its concentration in the MP. If this is true, then the effect of changing additive concentration on solute retention was a MP phenomenon and probably supports the concept of solute ionization suppression.

A subject needing further study involves solutes that can have several different structures (tautomers). Such compounds can be more difficult to elute and exhibit poorer peak shapes than much more polar solutes.

Changing Mobile Phase Polarity

This section logically belongs in Chapter 3, where MP phenomena are discussed. However, any discussion of the other effects of additives is incomplete without a discussion of their effect on MP solvent strength.

Solvatochromic dyes have been used to measure the solvent strength of tertiary mobile phases. Small concentrations of additive at least sometimes increase the apparent polarity or solvent strength of the MP.[28] One percent TFA increased the apparent solvent strength of $MeOH/CO_2$ mixtures marginally, as shown in Figure 4.13. However, decreasing the modifier polarity by replacing the MeOH with acetonitrile or methylene chloride results in dramatic increases in the apparent solvent strength of the MP. For example, a mixture of

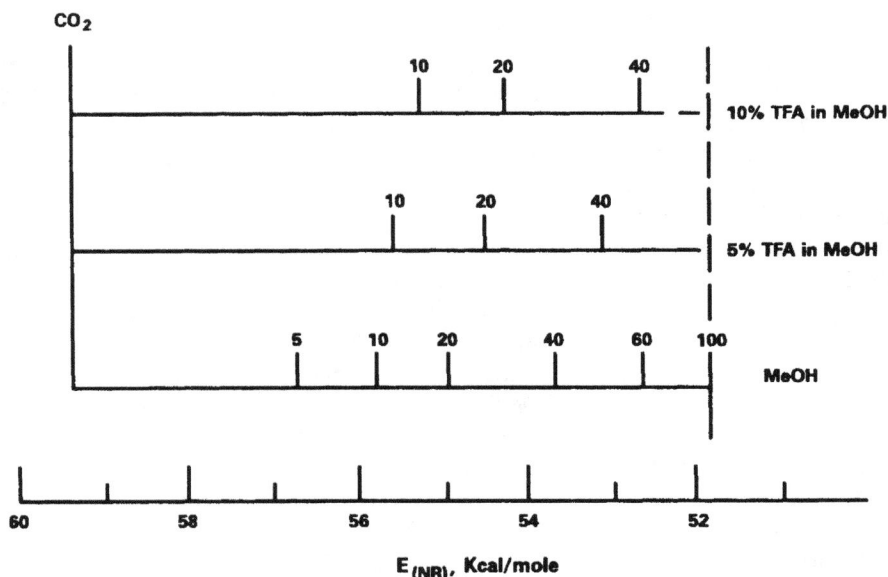

Figure 4.13 *Apparent increase in solvent strength of MeOH/CO$_2$ mixtures when trifluoroacetic acid was added*

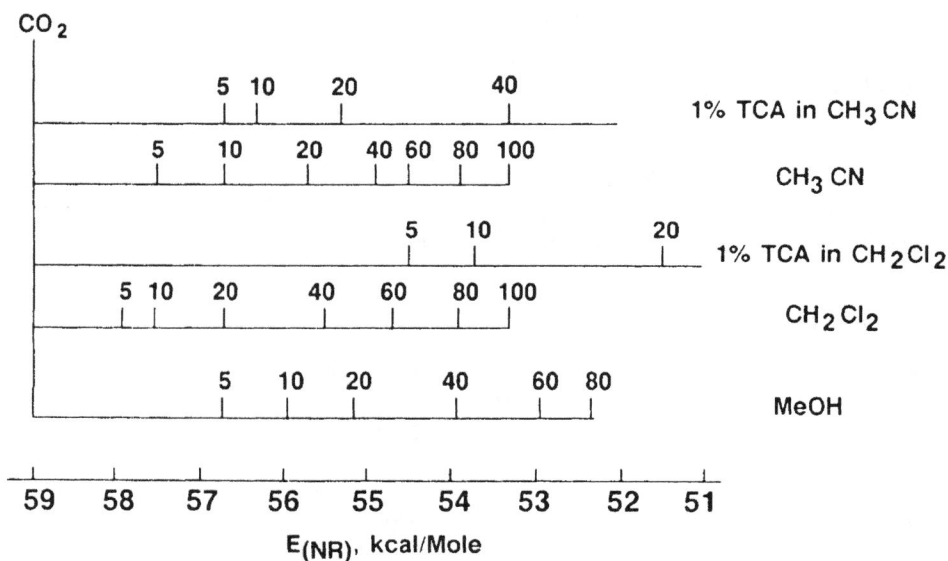

Figure 4.14 *Solvent strength enhancement when polar additive was added to acetonitrile/CO$_2$ and methylene chloride/CO$_2$ mixtures*
(Reproduced by permission from ref. 28)

0.2% TCA, 19.8% methylene chloride, and 80% CO_2 produced an **apparent** solvent strength substantially higher than that of pure MeOH! The effect of TCA on acetonitrile/CO_2 and methylene chloride/CO_2 mixtures[28] is shown in Figure 4.14.

These results suggest that additives are just an extreme form of a modifier. Even very low concentrations can have relatively large impact on MP solvent strength.

6 Choosing an Additive

If active sites are acidic, as is often assumed, basic compounds should tail worse than acidic solutes. However, solutes tail regardless of whether they are acidic or basic. They even tail on polymer based columns that contain no silica. Nevertheless, tailing is still almost universally attributed to the presence of active sites. Such designation offers no insight toward improving the chromatography.

Acidic additives are needed to improve the peak shapes of acidic solutes. Basic additives are needed to improve the peak shapes of basic solutes.[22,23] So far we have found no instance where a basic additive has improved the peak shapes of acidic solutes or *vice versa*. Although there have been several papers claiming ion pairing in SFC, only one appears unequivocally to support ion pairing[29] without a possible alternative explanation.

One difference between acidic and basic additives has been noted. Multifunctional acidic additives have been effective in suppressing tailing, even when the solutes are similar or slightly more acidic. Multifunctional bases, however, are ineffective or even detrimental to improving peak shapes.[22] Adding such compounds as ethylenediamine or diethylenetriamine as additive degraded peak shapes of such compounds as aliphatic amines. Such action is equivalent to the creation of 'active sites'.

Solutes Can Act as Additives

A behaviour which has not been discussed in the literature involves solute–solute interactions. The most polar entity present in a chromatographic system affects the retention characteristics of all other entities. If solutes are more polar than the phases, solute-solute interactions can determine retention and peak shapes.

With methanol modified mobile phases (NO additive), resorsinol and catechol (two hydroxyl groups on a benzene ring) elute with reasonable peak shapes. However, retention is a function of solute concentration. If several polar solutes are present, the concentration of each may affect the retention of the others. When a more polar additive is included in the mobile phase this behaviour ceases. It appears that the most polar entity dictates the retention of all less polar solutes.

Similar behaviour occurs on capillary columns when pure fluids like CO_2 are used as the MP. An example is provided by an attempt to resolve all the mono-

and di-acids present in the synthesis of nylon. Initial experiments indicated that polar SPs were required to obtain both significant retention and good peak shapes. With polar SPs, individual acids could be eluted with symmetrical peaks. Methods could also be developed producing symmetrical peaks and reproducible retention times for specific mixtures containing relatively narrow ranges of molecular weight. However, significant changes in the concentration (*e.g.* two times) of any one component caused shifts in the retention times and peak shapes of all later eluting compounds. Using broader ranges of molecular weights also resulted in the distortion of some of the peaks. In the end it was concluded that neither the MP or the SP was polar enough to suppress solute–solute interactions.

Support Deactivation–Polymer Based Columns

Several manufacturers have coated silica particles with a deactivating polymer layer before application of the main bonded phase. The most widely known is Deltabond by Keystone Scientific. The intent is to improve peak shapes and allow the elution of more polar solutes using pure CO_2 as the MP. This approach has been only modestly effective in improving peak shapes.

While retention can be decreased and peak shapes improved somewhat, the degree of improvement, using pure fluids as the MP, is only modest. The main effect appears to be an improvement of the peak shapes of molecules that can be eluted from columns without the treatment. The treatment does not allow more polar solutes to be eluted.[23] However, with modified MPs, these columns can produce performance that differs from standard silica based columns. The effective SP on most columns includes the adsorbed film of MP components. More polar phases tend to adsorb more polar modifier. Adsorption and retention characteristics are likely to be different on a polymer deactivation layer, compared with bare silica. It is likely that these polymer deactivation layers do cover or suppress active sites but do nothing to overcome other sources of peak tailing.

Most workers have assumed that peak shape problems have been primarily due to the presence of relatively undefined 'active sites' on silica. Some further reasoned that the way to produce better peak shapes was to eliminate the silica. Subsequently, there has been considerable effort to develop polymer based particles. The use of such particles as a chromatographic support is a severe test of their physical strength. The pressure drop across columns, due to the flow of mobile phase, can collapse some polymer particles and dramatically swell others. Unfortunately, polymer based supports (no silica) have only been modestly successful in improving peak shapes.

Gere[30] was able to separate the homologues of polyethylene glycol derivatives such as Triton X-100 on a PRP-1 column with pure CO_2 but could not separate them on silica based C_{18} columns.

Asahipack materials, recently developed for reversed phase LC, have been somewhat successful in minimizing the required polarity of the MP but still produce tailed peaks.[23] A C_{18} bonded phase produced little retention but severe

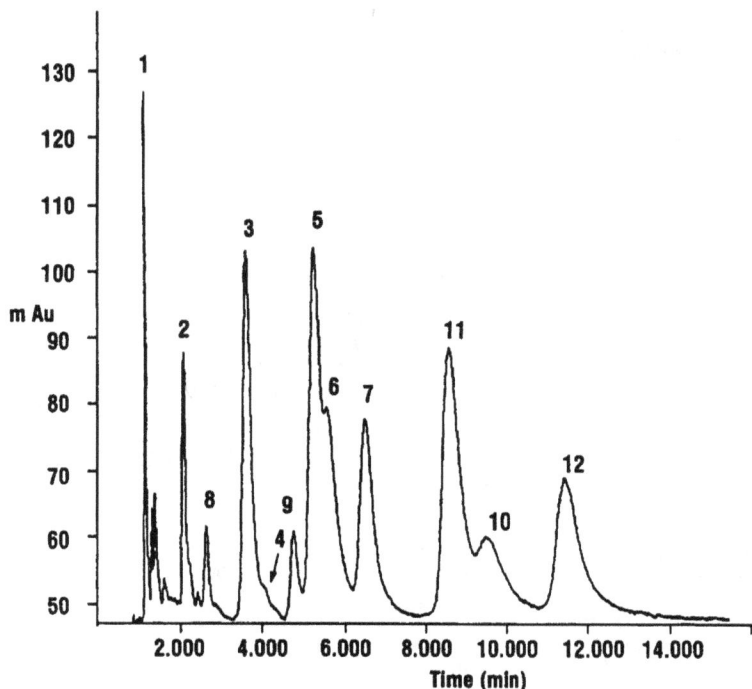

Figure 4.15 *Elution of a mixture of anilines, benzamides, benzylamines, and other strong bases from an Asahi Pac polymer based amino column using MeOH/CO₂. The stronger bases would not elute from silica based columns without an additive in the mobile phase (cf. also Figure 6.4)*
(Reproduced by permission from ref. 23

tailing for amines. However, an amino phase eluted even strong bases without an additive although the peaks were severely tailed, as shown in Figure 4.15. Despite the poor peak shapes, this column produced the best peak shapes for aliphatic amines without a strong base additive in the MP. This partial improvement suggests that active sites were suppressed or eliminated but other problems, leading to peak tailing were not overcome.

All these deactivation approaches share the central concept that peaks tail primarily because of the presence of active sites. It is probably true that these approaches do minimize the effect of active sites but fail to suppress other mechanisms which also lead to tailed peaks. In our opinion, the central concept that 'active sites' are the source of all peak tailing is flawed.

7 Conclusions

Stationary phase phenomena on packed columns appear to be dominated by the adsorption of MP components. This adsorption appears to change dynamically both SP volume and polarity. Since the adsorbed film is always more dense and a stronger solvent than the MP, SFC (with CO_2 based fluids) is almost always a normal phase technique. Descriptions of retention mechanism need to account for such adsorption.

8 References

1. J. Nawrocki and B. Buszewski, *J. Chromatogr.*, 1988, **449**, 1.
2. A. Nomura, J. Yamada, and K.I. Tsunoda, *J. Chromatogr.*, 1988, **448**, 87.
3. T.A. Berger and J.F. Deye, *J. Chromatogr.*, 1992, **594**, 291.
4. J.R. Strubinger, H. Song, and J.F. Parcher, *Anal. Chem.*, 1991, **63**, 98.
5. J.R. Strubinger, H. Song, and J.F. Parcher, *Anal. Chem.*, 1991, **63**, 104.
6. C.H. Lochmuller and L.P. Mink, *J. Chromatogr.*, 1989, **471**, 357.
7. B.W. Wright and R.D. Smith, *J. Chromatogr.*, 1986, **355**, 367.
8. J.M. Levy, PhD Thesis, Case Western Reserve University, 1986
9. C.R. Yonker and R.D. Smith, *J. Chromatogr.*, 1986, **361**, 25.
10. S. Schmidt, G.L. Blomberg, and E.R. Cambell, *Chromatographia*, 1988, **25**, 755.
11. K.D. Bartle and D. Leyendecker, in 'Supercritical Fluid Chromatography', ed. R.M. Smith, The Royal Society of Chemistry, London, 1988, Chapters 1 and 3.
12. S.M. Fields, K.E. Markides, and M.L. Lee, *J. Chromatogr.*, 1987, **406**, 223.
13. Ph. Morin, M. Caude, and R.J. Rosset, *J. Chromatogr.*, 1987, **407**, 87.
14. C.R. Yonker, D.G. McMinn, B.W. Wright, and R.D. Smith, *J. Chromatogr.*, 1988, **396**, 19.
15. A.L. Blilie and T. Greibrokk, *Anal. Chem.*, 1985, **57**, 2239.
16. T.A. Berger and J.F. Deye, *Anal. Chem.*, 1990, **62**, 1181.
17. M. Novotny, W. Bertsch, and A. Zlatkis, *J. Chromatogr.*, 1971, **61**, 17.
18. B.P. Semonian, L.B. Rogers, *J. Chromatogr. Sci.*, 1978, **16**, 49.
19. T.L. Chester and D.P. Innis, *J. High Resolut. Chromatogr.*, 1985, **8**, 561.
20. T.A. Berger, *Chromatographia*, 1993, **37**, 645.
21. T.A. Berger and J.F. Deye, *J. Chromatogr.*, 1991, **547**, 377.
22. T.A. Berger and J.F. Deye, *J. Chromatogr. Sci.*, 1991, **29**, 310.
23. T.A. Berger and W.H. Wilson, *J. Chromatogr. Sci.*, 1993, **31**, 127.
24. T.A. Berger and J.F. Deye, *J. Chromatogr. Sci.*, 1991, **29**, 26.
25. T.A. Berger and J.F. Deye, *J. Chromatogr. Sci.*, 1991, **29**, 141.
27. T.A. Berger and J.F. Deye, *J. Chromatogr. Sci.*, 1991, **29**, 280.
28. T.A. Berger and J.F. Deye, in 'Supercritical Fluid Technology', ed. F.V. Bright, and M.E.P. McNally, ACS Symposium Series No. 488, American Chemical Society, Washington, DC, 1992, Chapter 11.
29. W. Steuer, M. Schindler, G. Schill, and F. Erni, *J. Chromatogr.*, 1988, **447**, 287.
30. D.R. Gere, Hewlett-Packard Application Note AN 800–3, 1983.
31. T.A. Berger and J.F. Deye, *J. Chromatogr. Sci.*, 1991, **29**, 390.

CHAPTER 5

Effect of Instrumental Parameters on Retention, Selectivity, and Efficiency

1 Introduction

In Chapters 3 and 4 the physical chemistry of mobile (MP) and stationary phases (SP) were detailed. However, the effects of the instrumental variables on efficiency (N), retention (k' or t_R), and selectivity (α) were not addressed. This chapter attempts to relate the effects of modifier concentration, pressure, temperature, and flow to the quality of separations. Only items controlled by the instrument keyboard are considered. Such items as the identity of the MP or SP are excluded. Obviously this is an arbitrary distinction since switching valves could be controlled by the keyboard to change either phase on demand. Phase selection is discussed elsewhere (see Chapters 6 and 7).

The material in this chapter is presented in a somewhat unusual manner in that the summary is presented first. This is done to try to avoid confusion. By presenting the summary first, an overview is obtained. The reader can then find a deeper discussion of any specific topic in later sections. Sections 3, 4, and 5 deal with modified fluids and polar solutes. Section 3 deals with retention, Section 4 with selectivity, and Section 5 with efficiency. Section 6 discusses pure fluids and lower polarity solutes.

2 Summarizing the Effects of Physical Parameters

Essentially all the control parameters available in both GC and LC are available and useful in packed column SFC. Mobile phase composition and identity, temperature, pressure, flow, and SP identity all impact k', α, and N. Such a large selection of parameters has both positive and negative aspects. Although it is always helpful to have extra options for optimizing a separation, a large number of variables can also be quite confusing. If half a dozen options are available, which should one try first? Overall, there are logical step by step approaches that will solve most problems in a minimum number of steps. Step by step method development is outlined in Chapter 7. In this chapter, the effect of keyboard controlled physical parameters is summarized, providing background for Chapter 7.

Table 5.1 *Importance of physical parameters in adjusting performance*

	Retention	Selectivity	Efficiency
Most important	percentage modifier		flow
		temperature	
	pressure	pressure	
	temperature	percentage modifier	pressure
Least important	flow		temperature
		flow	percentage modifier

Binary and tertiary MPs are used in the majority of packed column applications. Binary and tertiary MPs are most useful with polar solutes. With binary fluids, modifier concentration is the most effective means of adjusting k'. Pressure is the next most effective parameter, while temperature is the least effective in adjusting k'.

This order follows the effect of each parameter in changing the MP solvent strength. Modifier concentration is far more important in determining solvent strength than any of the other parameters. Pressure or density, although important, are less effective in changing solvent strength. Solvent strength increases due to temperature are limited. Increasing temperature at even high pressure causes a decrease in density. Consequently, $T > 200\ °C$ is usually avoided in SFC.

The effect of physical parameters on α follows a different order, as indicated in Table 5.1. To adjust α, temperature is often the most important parameter. This is in part due to the adsorption–desorption characteristics of MP components on the SP in response to changes in temperature. Both the thickness (volume) and polarity of the adsorbed film change rather dramatically with small changes in temperature (between T_c and $1.2 \times T_c$, in K). Pressure tends to be the next most important parameter for adjusting α, again likely owing to its effect on MP adsorption. Modifier concentration, although producing the largest changes in k', tends to have the least effect on α.

The MP flow rate tends only to affect N and analysis time. All the other parameters tend primarily to affect k' and/or α.

Pure fluids as the MP are most effective with non-polar solutes. Pressure tends to be the primary means of changing k', although negative temperature programming can also be useful. There has been little written about α adjustment in such systems other than about differences in bonded SPs.

3 The Effect of Physical Parameters in Controlling Retention of Polar Solutes

Modifier Concentration–Retention Effects

The instrumental parameter that has the largest effect on the k' of polar solutes

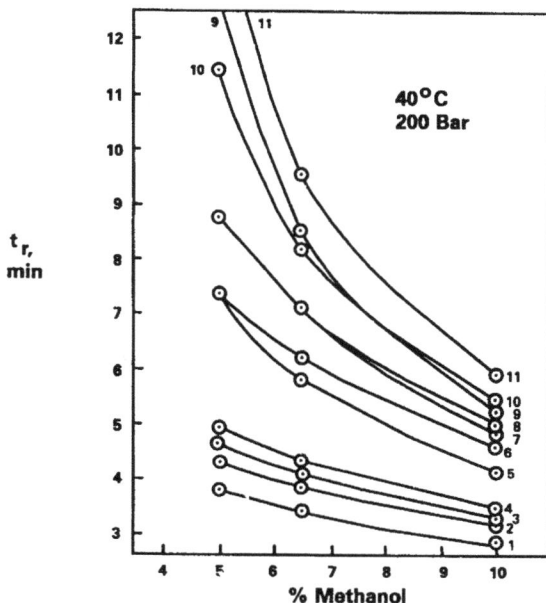

Figure 5.1 *The effect of modifier concentration on the retention of the eleven solutes covered by EPA Method 531. Seven of the solutes are commercial carbamate pesticides. They include: Aldicarb, Baygon, Barban, Carbofuran, Carbomyl, Methomyl, and Oxamyl. The rest are breakdown products of some of the pesticides. They included 1–naphthol, aldicarb sulfoxide, aldicarb sulfone, and 3–hydroxycarbofuran. Column: 4.6 × 250 mm, 5 μm Lichrosphere diol, flow 2.5 ml min⁻¹, 40 °C, 200 bar*
(Reproduced by permission from ref. 3)

is the presence and concentration of polar modifiers.[1] Small changes in modifier concentration tend to produce large changes in k'. The most useful range of modifier concentration is 1–30%, where the largest changes in solvent strength occur.[2]

As a rule of thumb, doubling the modifier concentration results in a halving of k'. For example, the effect of modifier concentration on the separation of carbamate pesticides[3] is illustrated in Figure 5.1.

Pressure and Retention

When polar modifiers, such as MeOH, are added to CO_2, solvent strength increases dramatically.[2] Large changes in solvent strength accompany small changes in modifier concentration (see Chapter 3). Changing the pressure of modified fluids changes the density, similar to pure fluids. A change in density produces similar changes in relative retention when either pure or modified fluids are used. The only difference is that the pure fluid has a solvent strength appropriate for low polarity compounds whereas the binary fluid solvent strength is more appropriate for more polar compounds. In either case, the

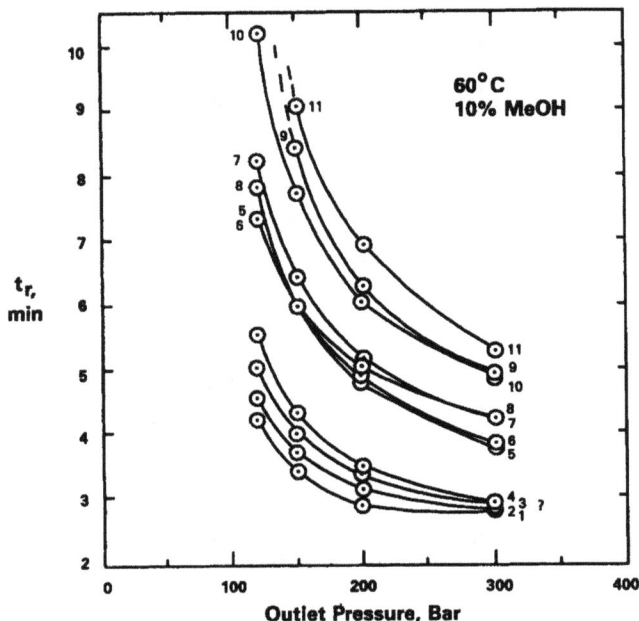

Figure 5.2 *The effect of pressure on the retention of the carbamate pesticides in Figure 5.1. Conditions the same except: 60 °C, 10% modifier, and pressure varied*
(Reproduced by permission from ref. 3

magnitude of the changes due to variations in pressure is small compared with the changes due to modifier concentration.[1] The effect of pressure can also be shown using the same carbamate pesticides used as examples of retention *vs.* modifier concentration, above. The retention times of the 11 carbamates as a function of pressure are shown[3] in Figure 5.2.

The range of densities available with binary fluids drops dramatically compared with pure fluids, particularly at low temperature. With MeOH/CO_2, most compositions are unstable[4] below densities of *ca.* 0.5 g cm^{-3}, breaking down into two phases (see Chapter 3). The usable range is, therefore, 0.5–1.0 g cm^{-3}, but it varies with temperature.

Since density is both a less effective control variable and the range of densities is curtailed, pressure is a secondary control variable compared with modifier concentration when binary fluids are used with polar solutes.

Temperature and Retention

Increasing the temperature of binary fluids, at constant pressure, decreases the density of the fluid,[4] but may or may not increase k'. At high temperature and low pressure, greater care is required to avoid two-phase formation. At high temperature, the minimum pressure that must be maintained to avoid two-

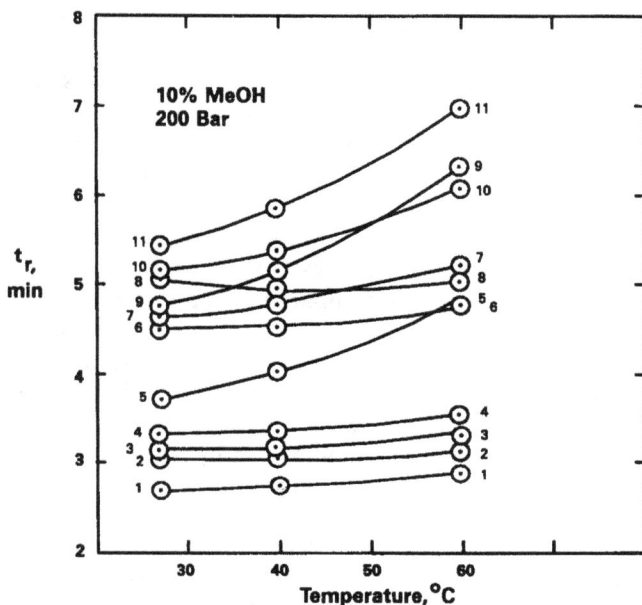

Figure 5.3 *The effect of temperature on retention of the carbamate pesticides. Conditions as in Figure 5.2 except: 10% methanol in carbon dioxide, temperature varied* (Reproduced by permission from ref. 3

phase formation increases[5] (see Chapter 3). With MeOH/CO_2 mixtures, low temperatures require only modest pressure. For example, at 40 °C, $P > 80$ bar always produces a single phase. However, at 150 °C, a pressure of up to 150 bar may be required to avoid two-phase formation. At each temperature, every composition has its own minimum pressure. Above 150 bar all compositions are stable at all reasonable temperatures.

Increased temperature also causes desorption of both CO_2 and MeOH from SPs[6] (see Chapter 4). This tends to decrease both the volume and the polarity of the effective SP, which tends to decrease k'. Note that this opposes the effect of increasing temperature causing a decrease in mobile phase density which would increase k'. In fact, there are numerous cases where higher temperature increases, decreases, or has no effect on k'. All three effects are obvious using the same carbamate mix discussed previously.

The effect of temperature on t_R[3] of the 11 compounds in EPA method 531 for carbamates is shown in Figure 5.3. Some compounds show a dramatic increase in t_R, some a modest increase, and one a slight decrease in t_R in response to a temperature increase. Such a wide range of response makes temperature an unpredictable control variable. Further, compared with pressure or modifier concentration, temperature changes have the least effect on k', and it is used as a tertiary control.

Mobile Phase Flow Rate and Retention

In LC, N is almost always low ($N < 20\,000$ plates is typical) so significant losses in N usually cause problems. Consequently, most analyses are conducted at near optimum flow rates. Since viscosities tend to be high, pressure drops are also high. This further discourages the use of much higher than optimum flow rates.

In packed column SFC, using pumps in the flow control mode with independent control of p_0, it is both simple and desirable to use flow as a routine control variable. Since D_{MS} tend to be high compared with those in LC (see Chapter 3), the μ_{opt} of the fluids is higher. Multiples of μ_{opt} in SFC produce much higher (absolute) μ than the same multiple in LC. For example a 4.6 mm column with 5 μm particles might have an optimum LC flow rate between 0.7 and 0.9 ml min^{-1}, but an optimum SFC flow rate between 2.5 and 3 ml min^{-1}. Two times optimum flow is as low as 1.4 ml min^{-1} in LC and as high as 6 ml min^{-1} in SFC. Such high SFC flows produce only a small loss in N compared with optimum.

Since viscosity in SFC is much lower than in normal liquids (see Chapter 3), pressure drops even at much higher velocities are still lower than in LC using much lower flows. This means that it is easy either to make the column substantially longer (more plates) or to make μ much higher (more plates s^{-1}), including using smaller particle packings.

We have often used smaller diameter columns, such as 2 mm i.d., operated at many times optimum (5–10 times!!) to scout out optimum conditions for a separation. Others[7,8] have followed a similar approach. Such high speed scouting can dramatically decrease the time required for method development.

4 The Effect of Physical Parameters on Selectivity between Polar Solutes

The primary purpose of chromatography is, of course, to separate solutes in the shortest time so that each can be quantitatively and/or qualitatively analysed. There are three ways to improve resolution between any two solutes as indicated by the universal resolution equation:

$$R_s = (\text{constant})[(\alpha - 1)/\alpha][k/(k + 1)](N)^{0.5} \qquad (5.1)$$

where R_s is the resolution between a pair of solutes, α is selectivity = k_2/k_1, k is partition ratio = $(t_R - t_0)/t_0$, t_R is retention time, t_0 is hold up time (or the transit time of an unretained peak), and N is efficiency expressed as theoretical plates [$N = 5.54(t_R/W_H)^2$, where W_H is peak width at half height].

Simply increasing retention (increased k) tends to increase resolution, particularly for initially little retained peaks. Similarly, increasing the column length (L) increases R_s, although slowly ($R_s \propto N^{0.5}$, but $N \propto$ length). Finally, the most powerful way to change R_s is to change the chemical interactions between the solutes and the chromatographic phases (α).

Since it is easy to generate high N in GC, long capillaries have become the columns of choice. However, as seen from Equation (5.1), higher N is the most expensive way to increase R_s. Resolution only increases with the square root of N whereas analysis time increases linearly with N.

In GC, after choosing a column with an inherently high N, R_s is further enhanced by increasing retention (lower temperature). However, above a k' of *ca.* 7, there is little further improvement in R_s while analysis time continues to increase. The most powerful way to improve R_s is to change α. In GC, changing α largely means changing the SP. Solute interactions with the MP are generally ignored. Peak reversals due to changes in temperature are minor.

In LC, it is difficult to generate high N. However, it is easy to change either the MP or SP. Changing the modifier changes both k' and α, and is the approach generally taken for increasing R_s. If changing the modifier doesn't work then the column is changed to one with a different SP.

Representing Selectivity Changes

To indicate a change in selectivity, α for each adjacent pair of solutes can be plotted against some parameter such as modifier concentration. However, packed column SFC probably produces more major peak reversals than any other chromatographic technique. One peak can shift past several others. In order to make it obvious that one peak is shifting past several others, at least several α values are required to represent the behaviour of each peak. When there are many such reversals, the values for α can be difficult to interpret and plots of α *vs.* some parameter can become extremely complex.

In this chapter, plots of k' and even t_R *vs.* the values of various control parameters are preferred over plots of α *vs.* the same variables. This produces simpler, more informative, and more understandable plots since only a single line represents each solute, and peak reversals remain obvious.

Modifier Concentration and Selectivity

The effect of modifier identity is discussed in Chapter3. The modifier concentration tends to have a large impact on k', but much less impact on α. As indicated previously, an increase in k' tends to increase R_s, although slowly. A typical increase in k' with decreasing modifier concentration was shown[3] in Figure 5.1 for carbamate pesticides using MeOH/CO_2 on a diol column. Note that there are few peak reversals (few of the lines cross each other) over the range of modifier concentration indicated. At low modifier concentration, R_s is better than at high modifier concentration but this can be largely attributed to increases in k'. **Relative** retention (solutes compared with other solutes) apparently did not change significantly.

Pressure and Selectivity

In capillary SFC, the most popular means of controlling k' involves changing the pressure, in order to vary density. As pointed out previously, changing the pressure of binary fluids is a far less effective way to change k' than changing modifier concentration. Over the full range of densities available, k' changes far less than over the full range of modifier concentration available. With most binary mixtures at low temperature, low densities cannot be made. Instead, the fluid tends to break up into two phases, making chromatography impossible. Two-phase formation severely limits the range of densities available with binary fluids. Thus, modifier concentration is the primary method for controlling retention in packed column SFC.

However, pressure changes tends to produce larger changes in α than do modifier concentration adjustments, although the differences are subtle. An example[3] is presented in Figure 5.2. In examining many such plots of k' or t_R vs. P for different families of solutes it becomes clear that peak reversals are more common with changes in pressure than with changes in modifier concentration.

Temperature and Selectivity

Chromatograms[4] of two sulfonamide antibiotics are presented in Figure 5.4. Notice that, initially the retention of both drugs decreased as temperature increased, until a minimum was reached. Further increases in temperature produced increased retention. Thus, plots of retention vs. temperature produce a minimum. This is opposite to the behaviour observed in capillary SFC, where maxima in retention vs. temperature plots occur. These changes in retention were accompanied by dramatic changes in α, resulting in the peak reversal shown in the chromatograms. Such a significant change in α is not an unusual occurrence in packed column SFC.

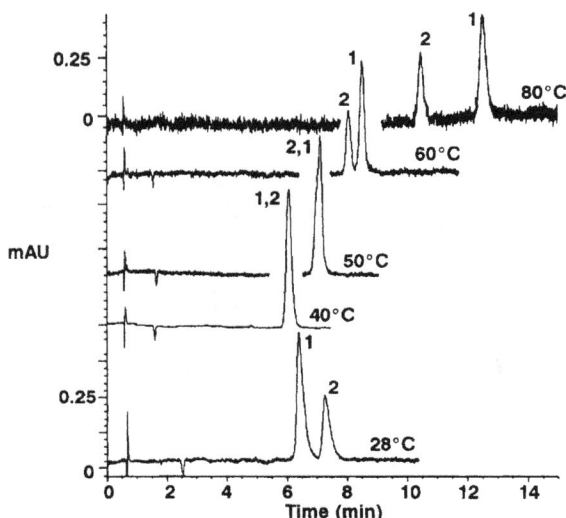

Figure 5.4 *Chromatograms of two sulfa drugs at different temperatures indicating selectivity changes. Key: 1 = sulfamethoxypyridazine, 2 = sulfisomidine. 5% methanol in carbon dioxide, 182 bar outlet pressure, 1 ml min^{-1}, Column: 2 × 150 mm, 3 μm Zorbax CN*
(Reproduced by permission from ref. 4)

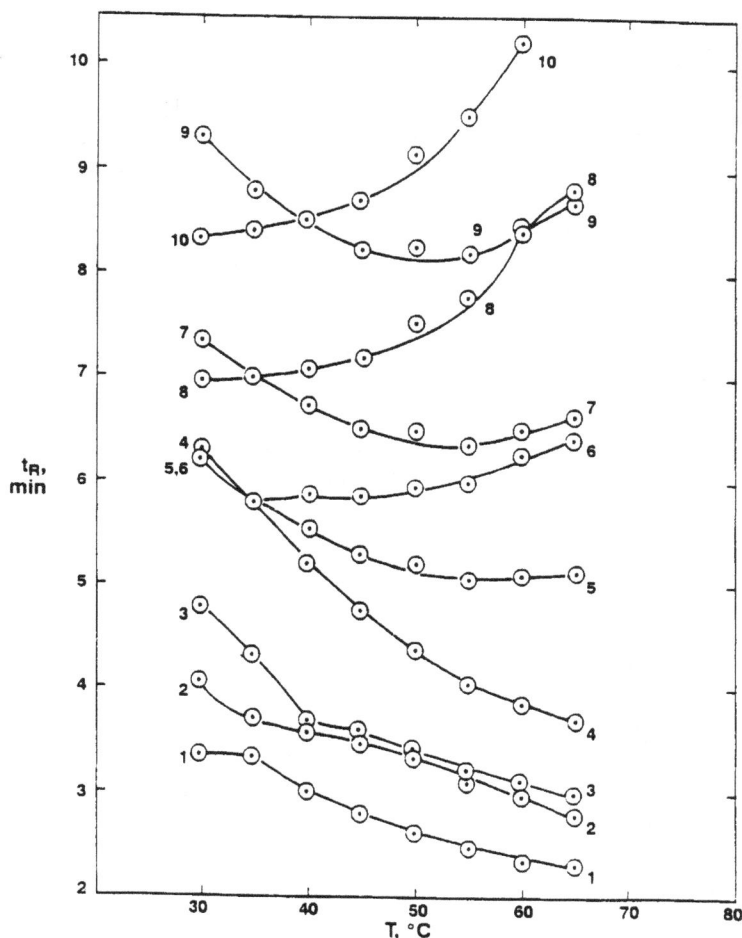

Figure 5.5   *The effect of temperature on the selectivity of antipsycotic drugs. Note in particular compounds 4, 5, and 6. Key: 1. Triflupromazine, 2. Carphenazine, 3. Methotrimeprazine, 4. Promazine, 5. Molindone, 6. Perphenazine, 7. Chlorprothixene, 8. Deserpidine, 9. Thiothixene, 10. Reserpine. 3 ml min^{-1} of 6% (methanol containing 0.5% isopropylamine) in carbon dioxide, 200 bar. Column: 4.6 × 250 mm, 5 μm Lichrosphere cyanopropyl (Reproduced by permission from ref. 9)*

A plot of retention *vs.* temperature using the same carbamates, in Figure 5.3,[3] produced another typical result for packed column SFC. The results divided into three types of behaviour. The retention of one compound decreased as temperature increased. The retention of seven of the compounds increased slightly, while the retention of three of the compounds increased much more noticeably. **Within** each of the latter two groups, plots of retention *vs.* temperature produced nearly parallel lines. However, the slopes of the lines of the two groups were quite different.

Not surprisingly, the compounds in the three groups belonged to three different families of compounds. The first compound was a naphthol, a non-nitrogen containing breakdown product of one of the carbamates. The second major group consisted of the parent pesticides. The third group was another set of breakdown products including a sulfoxide and a sulfone. Thus, each family of compounds produced a unique response to changes in temperature.

A larger change in α with small changes in temperature was observed in studies[9] of antipsycotic drugs. A plot of t_R *vs.* T for 10 members of this group is presented in Figure 5.5. Again, some compounds showed an increase, others a decrease, and still others almost no change in t_R when temperature was increased. Notice that the plots of many of the compounds appear to have minima, which is, again, directly opposite to the maxima found using capillaries.

The compounds numbered 4 (promazine), 5 (molindone), and 6 (perphenazine) in Figure 5.5 provide a startling change in selectivity with a temperature change of only 15 °C. At 30 °C, all three compounds co-elute ($\alpha = 1.0$). On raising temperature, the retention of perphenazine increases, the retention of molindone decreases modestly, and the retention of promazine decreases dramatically. At 45 °C, the α between promazine, and perphenazine reaches 1.35. As with the carbamates, the presence of various functional groups or structures correlates with differences in behaviour. All the tertiary amines behaved similarly. Each of the secondary amines behaved like all the other secondary amines and differently from the tertiary amines.

Small changes in temperature do not always produce large variations in α or even t_R. Some families appear to react to changes in temperature in exactly the same way. In other cases, temperature appears to have almost no effect on either retention or α.

Flow Rate and Selectivity

Flow has no direct influence on α. However, significant changes in flow through a packed column containing small particles generally cause changes in the pressure drop. This means that the inlet pressure and the average pressure both change. This change in pressure can alter α, but the effect is usually small.

There is no direct evidence that high flows cause heating of the fluid, which could also change α, due to thermal gradients. The possibility of such heating has been mentioned as a possible source of efficiency losses.

5 The Effect of Physical Parameters on Efficiency (with Polar Solutes)

Pressure and Efficiency

Changing the pressure (while holding temperature constant) changes the mobile phase density. Solute D_Ms are affected by the density of the fluid. In high density fluids, solute D_Ms in the MP are lower than in low density fluids. This means that the μ_{0pt} of the MP drops when the density increases.

In situations where the density is programmed, it is important to use instrumental schemes which independently control pressure and flow. If the MP mass flow is fixed, while the density is increased, both the solute D_M and μ decrease. In fact both parameters decrease in the same way when density is changed by changing the pressure at constant (mass) flow. The ratio D_M/μ is unchanged so N is unaffected by pressure programs with constant mass flow.

Previous generations of chromatographs which used the pump (usually a syringe pump) as a pressure source are incapable of independent control of P and flow. Such systems rely on fixed restrictors to control flow, which allow dramatic increases in mass flow when the pressure is increased. The MP μ generally increases at the same time that D_Ms decrease. A ten-fold increase in mass flow, occurring over a range of densities that decrease D_M more than five times, is common. The ratio D_M/μ, thus, can change more than 50 times as can N, during an extensive pressure program with fixed restrictors.

At low temperature and pressure there is a narrow range of conditions that can cause excessive adsorption of MP onto the SP. In this narrow range of conditions, N can severely degrade.[10]. Raising the pressure to 10–20 bar or increasing the temperature above 60 °C eliminates the problem.

Temperature and Efficiency

Changing the temperature (at constant pressure) also changes the density. However, temperature has an effect on solute D_Ms independent of any changes in density. Increasing temperature, even at constant density, causes an increase in D_M (increasing kinetic energy of the solute molecules). Therefore, both decreasing density and increasing temperature increase D_M. The μ of the MP only changes in proportion to the density change. During a negative temperature program (at constant pressure, and constant mass flow), D_M/μ still changes. Thus even with constant pressure and mass flow, μ_{opt} and N still change with a negative temperature program. Such behaviour is different from constant mass flow, constant temperature operation where N is not a function of pressure or density.

Mobile Phase Composition and Efficiency

At constant pressure and temperature, the density of $MeOH/CO_2$ mixtures increases with increasing MeOH concentration. However, both the solute D_Ms and μ change in a similar manner, minimizing any effect on N.

Making accurate, reproducible binary fluid compositions requires accurate mass flow control. Without such control, retention and variations in μ make the chromatography virtually useless for routine analysis. Therefore, with an adequately designed system, reproducible results effectively guarantee that the system efficiency changes little with composition programming.

Flow Rate and Efficiency

As with any other chromatographic technique, packed column SFC obeys a form of the Van Deemter equation. There is a flow rate that produces maximum N for a particular column. Either lower or higher flow rates cause a loss in N. Naturally, an analyst desires adequate separation in the shortest time. Because the μ_{opt} in SFC is significantly higher than μ_{opt} in LC, multiples of μ_{opt} in SFC are much higher than the typical values used in LC. Thus, a 4.6 mm i.d. column exhibits its μ_{opt} with a pump flow rate of 2.5–3.0 ml min^{-1}. Such a column can also be operated at 5 ml min^{-1} in SFC yet produce as much as 70% of the efficiency the same column produces at 0.7–0.9 ml min^{-1} in LC.

Many authors (see *e.g.* ref. 11) have observed what appears to be an efficiency loss sometimes associated with large pressure drops across columns. The losses only occur under a small subset of common conditions and appear to be due to retention gradients (due to the change in density which was due to the pressure drops) down the column. Therefore, high flow rates, which is one means of generating steep gradients in pressure, appear at least sometimes to degrade N by a mechanism other than those embodied in the Van Deemter equation. In a recent study,[12] these effects were quantified. Losses from this mechanism were extremely difficult to produce. It should be remembered that several groups,[7,8] including this one, have proposed using very high flow rates (*e.g.* 10 × optimum) in screening. Results from those groups show no hint of the problems suggested. Although the losses are real, it is trivial to avoid the limited sets of conditions that produce the problems. Therefore, such losses have been over emphasized. In general, high flow rates have only modest, expected impacts on efficiency.

6 The Effects of Physical Parameters Using Pure Fluids on Non-polar Solutes

Packed column SFC using pure fluids is both the form most similar to capillary SFC and the least useful form of packed column operation. The similarity makes it appropriate to compare packed columns using pure fluids with capillaries.

As discussed in Chapter 4, packing materials are inherently up to hundreds of times more retentive than capillaries. This makes it up to hundreds of times more difficult to elute solutes from packed columns than from capillary columns with the same fluid, at the same temperature and pressure. Typical pure SFs, such as CO_2, are weak, non-polar solvents only capable of displacing relatively non-polar solutes from packings.

With pure fluids, changing the pressure or temperature tends to change only the intensity, not the nature of interactions between solute and fluid. It is not surprising, therefore, that α is seldom mentioned or addressed in either packed column SFC using pure fluids or in capillary SFC other than by changing the SP or perhaps the MP.

In SFC with pure fluids, the density of the MP is the primary determiner of retention.[13–17] However, density is not a control variable (there is no 'density' control knob). To change the density of the fluid, either the temperature or pressure or both are changed. The most common approach to increase density is to increase pressure at constant temperature. There are, also, numerous examples of inverse temperature programming[18–21] where temperature is decreased at constant pressure to produce increasing density. Both temperature and pressure or temperature and density can be programmed simultaneously.[22]

In capillary SFC, solutes exhibit maxima in plots of k' vs. T at constant pressure.[18,23–29] This is often explained in terms of competition between solvation and volatility. However, in supercritical fluids only a single, compressible phase exists. Discussions of volatility are meaningless. When the data are replotted as $\log k'$ vs. **density**, monotonic, nearly straight lines are obtained. Thus, with the non-linearity explained, retention appears to be a simple function of solute interactions with the MP and SP.

The single largest problem with capillaries is the necessity of using fixed restrictors as passive control devices. Subsequently, the user can only actively control pressure or flow but not both. Since flow through restrictors varies with temperature and pressure[11,30,31] in a not easily visualized way, retention times vary due to both actual changes in chemical interactions (represented by k') and simple physical changes in flow. The observer cannot readily deconvolute these two effects. The much larger flow rates used with packed columns provide an advantage over capillaries in that a back pressure regulator can be used actively to control the pressure. The pump can then control flow. Independent, accurate, simultaneous control of pressure and flow allows simplified interpretation of t_R during pressure programs.

Pressure Control

With pure CO_2 and packed columns, changing density from air-like (density *ca.* 0.002 g cm^{-3}) to liquid-like (density *ca.* 1.0 g cm^{-3}) can result[10] in a change in retention of nearly 10^4 times. The number suggested is actually a maximum change accessible only with difficulty and covering the entire available range of densities possible. Initial densities are seldom below 0.05 g cm^{-3} and, especially at higher temperature, seldom exceed 0.85 g cm^{-3}. In LC, changing the MP composition can theoretically change the retention of some solutes by $> 10^{10}$ times.

The retention of fluoranthene[10] on Hypersil silica is shown in Figure 5.6. The data are presented as $\log k'$ vs. density at six temperatures. Pressure was varied from below 70 bar to nearly 400 bar at temperature from 40 to 150 °C. Retention varied nearly 10^4 times. To collect these data, short columns (10 cm) packed with relatively large particles (10 μm) were operated at low flow rates to produce pressure drops of < 1–2 bar.

Packed columns are usually longer, operated at higher flow rates, and packed with smaller particles compared with the columns used to collect the data in Figure 5.6. Common analytical flow rates tend to produce pressure drops greater than 10 bar. Under such conditions, no two locations in the column will

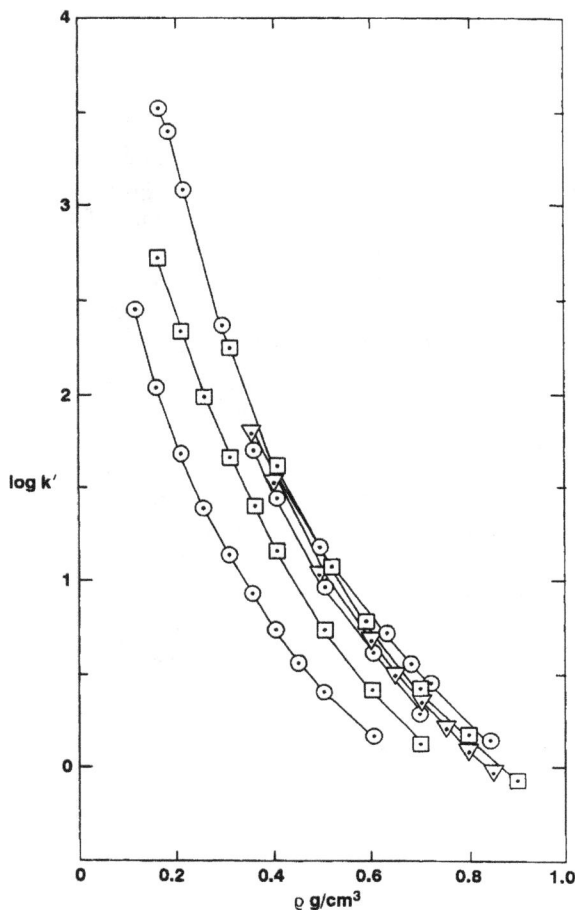

Figure 5.6 *Log k' vs. density of fluoranthene on a Hypersil silica column using pure carbon dioxide as mobile phase at various temperatures. Columns were 2.1 × 100 mm, packed with 10 μm particles. Flow was minimized to maintain pressure drops less than 2 bar in all cases. From bottom to top: 150, 100, 70, 60, 50, and 40 °C*
(Reproduced by permission from ref. 10

have the same density. This means that the association between retention and density is more nebulous on packed columns than on capillaries, since retention becomes a function of location. Changes in the pressure drop across the column cause changes in retention, even if the outlet density remains constant.

The pressure drop across the column is a relatively uncontrolled phenomenon. Changes in temperature, pressure, MP, the accumulation of dirt, or the break up of particles with sudden pressure changes (depressurization) can all cause the pressure drop across a column to change. If the column inlet pressure is controlled by a feed-back loop, changes in pressure drops tend to cause the p_o to decrease. If pressure drops become larger, p_o can become too low, the outlet density may become inadequate, and the solutes may not elute or elute with poor peak shapes. Such considerations make it clear that one should control p_o, instead of the inlet pressure, since this ensures that the location with the poorest solvent strength is controlled.

Temperature Control

It is also feasible to control density by varying temperature.[25-29] To increase the density at constant pressure, the temperature must decrease. Such inverse temperature programming has several advantages.

Decreasing temperature (at constant pressure) increases MP density without increasing mass flow, even using fixed restrictors.[32,33] With constant mass flow, μ actually decreases in proportion to the increase in density. D_Ms are also proportional to density. Since D_M/μ remains fairly constant when density changes at constant mass flow, efficiency does not degrade. The same is true using downstream control, since the mass flow is fixed by the pump. Changing either temperature or pressure or both causes no problems with N using packed columns and a back pressure regulator.

Negative temperature programming also has several drawbacks. With packed columns, the thick column wall can potentially delay thermal equilibration and help create a radial temperature gradient. More importantly, large molecules often require relatively elevated temperature and pressure to elute with good peak shape. Negative temperature programs produce the lowest temperature near the end of the program, where the largest molecules elute.

7 References

1. T.A. Berger and J.F. Deye, *Anal. Chem.*, 1990, **62**, 1181.
2. J.F. Deye, T.A. Berger, and A.G. Anderson, *Anal. Chem.*, 1990, **62**, 615.
3. T.A. Berger, J.F. Deye, and W.H. Wilson, *J. Chromatogr. Sci.*, 1994, **32**, 179.
4. T.A. Berger, *J. High Resolut. Chromatogr.*, 1991, **14**, 312.
5. T.A. Berger and J.F. Deye, *Chromatographia*, 1991, **31**, 529.
6. J.R. Strubinger, H. Song, and J.F. Parcher, *Anal. Chem.*, 1991, **63**, 104.
7. M.A. Morrissey, A. Georgetti, M. Polasek, N. Pericles, and H.M. Widmer, *J. Chromatogr. Sci.*, 1991, **29**, 237.
8. A. Georgetti, N. Pericles, H.M. Widmer, K. Anton, and P. Datwyler, *J.Chromatogr. Sci.*, 1989, **27**, 318.
9. T.A. Berger and W.H. Wilson, *J. Pharm. Sci.*, 1994, **83**, 281.
10. T.A. Berger, *Chromatographia*, 1993, **37**, 645.
11. T.A. Berger, *Anal. Chem.*, 1989, **61**, 356.
12. T.A. Berger and C. Toney, *J. Chromatogr.*, 1989, **465**, 157.
13. J.C. Giddings, M.N. Meyer, L. McLaren, and R.A. Keller, *Science*, 1968, **162**, 67.
14. R.E. Jentoft and T.H. Gouw, *J.Chromatogr. Sci.*, 1970, **8**, 138.
15. J.A. Graham and L.B. Rogers, *J. Chromatogr. Sci.*, 1980, **18**, 75.
16. J.C. Fjeldsted, W.P. Jackson, P.A. Peaden, and M.L. Lee, *J. Chromatogr. Sci.*, 1893, **21**, 222.
17. S.M. Fields and M.L. Lee, *J. Chromatogr.*, 1985, **349**, 305.
18. M. Novotny, W. Bertsch, and A. Zlatkis, *J. Chromatogr.*, 1971, **61**, 17.
19. B. Wenclawiak, *Fresenius' Z. Anal. Chem.*, 1986, **323**, 492.
20. Y. Hirata, F. Nakata, and S. Murata, *Chromatographia*, 1987, **23**, 663.
21. B. Wenclawiak, *Fresenius' Z. Anal. Chem.*, 1988, **330**, 218.
22. D.W. Later, E.R. Cambell, and B.E. Richter, *J. High Resolut. Chromatogr.*, 1988, **11**, 65.
23. S.T. Sie and G.W.A. Rijnders, *Sep. Sci.*, 1967, **2**, 729.

24. S.T. Sie and G.W.A. Rijnders, *Sep. Sci.*, 1967, **2**, 755.
25. B.P. Semonian and L.B. Rogers, *J.Chromatogr. Sci.*, 1978, **16**, 49.
26. D. Leyendecker, F.P. Schmitz, and E. Klesper, *J. Chromatogr.*, 1984, **315**, 19.
27. F.P. Schmitz, D. Leyendecker, and E. Klesper, *Ber. Bunsenges. Phys. Chem.*, 1984, **88**, 912.
28. T. Takeuchi, K. Ohta and D. Ishii, *Chromatographia*, 1988, **25**, 125.
29. T.A. Berger, *J. Chromatogr.*, 1989, **478**, 311.
30. R.W. Bally and C.A. Cramers, *J. High Resolut. Chromatogr.*, 1986, **9**, 626.
31. T.A. Berger, *J. High Resolut. Chromatogr.*, 1989, **12**, 96.
32. P.J. Schoenmakers, in 'Supercritical Fluid Chromatography', ed. R.M. Smith, Royal Society of Chemistry, London, 1988, Chapter 4.
33. T.A. Berger and L.M. Blumberg, *Chromatographia*, 1994, **38**, 5–11

CHAPTER 6

Concepts That Simplify Phase Selection

1 Introduction

In SFC, neither phase is obvious. No one column type or stationary phase (SP) is appropriate for more than a fraction of likely applications. In addition, the mobile phase (MP) fluids used cover a wide range of solvent strength. Matching the two most appropriate phases for a separation is not, necessarily, a trivial undertaking. Most chromatographers are still unfamiliar with SFC, and do not know what trade-offs are appropriate. The 20–30 years experience by a large number of practitioners is simply not there. Fortunately there are many 'rules of thumb' or practical guidelines that allow prediction of near optimum phases. This chapter is an attempt to establish concepts that form the background for phase selection guidelines.

2 'Polarity Windows' Concept Helps Choice of Phases

Choosing appropriate chromatographic phases for packed column SFC is simplified using the concept of 'polarity windows'. This concept was empirically developed over a number of years. Simply stated, the polarity of the chromatographic phases should bracket (or form a window around) the polarity of the solutes.

The concept of polarity windows is an alternative way of thinking about distribution coefficients (K_D) (see Chapter 4). Such coefficients express the ratio of the equilibrium concentrations of the solute in the SP ($[C_s]$) divided by the concentration in the MP ($[C_m]$), which forms the basis of retention and separation. In capillary columns it is easy to envision the partition coefficient of a solute as the product ($K_D = k'\beta$) of a partition ratio, k', and a phase ratio, β. Similar concepts work for packed columns although the SP volume is poorly defined.

The β on packed columns is ≤ 10. When k' is between 1 and 10, K_D must be $\leq 10–100$. The C_s must be $\leq 10–100$ times C_m. Obviously, stationary phase–solute (SP–S) interactions must be 10–100 times more intense than the mobile phase–solute (MP–S) interaction.

Most schemes for optimizing LC mobile phase composition attempt to 'match' the solvent polarity to the solute polarity. Such 'matching' does not mean that the solvent and solute should have the **same** polarity. The solvent polarity must actually be **significantly different** from the solute polarity. If a solvent were chosen that exhibited the maximum interaction with the solute (the same polarity), it would be extremely difficult to find a SP with 10–100 times more intense interaction to produce $K_D \leq 10$–100. A solvent producing optimum solubility would likely produce negligible retention. Therefore, 'matching' solvents to solutes means finding an **appropriate** but, in normal phase chromatography, a significantly less polar solvent than the solutes.

SFC Is Always Normal Phase

The strong adsorption of MP components, described in Chapter 4, along with extensive experience with a wide range of solutes leads us to state that SFC (at least with CO_2 based fluids) is virtually never a reversed phase technique. However, it must also be immediately stated that SFC is NOT adsorption chromatography.

At low temperature and moderate pressure, adsorbed MP forms a thick, condensed film on the SP, which is a more dense version of the MP. When modifiers are present, the concentration of modifier in the condensed film is much higher than its concentration in the MP. The adsorbed film is always more dense, and contains a higher concentration of modifier (if present) than the MP. Both these effects mean that the adsorbed film is a stronger solvent than the MP, and the technique is virtually always normal phase.

Contrary Views

Some workers have suggested that the polarity of the phases is not relevant, rejecting the notion that 'polarity windows' are necessary. They proposed that even very weak solvents can be used to elute polar solutes at low concentrations, provided that the SP is also weak. The combination of weak MP–S and weak SP–S interactions would, theoretically, allow the separation of polar compounds using pure CO_2 and low polarity SPs.

Others maintain that CO_2 can be more polar than SPs like C_{18}. After all, there is no real consensus on the meaning of various 'polarity' scales. If true, SFC could sometimes be a reversed phase technique and the 'polarity window' could form the other way (solutes would be more polar than the SP but less polar than the MP). This concept will be extensively addressed below.

Active Sites

When low polarity phases are used with polar solutes, peak shapes are generally poor. Poor peak shapes can be (and often are) interpreted as a problem with 'active sites'. This has led to extensive efforts to deactivate or eliminate active sites. However, such attempts have been largely unsuccessful. If solute molecules are more like 'active sites' than they are like the SP then it is extremely difficult to obtain good peak shapes. A few intense interactions with 'active sites' can produce a significant effect that distorts the much larger number of weak interactions with the intended SP. In the extreme, peaks become symmetrical but the separation takes place on the uncontrolled 'active sites' and the SP only serves to 'deactivate' most of the surface. This is equivalent to a polarity window forming where the solvent is on one side of the solute (less polar) and the active sites are on the other (more polar). The purchased SP is irrelevant and not in the window.

Avoiding Unwanted Interactions

Both phases should be chosen to produce intense MP–S and SP–S interactions (an extension of the concept: 'like dissolves like'). If the solutes strongly interact with the chosen phase, other, unintended interactions will have less effect. This minimizes extraneous interactions such as between solutes and 'active sites'.

Of course, avoiding unwanted interactions is only a goal. Many molecules contain large non-polar sections but also possess one or more small, polar functional groups. Different parts of the same solute molecule may be better matched to different local environments.

Subtle *vs.* Gross Separations

There is a relationship between the similarity of the phases and solutes and the subtlety of separations. If the two chromatographic phases are very different from each other, dissimilar solutes exhibit enormous differences in relative retention. However, under the same conditions, similar solutes often cannot be resolved. On the other hand, if both solutes and phases possess similar polarity, subtle differences between solutes can often be exploited to separate them. Consequently, the range of solute polarity within a sample is extremely important in choosing appropriate phases for their resolution.

A simple analogy clarifies the point. Standing on a hill, it is easy to see panoramic overviews of the surrounding country, but it can be difficult to observe subtle details in a valley below. Looking through a telescope, one can focus on a much smaller area with much greater clarity. Increasing the magnification of the telescope makes subtle details even more apparent. However, increased magnification also causes a decrease in the field of view. One must choose between the panorama and the detail.

In chromatography, a large difference in the phases allows the equivalent of a wide field of view (or a wide window) but limits the resolution of details. Choosing similar phases is equivalent to a narrow field of view (a narrow polarity window). It allows subtle differences in solutes to be exploited, but only a narrow range of solutes can be conveniently resolved.

The Most Polar Entities Present Dictate Retention

If one attempts to separate mixtures of polar solutes using much lower polarity phases, changing the concentration of one **solute** can change the retention times of all the other solutes. If the SP is made more polar than the solutes, this behaviour ceases. Polar additives included in the MP also prevent interactions between less polar solutes.[1].

Examples of Polarity Windows

The most common mistake with SP selection in packed column SFC involves choosing a phase that is too non-polar for the solutes of interest. For instance, there have been many attempts to separate polar molecules with CO_2 and a C_{18} bonded phase. Since even dense CO_2 has a solvent strength similar to that of straight chain hydrocarbons, CO_2 and C_{18} form a narrow polarity window, appropriate only for low polarity solutes.

Hydroxysteroids

Pure CO_2 and C_{18} produce little retained but tailed peaks for polar solutes. This can be demonstrated using a test mix of four hydroxysteroids[2] as shown previously in Figure 4.9. The solutes contained up to three hydroxyl groups and a keto group on a four ring, steroidal hydrocarbon backbone. In Figure 4.9, pure CO_2 eluted only one of four solutes. The single peak was broad and badly tailed. Three of the solutes did not elute. The addition of 2% methanol (MeOH) to the mobile phase caused all the solutes to elute but several remained tailed badly even though they were barely retained.

Increasing the polarity of the SP both increased retention and improved peak shapes. A phenyl column (see Figure 4.10a) produced reasonable peak shapes for the first two compounds (testosterone and estradiol), but the last two (the more polar hydroxycortisone and estriol) continued to tail. Increasing MeOH concentration decreased retention but did not improve the peak shapes of the later eluting peaks.

On still more polar SPs with MeOH/CO_2, all the solutes eluted with symmetrical peaks and exhibited near theoretical efficiencies! Typical chromatograms obtained with four different polar SPs are presented in Figure 6.1. The selectivity (α) is ideal on the cyanopropyl column, with virtually even spacing between peaks. Selectivity is slightly worse on the diol column. The middle two peaks are just baseline resolved but the analysis time is substantially longer. Peak spacing improved on a sulfonic acid (SA) column, but retention was excessive (the column was too polar).

Benzylamines

The benzylamines are strong bases with aqueous pK_as in the vicinity of 11. In reversed phase LC, one would consider using a MP with a pH greater than 11 to prevent dissociation of the amine. However, silica based packing materials are dissolved by strong bases. Modern LC SPs, intended for such separations, are either especially deactivated and end capped to prevent basic attack on the silica (which tend to be less than completely effective), or use other, non-silica based supports that are not subject to attack by bases.

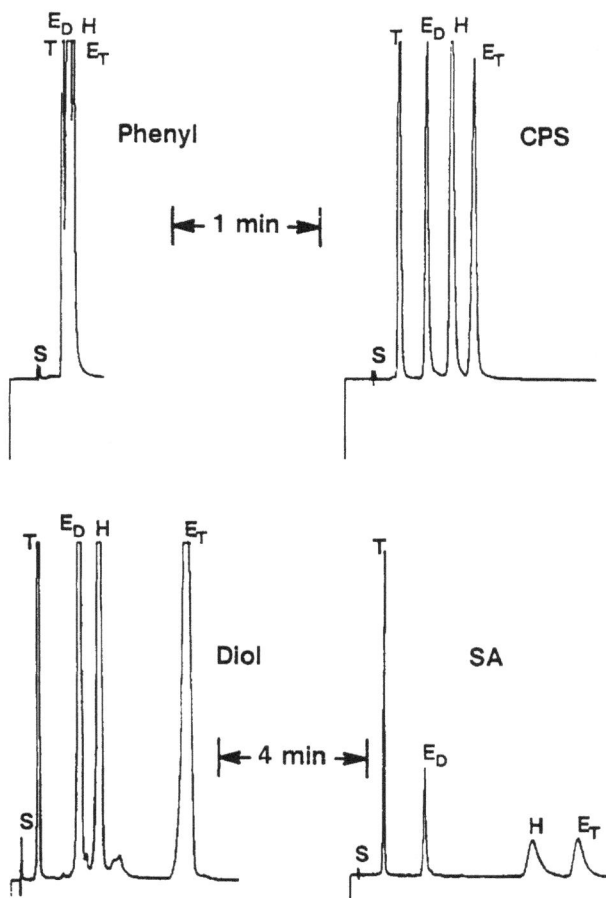

Figure 6.1 *The hydroxysteroids on four different polar stationary phases, phenyl cyano-propyl, diol, and SA (sulfonic acid). Note that all the peaks are symmetrical with good peak shapes on adequately polar stationary phases. Intermediate polarity CN is 'best'*
(Reproduced by permission from ref. 2)

Tribenzylamine, dibenzylamine, and benzylamine were separated[3] on six different silica based stationary phases using MeOH/CO_2, with isopropylamine (IPAm) as an additive. The results, in Figure 6.2, show that low polarity phases produced almost no retention, yet the peaks tended to tail. Better peak shapes were obtained on more polar phases. These chromatograms make the same point as the hydroxysteroid example. Polar solutes are best separated on a polar SPs.

The above examples do NOT mean that C_{18} is never useful in SFC. Instead, the examples were intended to show that it is important to use phases of the **appropriate** polarity. Most polar solutes that require polar modifiers exhibit

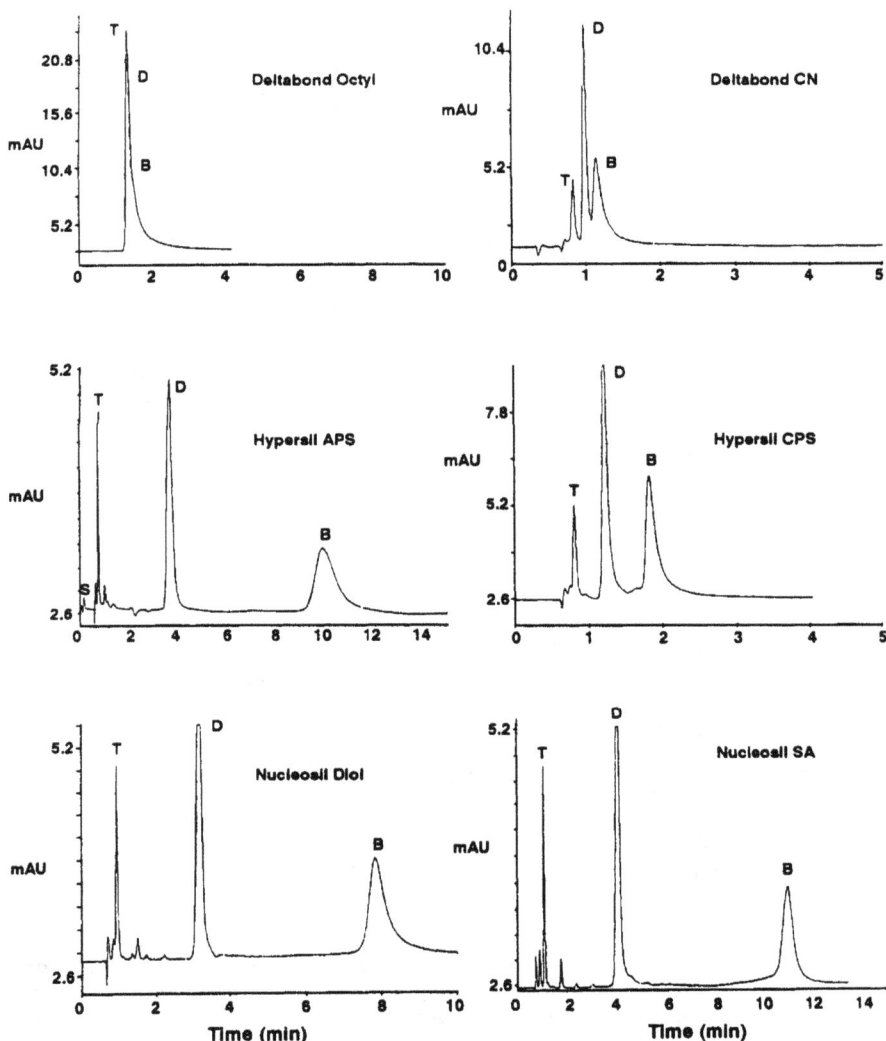

Figure 6.2 *The separation of tribenzylamine, dibenzylamine and benzylamine on six different stationary phases, using methanol modified carbon dioxide with isopropylamine as additive*
(Reproduced by permission from ref. 3)

increasing retention with increasing SP polarity. For low to moderate polarity solutes it **is** appropriate to use low to moderate polarity phases, such as C_{18}. For such solutes it is possible to choose a phase that is too polar. On polar phases, the retention of such solutes can actually decrease compared with the level of retention on less polar phases. On the other hand, low molecular weight, low polarity solutes are generally inappropriate for SFC since they can usually be separated by GC (see Chapter 1).

3 Previous Polarity Scales in GC, LC, and SFC

The concept of polarity windows requires a scale of increasing chemical 'polarity' on which all compounds and phases can be compared. The term 'polarity' is arbitrary. One could call such a scale an energy scale, an elution scale, or perhaps a retentivity scale and mean nearly the same thing. The intent is to express the appropriateness of specific stationary phases and mobile phases for the separation of specific types of solutes.

There are many precedents for single axis polarity scales in chromatography. Some are based on the actual retention of various model solutes on a given column type using various binary MPs. Some are based on a measure of the solvent strength of the phases as measured by some method other than chromatography.

To create a single 'polarity' axis for all solutes (that can be related to chromatographic phases), several aspects of molecules such as van der Waals forces, inducible dipoles, permanent dipoles, hydrogen bonding, acid/base interactions, and others must be represented. Obviously, it would be more accurate to visualize these very different aspects of polarity in multiple dimensions. However, this sacrifices the simplicity of lumping all forms of polarity into a single axis.

In the following sections, the history of solvent strength and elution scales in LC and GC is discussed in an attempt to provide a framework for the SFC polarity scale proposed in Section 4, and used in Chapter 7.

The ε^0 Elution Scale from Normal Phase LC

'Polarity' scales based at least partially on chromatographic retention, have been somewhat successful. Snyder[4] developed a linear elution scale, called ε^0, for normal phase adsorption chromatography also called liquid–solid chromatography (LSC). Each increase in ε^0 of 0.05 decreased retention by 2–3 times. Retention was not a linear function of modifier concentration. Instead, retention appeared to decrease logarithmically with increasing modifier concentration.

Solutes were eluted from a single column type. The modifier concentration was varied until each solute eluted with $k' = 3$. The solutes were then ranked in order by assigning them the ε^0 value of the MP producing $k' = 3$. Different solvent mixtures with the same ε^0 value would produce approximately the same solute retention on the same support material.

Saunders[5] produced a graphical version of Snyder's data. In one graph the range of retention of solute families was detailed, showing both the relative retention of different families, and the range of retention within each family. Some rules evolved to indicate, *a priori*, how various structures and substitutions should change retention, compared with a model compound.

In a second graph, both the ranges and non-linear behaviour of solvent mixtures were detailed. The user could use the first graph to determine an approximate ε^0 value for the solutes of interest, then from the second graph find one or more solvent mixtures producing that same ε^0 value, and, therefore, providing appropriate elution conditions.

The non-linear relation between composition and elution strength was interpreted to follow the adsorption of the modifier onto the stationary phase. Surprisingly, Reichardt's E_t scale,[6] obtained using a solvatochromic dye, has been found[7] to correlate with the ε^0 scale. Since the solvatochromic dye has nothing to do with adsorption, such a correlation is curious, and implies that the underlying assumptions about mechanism in LSC are flawed. The LSC experience with Reichardt's dye also mirrors the experience found in SFC, using other solvatochromic dyes, described in Chapter 3. There, the solvatochromic dye indicates that non-linear retention phenomena are due to non-linear solvent strength *vs.* mole percent modifier concentration (a MP phenomenon) and not adsorption (a surface phenomenon). Is it possible that retention in normal phase adsorption chromatography does not actually follow the adsorption isotherm of the modifier?

The procedure for finding an appropriate solvent for LSC involves 'matching' the ε^0 value of the solute with the ε^0 value of the solvent. Such an approach is, at least on the surface, philosophically different from the concept of polarity windows. However, since a successful match between the ε^0 value of the solvent and the ε^0 value of the solute produces a partition ratio (k') of 3, there is actually no conflict between the two concepts. With 'matching' ε^0 values, the solute spends a majority of the time on the SP ($k' = 3 = K_D/\beta$; since $\beta \leq 10$, $K_D \leq 30$, and $C_s \leq 30 \times C_m$), which means that the solutes significantly 'prefer' the SP over the MP. If the solvent was 'ideal' for the solutes (maximum solubility, maximum interaction between the solutes and the solvent) then solute retention would probably be negligible. So, in reality, 'matching' the ε^0 values of the solutes and the solvents requires a weaker (less polar) MP than the SP, and the phases create a window around the solutes.

Gas Chromatographic Retention

In another attempt to build a 'polarity' scale based on relative chromatographic retention, Rohrschneider[8] developed a series of constants to describe SPs used in GC. The initial work used differences in retention indices between a 'standard' phase (squalene) and other SPs. The differences in retention were divided into factors characteristic of the solute and others characteristic of the solvent. This approach was extended to the use of other 'standard' phases[9] which allowed higher temperature to be used. Although most of these phases are no longer used, the concepts are very much alive and an underpinning of most attempts at correlating retention indices.

The P' Scale of Solvent Polarity and Selectivity

Rohrschneider also published data on the gas–liquid distribution coefficients of six carefully chosen solutes in a wide range of normal liquids (initially 80 liquids), sometimes used as LC solvents. Snyder[10] attempted to differentiate between solubility and selectivity using Rohrschneider's data. He corrected for molecular weight, and molecular volume, then normalized Rohrschneider's data to n-hexane. He then devised a P' scale based on interactions with the polar solutes ethanol, dioxane, and nitromethane with the various solvents. These three solutes are meant to represent contributions by proton donors (X_e, where e represents ethanol), proton acceptors (X_d, where d represents dioxane), or dipoles (X_n, where n represents nitromethane).

The overall P' value provides a measure of polarity, while the specific interactions (X_e, X_d, and X_n) with polar solutes provide a measure of selectivity. The selectivity values can be used to separate Rohrschneider's solvents into eight families. In general, any one of the solutes within a family can be thought to represent the whole family, so the total number of possible solvents required can be reduced.

There are several problems with these scales. They are based on interactions involving a few model compounds, so interactions with different compounds in the same family might vary substantially. The values depend heavily on multiple mathematical manipulations. The different interactions of the model compounds are assumed to be entirely due to the single designated attribute of each solute. There are several overlaps between the families.

The proposed mixing rules are simplistic, relying on all the solvents to form ideal mixtures. Despite these inadequacies, the P' scale is still the most widely used solvent strength scale for reversed phase LC, because it gets the user close to optimum solvent choices in minimal time and predicts alternatives if the first choice is either unsuccessful or can't be used for some other reason.

The P' scale is largely calculated and is based on solubility. Some care needs to be exercised in comparing the solvent and solute polarity because the solvent producing the 'ideal' solubility is probably not the appropriate elution solvent for that solute (see above).

The proposed means of determining the solvent strength of mixtures uses the equation: $P' = P_A\Phi_A \, P_B\Phi_B$, where P_A and P_B represent the solvent strengths of each of the pure solvents, P' represents the mixture, and Φ_A and Φ_B represent the volume fractions of the components. This mixing equation assumes that the solvents produce ideal solutions (the volumes are additive), and solvent strength is a linear function of volume percentage. Neither assumption is true in normal liquids, but the deviations are generally small enough for the equation to be of practical value. Generally, the sum of the volumes of two solvents is significantly different from the volume of the mixture. Similarly, the solvent strength of mixtures is often NOT a linear function of composition.

In SFC, the deviations from ideality are so overwhelming that the equation in the previous paragraph would completely obscure the most powerful aspects of the technique.

Solvatochromic Dyes

Solvatochromism is becoming the accepted approach to elucidating actual interactions between solutes and solvents (see also Chapter 3). Dorsey[11–13] has mostly used Reichardt's E_{t30} dye as a probe of retention in reversed phase LC.

Kamlet and Taft[14-18] and co-workers developed a series of scales based on the interactions of numerous solvatochromic dyes with solvents. These dyes were carefully chosen to 'independently' measure polarizability, ion–dipole, acid–base, and hydrogen bonding characteristics (hydrogen bond donor, *vs.* hydrogen bond acceptor) of solvents. This is similar to the P' scale based on interactions with a few solutes, each meant to be the model for one of these sorts of interactions. However, the solvatochromic scales are each determined with a series of measurements using numerous solutes.

Carr and co-workers[19] used a series of Kamlet and Taft's dyes to study methyl/phenyl siloxanes which are widely used as SPs in GLC. This was perhaps the first study to probe directly the polarity of the SP by non-chromatographic methods. Carr and co-workers[20-23] have also extensively studied retention in reversed phase LC using the same series of solvatochromic compounds.

Previous Polarity Scales in SFC

There has been surprisingly little published work on relative retention or elution strength in packed column SFC. Most reports dealing with retention have been focused on practical separations, not patterns of retention. A fairly substantial number of solvatochromic dye studies have been performed on SFs in attempts to elucidate the nature of solvation in such media (see Chapter 3). However, only a few have been directed at correlating MP solvent strength with chromatographic elution. There have been no universally accepted polarity or elution scales in packed column SFC. There have been a few attempts to systematize solvent strength and elution strength in SFC.

It is actually difficult to retain light hydrocarbons and other low polarity solutes on packed columns using CO_2 as the MP. Among hydrocarbons, ethers, esters, aldehydes, and ketones, there is little to recommend SFC unless the molecules are very large, or contain multiple functional groups. In spite of these facts, most efforts to generate a coherent picture of retention in SFC have used light hydrocarbons and other light, low polarity molecules.

Packed column SFC is primarily useful for the separation of solutes that are thermally labile, polar, or large and moderately polar. Using probe molecules without these characteristics fails to bring out the real trade-offs and strengths of the technique. The most logical place to start in modelling SFC is to collect empirical retention measurements using the kinds of solutes the technique is useful for.

Hildebrand Solubility Parameter

Some of the earliest attempts to devise a single 'polarity' scale involved the use of the Hildebrand solubility parameter, δ.[24] The δ scale is normally calculated from the heat of vapourization of the solvent and depends on the change in its molar volume with pressure.

Giddings[25] attempted to relate this parameter to retention in SFC. However, with SFs, the 'heat of vapourization' is meaningless (there can only be one phase and it is not defined as either a gas or a liquid) so an alternative formulation was tried. With this alternative formulation, Giddings predicted a solvent strength for CO_2 similar to that of isopropyl alcohol or pyridine. A few[26] results were obtained which at least partially substantiated the prediction by reporting the elution of some very polar biological compounds. Unfortunately, these findings could not be independently repeated. At present, most workers consider the solvent strength of CO_2 to be more nearly equal to that of pentane or hexane than isopropyl alcohol or pyridine (see Chapter 3).

The P' Scale

Snyder's P' scale developed for LC, discussed above, has also been proposed as an initial framework for understanding solvent strength in SFs. It was proposed[27] mainly as a means of choosing modifiers (*e.g.* proton donor *vs.* proton acceptor, *etc.*). However, the problems with this scale have been discussed. Solvent strength in binary SFs is clearly a very non-linear function of composition. A model based on linear extrapolation is likely to cause significant errors. Further, retention is more than a function of the solvent strength of the MP.

In SFC, many bonded phases with substantially different selectivities and polarities are used, so any polarity scale needs to account for such differences. SP interactions could be broken down into dipole, proton donor and proton acceptor contributions. However, the strong adsorption of MP components tends to cover or conceal the differences between SPs. If $MeOH/CO_2$ mixtures are used with different SPs, the amount of MeOH adsorbed varies depending on the nature of the SP. Still, a substantial amount of MeOH adsorbs even on the least compatible SPs. This aspect needs to be systematically studied, before anything more definitive can be said.

It should be clear that standard descriptions from LC are inadequate to understand SFC. In fact, the SFC results tend to cast doubts on some of the cherished ideas from LC. In the future it is likely that the P' scale will be found helpful in choosing modifiers but it provides little help or insight at the present.

4 A Solvent Strength Scale for SFC

The most fruitful approach to understanding retention mechanisms in SFC appears to involve empirical measurement of solute retention using a variety of MPs and SPs at different temperatures and pressures. Initially, a single MP mixture and a single SP can be used to generate a relative retention scale.

The retention of various compounds on diol columns using $MeOH/CO_2$ (sometimes containing additives) is detailed in Table 6.1. A more extensive table, including other columns and other MPs, is included in Chapter 7. To simplify comparisons the linear E_{NR} energy scale is used. Each group in the Table is discussed briefly to try to provide some insight into the degree of difficulty involved in their elution.

Table 6.1   *Retention at* $k' = 3$ *of selected solutes on a diol stationary phase, using*
MeOH/CO$_2$, at 30–80 °C, 110–300 bar

E_{NR}, kcal mol^{-1}

| 60 | 59 | 58 | 57 | 56 | 55 | 54 | 53 | 52 | 51 |

Methanol concentration

| 0 | 1 | 2 | 3 | 4 | 5 | 10 | 20 | 40 | 60 | 80 | 100% |

PAHs	1————————2
Steroids	————————3————4
Alcohols	
Phenols	6——7,8——
Acids	
Aliphatic	9——
Aromatic	10—11^a————12^a
Bases	
Aliphatic	—13^{b}14^b——
Anilines	————15
Amides	——16————(17)
Sulfonamides	——18——

^{a} indicates that an acidic additive is required; ^{b} indicates that a basic additive is required.
Key:
1. Any of the 16 EPA PAH priority pollutants; 2. Up to 10 fused rings; 3. Hydrocortisone; 4. Estriol; 5, 6. 2-Nitrophenol; 7. Phenol; 8. 4-Nitrophenol; 9. C$_{40}$ fatty acid; 10. Benzoic acid; 11. 2,4,6-Trihydroxybenzoic acid; 12. 1,2,3,4,5,6-Benzenehexacarboxylic acid: 13. Benzylamine: 14. β-Phenylethylamine; 15. Aniline; 16. Benzamide; 17. 4-Aminobenzamide; 18. Sulfamethazine.

Hydrocarbons, Ethers, Esters, Aldehydes, and Ketones

SFC can be used to elute hydrocarbons up to at least C$_{120}$ but shows little advantage over GC. The only disadvantage for GC is the extreme temperature required ($T > 400$ °C) for the largest molecules. In SFC, the MP density is at least 50 and as much as 500 times higher, while the temperature tends to be much lower than in GC. Subsequently, SFC is typically 1000 times slower in producing the same efficiency.

Most polycyclic aromatic hydrocarbons (PAHs) can be separated by GC but the larger members require high temperature. The 16 priority PAHs elute from packed columns using pure CO$_2$ and modest temperature. With more rigorous conditions, at least some 10 ring compounds can be eluted.

Alcohols

Most alcohols and phenols can be eluted in GC, but tend to tail on low polarity phases. Phenol is often included in polarity test mixes used in GC. Many steroids are commonly separated by GC, but multiple hydroxy substitutions make elution of hydroxysteroids difficult.

Packed column SFC using $MeOH/CO_2$ allows separation of phenols faster than by GC.[28] Phenols elute from packed columns using $MeOH/CO_2$ with decent peak shapes. Without a modifier, only a few of the phenols will elute. Additives can improve peak shapes. Phenols can be separated on a CN column, although a diol column offers better a.

Of the 11 phenols studied using EPA Method 604, phenol elutes eighth, 2-nitrophenol elutes first, and 4-nitrophenol elutes last. Such differences in retention are probably due to steric effects but other explanations are also possible.

Hydroxysteroids[2] mostly require $MeOH/CO_2$ and a relatively polar SP (cyanopropyl or diol) to achieve good peak shapes.

Acids

In SFC, C_{40} fatty acid can be eluted without derivatization using pure CO_2 at $T < 100\,°C$. The appropriate SP is generally a deactivated or end capped C_8 or C_{18}. A long hydrocarbon tail tends to 'dilute' the polarity of the acidic group. The addition of a hydroxyl group prevents the elution of fatty acids using pure CO_2.

Aromatic acids are generally more difficult to elute than aliphatic acids. Benzoic acid can sometimes be eluted with pure CO_2 but is not representative of its family. Aromatic hydroxyacids[29] and polyacids[30] require a modifier and usually an additive[31] in the MP to elute with decent peak shapes. Even six acid groups on a benzene ring (mellitic acid) could be eluted with reasonable peak shapes, provided a strong acid was included in the MP. The best peak shapes are obtained with a hydrogen bonding SP, such as diol.

Bases

Anilines

Anilines[32,33] are among the weakest organic bases, with aqueous pK_as in the vicinity of 4–5. They did not elute or eluted with poor peak shapes[32] using pure CO_2. With $MeOH/CO_2$, the anilines eluted with good peak shapes from cyanopropyl columns, but retention was inadequate. Changing the main fluid from CO_2 to Freon-13 produced both good peak shapes and longer retention. Increasing the SP polarity to diol while retaining CO_2 also increases retention while retaining good peak shapes.

The hindered anilines (*e.g.* *N*-methyl- and *N,N*-dimethyl-) were less retained than aniline itself. Retention decreased with increasing hindrance. *N*-Ethylaniline was less retained than *N*-methylaniline, despite its larger size. Hindering substitutions on the ring (*e.g.* toluidines, methyl substituted on aniline) had much less effect on retention than substitutions on the nitrogen. *o*-Toluidine was less retained than aniline. It was much more retained than *N*-methylaniline, which has the same molecular weight. Retention did not follow aqueous pK_as.

Diamines

Phenylenediamines are also weak bases but were much more difficult to elute.[3] and were much more retained than the anilines. The main problem with these compounds was their reactivity with each other before injection. Peak shapes, particularly with *p*-phenylenediamine (1,4-diaminobenzene), were extremely poor using low MeOH concentrations. With higher MeOH concentrations, peak shapes improved somewhat but retention became inadequate. A strong base used as an additive dramatically improved peak shapes.

Aliphatic Amines

Aliphatic[3,33] amines are 5–7 orders of magnitude stronger (aqueous) bases than the anilines. Phenylethylamines and benzylamines both contain a benzene ring, which allows the compounds to be detected using a UV detector but does not diminish their basicity. They are eluted as extremely broad, square topped bands with sharp edges from packed columns using pure CO_2. Only polymer based columns (no silica) allowed a degree of separation with binary fluids (no additive) but peaks remained badly tailed.[33] However, with tertiary fluids, containing a strong base as additive, peak shapes are good and apparent efficiency was quite high, even on silica based columns. The hindered α-phenylethylamine eluted before the less hindered β-form and before the slightly shorter benzylamine.

Amides

Saunders indicates that aliphatic amides are among the most difficult solutes to elute in normal phase LC and are more retained than aromatic amides. This is NOT the case in SFC. Fatty amides, such as oleamide, eluted easily even from polar stationary phases, such as diol, with low modifier concentration. The polar column was needed for the aromatic amides but not for the aliphatic amides. Both C_{18} and C_{22} amides nearly co-eluted even when they were retained with $k' = 10$ on a diol column, using 3% MeOH in CO_2. A double bond in the long hydrocarbon chain had almost no impact on retention. Such compounds can probably be eluted with better selectivity using pure CO_2 from a less polar phase like C_{18}. Compared with benzamide, the aliphatic amides were much less retained, even though their molecular weight was much larger.

Aromatic amides were more retained than anilines, or phenols, but less retained than benzylamine.[33] They appear to be retained approximately as strongly as multifunctional acids. Substitution of bulky groups onto the amide nitrogen decrease retention. The addition of an aniline ring onto the amide dramatically increased retention, compared with an unsubstituted ring.

5 Factors Affecting Retention

The results outlined in the previous section allow several generalizations to be made. Retention in SFC is similar to that in normal phase LC but there are noticeable differences. Aliphatic members of families tend to be easier to elute than aromatic members of the same family.

Selectivity changes significantly with apparently subtle differences in physical parameters. In many ways, supercritical fluids appear to be exaggerated forms of liquids.

Functional Group Polarity

Retention follows 'polarity'. Increasing polarity tends to produce increasing retention. Intuitively, organic acids are more polar than alcohols, which are more polar than ketones or aldehydes. Esters and ethers are even less polar, approaching hydrocarbons. This order also closely follows dipole moment. The most polar (*e.g.* the strongest acid) usually elutes later than a less polar member of the same family (*e.g.* a weaker acid). All members of a more polar group do not necessarily elute later than all members of a nominally less polar group.

It is important to follow the idea of polarity windows. Phases should be appropriate for the solutes. If both phases are either too polar or not polar enough, peak shapes will degrade and it will become more difficult to use selectivity to separate similar solutes.

There are a number of mitigating circumstances which are also important in determining retention. These include molecular weight, the number of functional groups present, and steric hindrance.

Number of Functional Groups

Retention increases when polar functional groups are added to a molecule. Multiple substitutions can cause major increases in retention. Several weak functional groups can increase retention as much or more than a single more polar functional group.

An example of multiple substitutions using polyfunctional benzene carboxylic acids[30] is shown in Figure 6.3. The molecules contain one, two, three, four, and six acid groups substituted onto a single benzene ring. Note that the isomers of each level of substitution elute nearly as groups under the conditions used. Three di-substituted isomers were available and nearly co-eluted. Two tri-substituted acids coeluted.

Steric Hindrance

The presence of a very polar functional group does not guarantee that a solute will be strongly retained. Steric effects in packed column SFC can be dramatic, reversing expected elution orders based on, for example aqueous pK_as.

The family of weakest bases in a test mix (anilines) eluted first,[33] as shown in Figure 6.4. However, they eluted in order of decreasing steric hindrance. The benzamides which are not bases, but are larger and contain several polar groups, eluted after the anilines. The benzylamines are all strong bases, with aqueous pK_as in the vicinity of 11, which makes them up to seven orders of

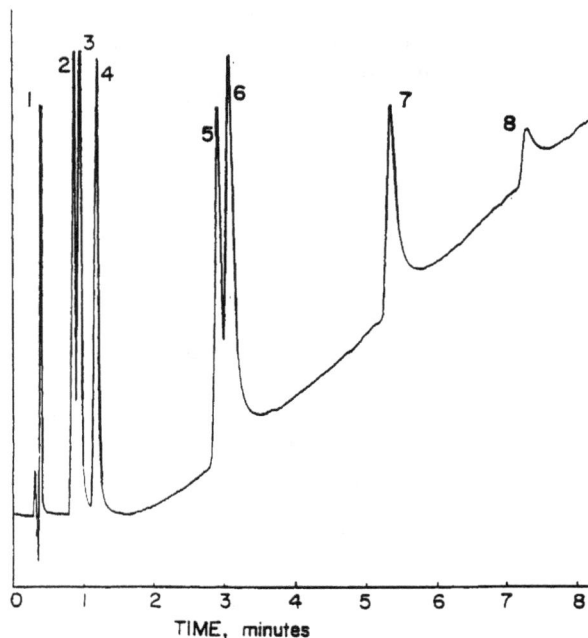

Figure 6.3 *The retention of polyfunctional aromatic acids increases with the number of acid groups. The solutes shown contain from 1 to 6 carboxylic acids on a benzene ring. The solutes elute as groups. The members of each group contain the same number of acid groups. 1 = benzoic; 2, 3, 4 = 1,4-, 1,3-, 1,2-diacids; 5, 6 = 1,3,5-, 1,2,4-triacids; 7 = 1,2,4,5-tetraacid; 8 = hexa-acid.*
(Reproduced by permission from ref. 30)

magnitude more basic (in water) than the anilines. However, this family eluted over the widest range of retention. Tribenzylamine, which is the most basic compound in the mix, but also the most hindered, eluted before aniline. Dibenzylamine, which is slightly less basic but also much less hindered, elutes later but before most benzamides. Benzylamine itself is the weakest base among its family, but it is the least hindered. Benzylamine is more basic and elutes much later than the anilines or benzamides. It elutes in the same general region as the phenylethylamines (other strong bases).

Obviously, steric effects were dramatic and more pronounced with the benzylamines than with the anilines. It appears that steric effects can overcome even gross differences in 'polarity'. Such steric effects also appear to be more dramatic in SFC than in LC.

Molecular Size

In most forms of chromatography, larger solutes are more difficult to elute than smaller solutes, provided both contain functional groups with similar polarity. In GC, retention generally increases with either molecular weight or boiling

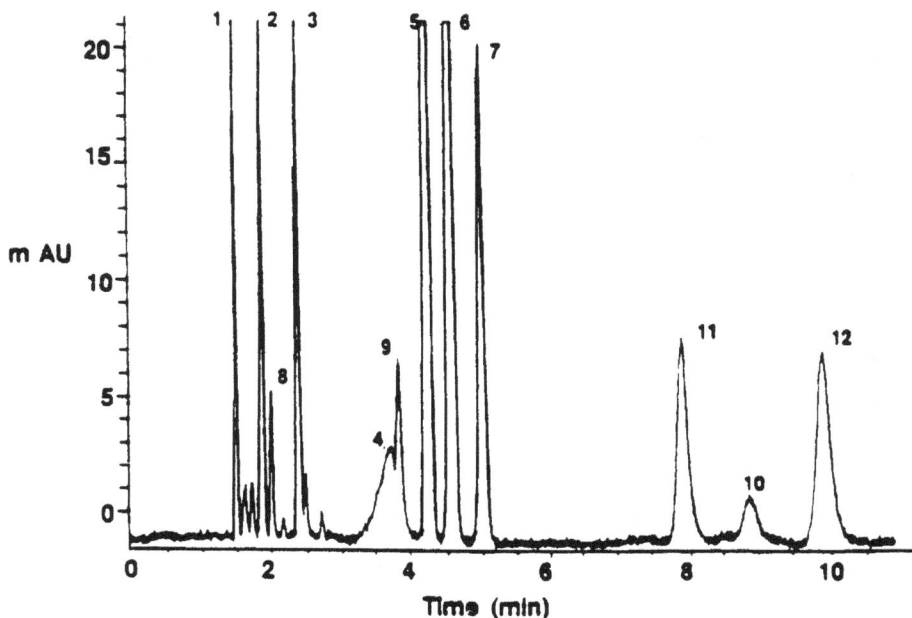

Figure 6.4 *Retention of bases and amides on a diol column. The results show that steric effects are at least as important as polarity or base strength.* 4.6 × 250 mm, 5 μm diol, 10% methanol modified carbon dioxide with 0.5% (in the methanol) isopropylamine additive: 40 °C, 200 bar, UV *detection* (Reproduced by permission from ref. 33)

point, although the two are linked. In SFC, larger compounds tend to elute later than small compounds containing the same functional groups.

Polywaxes are homologous mixtures of large, even numbered hydrocarbons that are often used as molecular weight markers in GC. Several such polywaxes, Polywax 500 and 655, are shown separated on a packed column in Figure 6.5. The designations 500, and 655 indicate the average molecular weight of the mix.

Stationary Phase Polarity

A wide range of SPs are used in SFC which offer significant variations in both retention and selectivity. Phases are chosen to be polar enough to interact with the most polar solutes. Phase selection attempts to exploit the physical and chemical differences between solutes to simplify separations.

On the other hand, the character of the bonded phase is muted by a covering of a thick film of MP components. Such a film tends to be an exaggerated form of the MP, which decreases the differences between phases. Both the quantity and composition of the adsorbed film can be varied by changing pressure and temperature. These possibilities are only beginning to be explored.

Phases range from hydrocarbons to strong acids and bases, with proton

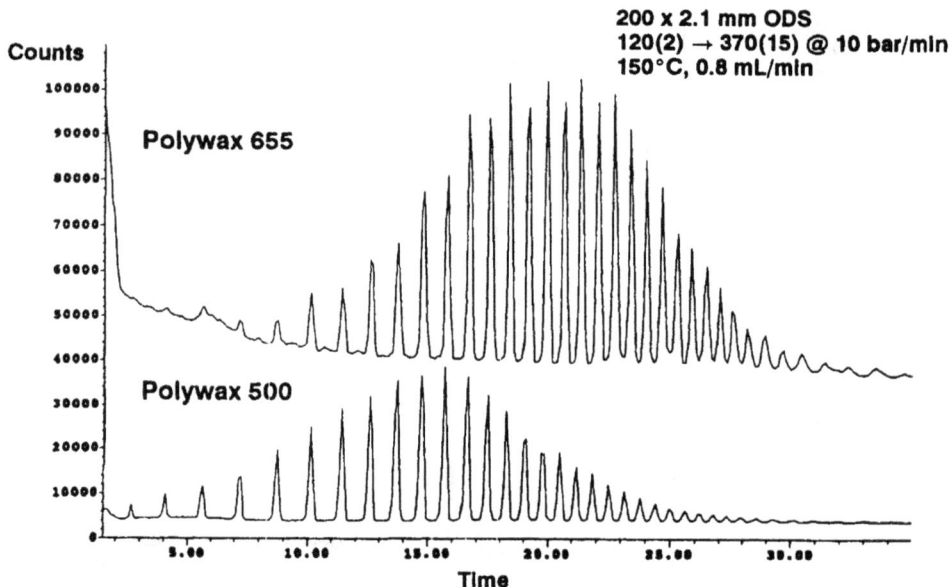

Figure 6.5   *The separation of Polywax 500 and 655 (av. mol. wt. = 500 and 655) on a C_{18} packed column using pure carbon dioxide as the mobile phase with FID detection*

donors, and acceptors, and permanent dipoles. In any solvent strength or retention scale, such variation must be accommodated.

A column can be too polar for a set of solutes. Such solutes can be more retained on a less polar column. However, low polarity solutes can usually be separated by GC so there is little incentive to use SFC. Most compounds requiring SFC with modifiers will show increasing retention with increasing SP polarity.

The general column order producing increasing retention for solutes follows the trend: C_{18} < C_8 < phenyl < CN < silica < diol < SA (sulfonic acid). Amino tends to fit somewhere between CN and SA, depending on the specific solutes. Modifiers can change the order. Some solutes will cause slight differences in order.

6   A Summary of Retention in Packed Column SFC

Plots of log k' *vs.* percentage modifier are non-linear. Most have assumed that this non-linearity makes SFC similar to liquid–solid (normal phase, or adsorption) chromatography. In LSC, the non-linearity is blamed on the adsorption of the modifier onto the SP.

Parcher[1] and Lochmuller[35] both measured the adsorption of MeOH onto packed columns from MeOH–CO_2 mixtures. At constant temperature and pressure, adsorption is complete below 2% MeOH. At higher modifier concentration, almost no additional adsorption occurs. Although the adsorption must modify both the volume and the polarity of the SP, these parameters become essentially constant above 1–2% MeOH in the MP. Adsorption can, thus, not explain non-linearities in the plots of log k' *vs.* percentage modifier. Further discussion of this topic can be found in Chapter 4.

Plotting log k' *vs.* E_{NR} produces straight lines. This suggests that the mechanism is not adsorption but fluid–fluid partition with non-linear MP solvent strength. Solvatochromic dyes also support the idea that the non-linearity in LSC is in the MP, not just adsorption of the modifier.

Eckert[36] implies that supercritical fluids do not form a rigid solvent cage around solutes but do strongly interact. Modifiers tend to form clusters (see Chapter 3) of high local polarity and high local density. Solutes from different families or even different parts of the same molecule appear to interact preferentially with different local environments. By choosing an appropriate SP, either the polar or non-polar parts of larger molecules can be used to drive a separation. The surfaces of both solute molecules and SPs appear to be more accessible in supercritical fluids than in normal liquids.

Using fluorescence, Eckert[36] observed that hydrogen bonding in supercritical fluids was different from that in normal liquids. They found no evidence for stoichiometric hydrogen bonding complexes but observed significant interactions between 0.5% MeOH in CO_2 and 7-azaindole. Those and other results indicate that the local composition around the solutes is substantially different from the bulk concentration of the modifier. The lack of semi-rigid hydrogen bonded structures is consistent with the low viscosities of binary mixtures evident from the low pressure drops across packed columns.

Nile Red is not the ideal dye for studying solvent strength in SFC. Olisek and co-workers[37] used Kamlet and Taft's dyes to measure solvent strength on a number of different 'polarity' axes. From a fundamental point of view, such an approach is to be preferred. Nile Red does provide a single dimension scale of modifier–carbon dioxide solvent strength that can be used as a polarity scale for acids and hydrogen bonding solutes.

7 References

1. J.R. Strubinger, H. Song, and J.F. Parcher, *Anal. Chem.*, 1991, **63**, 104.
2. T.A. Berger and J.F. Deye, *J. Chromatogr. Sci.*, 1991, **29**, 280.
3. T.A. Berger and J.F. Deye, *J. Chromatogr. Sci.*, 1991, **29**, 310.
4. L.R. Snyder, 'Principles of Adsorption Chromatography', Marcel Dekker, New York, 1968.
5. D.L. Saunders, *Anal. Chem.*, 1974, **46**, 470.
6. C. Reichardt, *Fortschr. Chem. Forsch.*, 1968, **11**, 1.
7. L. Rohrschneider, *Anal. Chem.*, 1973, **45**, 1241.
8. L. Rohrschneider, *J. Chromatogr.*, 1966, **22**, 6.
9. J.R. Ashes, J.K. Haken, and P. Souter, *J. Chromatogr.*, 1974, **92**, 273.
10. L.R. Snyder, *J. Chromatogr.*, 1974, **92**, 223.
11. B.P. Johnson, M.G. Khaledi, and J.G. Dorsey, *Anal. Chem.*, 1986, **58**, 2354.
12. J.G. Dorsey, *Chromatography*, 1987, **2**, 37.
13. J.J. Michels and J.G. Dorsey, *J. Chromatogr.*, 1988, **457**, 85.
14. M.J. Kamlet and R.W. Taft, *J. Am. Chem. Soc.*, 1976, **98**, 377.
15. R.W. Taft and M.J. Kamlet, *J. Am. Chem. Soc.*, 1976, **98**, 2886.
16. T. Yokoyama, R.W. Taft, and M.J. Kamlet, *J. Am. Chem. Soc.*, 1976, **98**, 3233.

17. M.J. Kamlet, J.L. Abboud, and R.W. Taft, *J. Am. Chem. Soc.*, 1976, **99**, 6027.
18. M.J. Kamlet and R.W. Taft, *Acta Chem. Scand., Ser B*, 1985, **39**, 611.
19. J.E. Brady, D. Bjorkman, C.D. Herter, and P.W. Carr, *Anal. Chem.*, 1985, **56**, 278.
20. P.C. Sadek, P.W. Carr, R.M. Doherty, M.J. Kamlet, R.W. Taft, and M.H. Abraham, *Anal. Chem.*, 1985, **57**, 2971.
21. P.W. Carr, R.M. Doherty, M.J. Kamlet, R.W. Taft, W. Melander, and C. Horvath, *Anal. Chem.*, 1986, **58**, 2674.
22. D.E. Leahy, P.W. Carr, R.S. Pearlman, R.W. Taft, and M.J. Kamlet, *Chromatographia*, 1986, **21**, 473.
23. M.J. Kamlet, R.M. Doherty, P.W. Carr, D. MacKay, M.H. Abraham, and R.W. Taft, *Environ. Sci. Tech.*, 1988, **22**, 503.
24. J.H. Hildebrand and R.J. Scott, 'The Solubility of Non-Electrolytes', 3rd Edn., Reinhold Publishing Corp., New York, 1950.
25. J.C. Giddings, M.N. Myers, L. McLaren, and R.A. Keller, *Science*, 1968, **162**, 67.
26. J.C. Giddings, M.N. Myers, and J.W. King, *J. Chromatogr. Sci.*, 1969, **7**, 276.
27. L.G. Randall, in 'Ultra High Resolution Chromatography', ed. S. Ahuja, American Chemical Society, Washington, DC 1984, Chapter 11.
28. T.A. Berger and J.F. Deye, *J. Chromatogr. Sci.*, 1991, **29**, 54.
29. T.A. Berger and J.F. Deye, *J. Chromatogr. Sci.*, 1991, **29**, 26.
30. T.A. Berger and J.F. Deye, *J. Chromatogr. Sci.*, 1991, **29**, 141.
31. T.A. Berger and J.F. Deye, *J. Chromatogr.*, 1991, **547**, 377.
32. T.A. Berger and J.F. Deye, *J. Chromatogr. Sci.*, 1991, **29**, 390.
33. T.A. Berger and W.H. Wilson, *J. Chromatogr. Sci.*, 1993, **31**, 127.
34. S. Scalia and D.E. Games, *J. Pharm. Sci.*, 1993, **82**, 44.
35. C.H. Lochmuller and L.P. Mink, *J. Chromatogr.*, 1989, **471**, 357.
36. D.L. Tomasko, B.L. Knutson, J.M. Coppom, W.Windsor, B. West, and C.A. Eckert, in 'Supercritical Fluid Engineering Science', ed. E. Kiran, J.F. Brennecke, ACS Symposium Series, No. 514, American Chemical Society, Washington, DC, Chapter 17.
37. Y. Cui and S.V. Olesik, *Anal. Chem.*, 1991, **63**, 1812.

CHAPTER 7

Systematic Method Development

1 Introduction

This chapter is intended to stand alone and act as a 'cookbook' for method development. Several different method development strategies are presented which have been shown to be effective. The intent of all of them is to determine as rapidly as possible whether packed column SFC is appropriate for a particular analytical problem and, further, indicate near optimum separation conditions.

Equilibrium is rapidly achieved using highly compressible fluids. This allows a number of different steady state conditions to be evaluated in a relatively brief time. Users familiar with LC will find such speed refreshing. In most cases, following the steps outlined below will result in the discovery of near optimum conditions in less than a day. The material is presented in a form assuming that the reader is new to packed column SFC.

2 Initial Non-chromatographic Tests

In this section a number of questions are outlined, the answers to which are intended to screen out applications unlikely to be helped by the use of SFC. The first few questions attempt to establish whether the solutes are compatible with the mobile phases (MPs) used in SFC.

Starting with dilute samples containing potential interferences always results in questions about whether a particular peak represents one of the solutes of interest. Avoid uncertainty. Use concentrated, pure standards to determine that the fluids are appropriate in as short a time as possible.

Once the solutes are shown to be compatible with a supercritical MP, it is then appropriate to get a feel for the solute concentrations and the number and concentration of interferences in the real samples. With real samples there are additional questions that can rule out other potential problems.

Screening Question No. 1: Are the Solutes Soluble in a Desirable

Mobile Phase?

To observe solubility or miscibility directly a high pressure view chamber may be required. It is generally more convenient to scout solute polarity by trying to dissolve the solutes or samples in various normal liquids. Attempt to find normal liquids that are appropriate solvents for the sample. Generally:

(1) If the solute readily dissolves in methanol (MeOH) or a less polar solvent it can probably be chromatographed by SFC.
(2) If the sample ONLY dissolves in water or a buffer it is a poor candidate for SFC.
(3) Solubility in water does not prevent analysis by SFC.
(4) If the sample only dissolves in a low polarity solvent such as hexane or methylene chloride, there is a good chance it can be eluted with pure CO_2.
(5) If the sample readily dissolves in solvents with a wide range of polarity, such as hexane, toluene, acetonitrile, and methanol it will likely require a modified fluid.

If the sample is a gas, it is extremely unlikely that

SFC will provide a better analytical solution than GC. Consider why you are contemplating using SFC.

Screening Question No. 2: What is the Concentration of the Solute in the Real Sample?

SFC is likely to exhibit sensitivity similar to LC. Even using selective detectors sensitivity is not likely to be more than one or two orders better. Derivatization can enhance selectivity and/or sensitivity.

One way to cope with low concentrations is to increase the sample size. However, excessively large injections of solvents can distort peaks and degrade efficiency. For sample solvents like MeOH, the largest sample size should be no more than 20–30 μl. Less polar solvents can allow somewhat larger injection volumes.

If the largest injection fails to provide adequate sensitivity, a pre-concentration step may be required. Since SFC is a normal phase technique, it offers some interesting possibilities in trapping moderate to low polarity solutes from polar samples.

Screening Question No. 3: Is the Sample Matrix Compatible with Supercritical Fluids?

It is important to keep particulates out of the column to avoid plugging. Most samples should be filtered before injection. The user should also consider using guard columns to protect the analytical column.

Many oils and other low polarity compounds will elute with pure carbon dioxide whereas most polar functional groups will not. One can often start an analysis with no modifier then make a step change after low polarity matrix elements elute.

SFs tend to be weaker solvents than those used in reversed phase LC. They are unlikely to transport high concentrations of ionic species such as inorganic salts. Such salts can drop out of solution when a sample is injected into a SF MP. As such compounds build up, the nature of the column changes. Pure fluids are the least polar and are particularly susceptible to ionic contaminants dropping out of solution and building up on the column.

Modified fluids can often transport weakly ionic salts. Many organic solutes are nominally salts. Such compounds are usually not difficult to elute. Very strong, organic acids and bases are miscible with modified fluids. Packed column SFC, using modified fluids, requires less sample preparation than GC but more than LC. Determine whether the MP compatible with the solutes prevents more polar solutes from being left behind on the column. If not, interferences may need to be removed by a sample preparation step. Alternatively, the method could be modified to wash off offending compounds during runs (*e.g.* program to a higher than normal modifier concentration), if the available solvent strength (*e.g.* pure MeOH) can remove the problem compounds. The least desirable, but effective option is to wash the column between runs.

Another alternative for dealing with interferences is to find a selective detector that responds to the solutes but not the interferences.

Although the occurrence is unusual, some solvents are not very miscible or require time to mix with the MP. Injecting such solvents can result in a momentary plugging or over-pressurization of the column. Such solvents can smear out, much like oil on water, spreading the sample and severely broadening peaks. If the sample solvent is not miscible with the MP, change the solvent, modify the injection, or the chromatographic conditions (*e.g.* increase the temperature).

Screening Question No. 4: Is a Universal, Near Constant Response Factor Critical?

Some situations require either universal or constant response factor detection. If such is the case, try to develop a method without modifier. None of the modifiers compatible with the FID are very effective. Consider developing a method using a capillary column or a packed column with low surface area.

3 Choosing between Packed and Capillary Columns

A more complete discussion of actual trade-offs between packed and capillary columns is contained in Chapter 1. More often, choosing a column type reduces to personal preference or to the type of instrumentation available. This is a book about packed column method development, so the reader should expect bias in that direction.

Packed column SFC does not compete with capillary SFC. Both packed and capillary SFC compete with LC, and to a lesser extent, GC. If SFC cannot outperform those other techniques, there is no place for it in separation science.

Capillary Columns

There is little incentive to evaluate capillary SFC as a replacement for an LC method, or even as a confirmatory test to complement an LC method. Capillary SFC cannot compete with LC in terms of speed of analysis, detection limits, reproducibility, or selectivity adjustments. The most important niche of capillary

SFC is the elution of larger, moderately polar solutes using pure CO_2 and detection with the FID. Capillaries tend to work best as an extension of GC into higher molecular weights.

Choose a capillary column if ALL these statements apply . . .

(1) Major/minor component analysis;
(2) If the FID (or other GC detector such as ECD, NPD, *etc.*) is the only detector producing the 'right' output;
(3) If pressure programming of pure fluids provides adequate range of solvent strength;
(4) If the available instrumentation only allows pressure control (no flow control) capillaries may be easier to deal with.

Packed Columns

Packed column SFC is superior to LC in terms of speed, efficiency, selectivity, and detection options. In addition it offers a complementary selectivity to reversed phase LC. Thus, there should be at least some incentive to evaluate packed column SFC as either a replacement for or a complement to reversed phase LC. Packed column SFC is already replacing normal phase LC.

Packed column SFC offers some potential for eluting polar solutes with pure solvents and detection with the FID. However, this is a secondary aspect of packed column performance.

The most important area for packed column use involves modified MPs. Many GC detectors, like the ECD, NPD, or sulfur chemilumenesence detectors, are compatible with at least limited ranges of modifier concentrations. These detectors allow selective and sensitive detection of compounds difficult or impossible to detect using traditional LC detectors.

Most polar compounds should be separated on packed columns. If it is of interest to analyse both a polar compound and the creme or suppository it is delivered in, packed columns probably have the advantage.

Choose packed columns . . .

(1) for polar to very polar solutes;
(2) If solutes contain multiple polar functional groups;
(3) If modifiers are required;
(4) If trace analysis is desired;
(5) If high precision is required;
(6) If high speed is required;
(7) Packed columns can also be used for major/minor component analysis, can be used with the FID, and with pressure programming of pure fluids.

4 Guidelines for Simplifying Method Development

There are important rules of thumb that aid in producing an optimized method in minimal time. They are often obvious, but it is helpful to list them together. Some which have proven helpful are summarized in the next three subsections.

Solute Characteristics That Affect Retention

(1) Increasing molecular weight increases retention.
(2) Addition of a polar functional group increases retention.
(3) The second polar functional group dramatically increases retention.
(4) Two weak polar functional groups can cause greater retention than one very polar group.
(5) Steric hindrance of polar functional groups can dramatically decrease retention.
(6) Hydrocarbons, ethers, esters, and some aldehydes and ketones can be eluted with CO_2.
(7) Acids and bases usually require modifiers.
(8) Strong or multifunctional acids and bases require modifiers and might additionally require polar additives (tertiary mobile phases).

Phase Selection Guidelines

(1) For most solutes, retention is directly proportional to SP polarity and inversely related to MP polarity.
(2) Solutes need a more polar SP and a less polar MP (SFC is normal phase).
(3) Try to bracket the polarity of the solutes with the phases. Form a 'polarity window' (see Chapter 6) around the solutes.
(4) The more similar the phases, the more subtle the separation. Very different phases tend to separate grossly different solutes but not subtly different ones.
(5) MP components adsorb onto SPs. The adsorbed film makes the SP more like an exaggerated form of the MP.

Instrumental Strategy

(1) Flow control pumps and back pressure regulators are easier to work with, and less confusing to use than pumps that control pressure, used with fixed restrictors.
(2) Larger diameter columns are better packed, have higher capacity, less extra-column band broadening problems, and allow more precise control of instrumental variables. Unless there is limited sample available, attempt to work with 4 or 4.6 mm i.d. columns (at least initially).
(3) Pick initial conditions that produce little or no solute retention. Avoid uncertainty. If the solutes don't elute with good peak shapes, a different phase may be required.
(4) Avoid phase problems. Initial temperature should be low and pressure high to avoid any possible MP breakdown.
(5) Don't waste much time trying to optimize the separation using a single variable. First, adjust k' to between 1 and 10 using modifier concentration, then pressure, then temperature.
(6) Focus on resolving difficult pairs through changing α. Temperature usually,

but not always, offers the greatest potential for α adjustment. Pressure provides less α. Modifier concentration tends to be least effective with α.

(7) Detector optimization can require significant development time. Try to develop the separation independent of detector optimization. With pure fluids try to work with the FID. With modified fluids, try to work with the UV detector.

5 Mobile and Stationary Phase Selection Guide for Packed Column SFC

Table 7.1 is intended to indicate appropriate initial conditions for SFC method development by matching relative retention to appropriate MP and SPs. Solutes on the left elute easily with mild SFC conditions. Solutes on the right are difficult to elute under the most extreme SFC conditions.

The most polar functional group in the solutes should define the compound type. Find a compound type in Table 7.1 that at least approximately describes the solutes of interest. Each compound type is represented by a horizontal line, the length of which is meant to provide an indication of the range of retention within the type.

Using the Solute Characteristics Guidelines from the previous section, determine where the solutes fit within the range. For example, a large, multifunctional polar solute should be much more strongly retained than a smaller, monofunctional, weaker solute in the same family, so the former should be located toward the right extreme of that sample type in the table.

After determining the relative location of the solutes, drop a vertical line to the bottom of the table, to find appropriate SPs and MPs. There are usually several choices for either phase. Use the Phase Selection Guidelines, above, to determine the most appropriate phases. All the ranges in the table are approximate. Column order can change.

There is little incentive to go through a speed–resolution optimization if the detection objectives cannot be met by the technique. Also see the bottom of the table, indicating the approximate range of conditions under which various detectors will work. If the detection option of choice does not appear to be compatible with the MP suggested by the table, it may be advisable to read through Section 9 to ensure that detection objectives can be met before proceeding further.

At this point, the user should have a reasonably good idea of solute polarity, and the relative ease of elution compared with other solutes. An initial MP and SP, plus a detector, should have been chosen. The next step is to go to Section 6 or Section 7, below, and follow the steps suggested. Section 6 deals with pure fluids, while Section 7 deals with modified fluids. Both outline method optimization schemes designed to determine feasibility in the fewest number of experiments. Each is a successive approximation approach, designed to find optimum conditions. An alternative optimization scheme is outlined in Section 8, which more systematically studies the effect of each variable, independent of the others. The reader might consider this alternative approach

Table 7.1   *Polarity ranges of solute families*

hydrocarbons
PAHs
ethers
esters
light polymers/oligomers
polyethylene glycols
aldehydes and ketones
alcohols
Phenols
acids
monofunctional fatty acids
hydroxy acids
polyacids
anilines
benzamides
benzylamines
sulfonamides

Pharmaceuticals
fat soluble vitamins
Stimulants
antipsychotics
antidepressants
diuretics
β-blockers
steroids
PTH-amino acids

Agricultural chemicals
phenylureas
sulfonylureas
triazines
carbamates
organochlorine
organophosphorus
phenoxy acids

Polarity range of each stationary phase
$C_{18} < C_8$
phenyl
CN
silica
NH_2
diol
SA

Appropriate range for several mobile phases
pure carbon dioxide
CO_2 + dioxane
CO_2 + THF
CO_2 + acetonitrile
CO_2 + ethanol
CO_2 + methanol
CO_2 + methanol + additive

Instrumental Considerations
FID
PID
NPD
ECD
SCD
UV
Fluorescence

for at least the first few times he or she attempts to use packed column SFC. Several other approaches are also briefly discussed in Section 8.

Whichever approach is tried, it is important to break down the problem into simple steps. The initial stages of method development should use relatively concentrated standards easily detected with a common detector. It may also be helpful to use small injection sizes, to avoid any possible distortions caused by sample solvents.

6 Step-by-Step Speed–Resolution Optimization Scheme for Molecules That Elute with Pure CO$_2$

In each of the following sub-sections a title describes a topic followed by a list. In sections such as **Initial Conditions**, a list of the set points to try first is provided. In sections which ask questions, usually about peak shape, a list of actions is provided. These lists suggest the actions most likely to improve performance. The most effective are listed first.

Initial Conditions

For low to moderately polar molecules likely to be eluted using pure CO$_2$, the Phase Selection guide typically indicates a non-polar column, such as C$_{18}$ with pure CO$_2$, FID.

Also use for initial conditions: 80 °C, $P = 80$ bar; for two column hold up times, program to 350 bar at 10 bar min^{-1}. Flow = 2.5 ml min^{-1} for 4.6 mm i.d., 2 ml min^{-1} for 4 mm i.d., 0.5 ml min^{-1} for 2 mm i.d., 0.125 ml min^{-1} for 1 mm i.d.

Inject a standard and observe peak shapes. Go to the appropriate section, below, to optimize speed and resolution. After the elution characteristics of the solutes are determined, inject real samples to see where the major interferences elute.

Non-polar Solute, Asymmetric Peaks

If no peaks emerge or if peaks are not symmetrical . . .
    (A) Make initial pressure = 200 bar, hold at P_{max} for 15 minutes. Reinject and observe peaks.
    (B) Increase temperature to 150 °C. If no peaks emerge, the separation probably requires a modifier.
If peaks emerge with only modest retention but are not symmetric, the SP polarity may be inadequate.
    (C) Increase SP polarity and repeat pressure/temperature optimization.
If the peaks remain asymmetric, use a modified mobile phase.

Non-polar Solute, Symmetric Peaks

If symmetric peaks elute . . .

(A) Adjust retention by changing initial pressure and hold time: higher pressure, less hold to decrease retention; lower pressure, longer hold to increase

If peaks are inadequately resolved . . .

(A) Decrease initial pressure. Don't be afraid to go below P_c (at $T \gg T_c$). increase initial hold time.
(B) Decrease program rate
(C) Vary temperature both up and down. Continue in the direction producing the better result.
(D) Change SP and start over

7 Step-by-Step Speed–Resolution Optimization Scheme for Molecules That Require Modifier

As in the previous section, in each of the following sub-sections a title describes a topic followed by a list. In sections such as **Initial Conditions,** a list of all the set points is provided. In sections which ask questions, usually about peak shape, lists of actions are provided. These lists suggest the actions most likely to improve performance. The most effective are listed first. The initial goal is to find steady state conditions (no programming) that resolve, or nearly resolve, the solutes.

The following scheme may appear complicated. However, for any one separation, the chromatographer is unlikely to go into more than a few of the sections. Method development tends to be rapid owing to fast equilibration between the MP and SP. In typical cases, it takes no more than 4–8 hours to determine whether SFC will perform a given separation and produce at least a preliminary method.

Initial Conditions for Polar Solutes

Choose appropriate SP and MP using the Phase Selection Guide in Table 7.1. For polar solutes this might typically suggest: A polar column (diol, aminopropyl, or cyanopropyl) with MeOH modified CO_2, UV detector.

Also use for initial conditions: 20–30% MeOH in CO_2, 40 °C, pressure = 200 bar; flow ≈ 2.5 ml min^{-1} on 4.6 mm i.d., 2 ml min^{-1} for 4 mm i.d., *ca.* 0.5 ml min^{-1} for 2 mm i.d., 0.125 ml min^{-1} for 1 mm i.d.; UV detector at low wavelength (*e.g.* 210 nm).

Observe Peak Shape

Run an initial chromatogram. The most important first goal is to obtain good peak shapes. Resolution of the solutes is initially a secondary issue. Observe peak shapes and go to the section, below, which best describes the worst peaks in the chromatogram. There is one exception. If peaks both tail and have virtually no retention, there is probably a severe mismatch requiring a different phase selection.

Tailing

If retained peaks tail . . .

    (A) Increase the modifier concentration further.

    (B) Decrease temperature, and/or increase pressure.

    (C) Dramatically increase temperature (*e.g.* 120 °C).

If peak shapes improve go to *Symmetric Peaks*.

If peaks fail to improve, go to *Distorted/No Peaks*.

Distorted/No Peaks

If peaks do not elute, or look like the Rock of Gibraltar . . .

    (A) Include an additive in the MP.

        (1) a strong base for basic solutes.

        (2) a strong acid for acidic solutes.

If Symmetric peaks elute, go to *Symmetric Peaks*. If tertiary fluids do not elute the solutes with good peak shapes . . .

    (A) Try a different additive.

If poor peak shapes persist, the solute is probably too polar for packed column SFC. Quit.

Symmetric Peaks

It is not necessary to baseline resolve solutes here. Observe retention. Choose from the following three sections. As soon as adequate retention is achieved go to *Adequate Retention, Adjust Selectivity*.

Excessive Retention. If peaks are symmetrical but retention is excessive (*e.g.* four solutes in > 30 minutes) . . .

    (A) Further increase modifier concentration (you can go all the way to 100%)

    (B) Increase pressure, decrease temperature.

    (C) Double the flow rate.

If retention improves, go to *Adequate Retention, Adjust Selectivity*. If retention remains excessive, there are further options but one should begin to consider other separation techniques. If it is desirable to proceed . . .

    (D) Decrease the polarity of the SP (SA > diol > silica > CN > phenyl > $C_8 > C_{18}$)

    (E) Decrease SP surface area (larger pores or pellicular, or shorter column)

If retention remains excessive (extremely unlikely), consider a different technique (not SFC, probably LC). Quit.

Inadequate Retention. As soon as adequate retention is achieved go to *Adequate Retention, Adjust Selectivity*.

    If peaks are symmetrical but even at low modifier concentration retention is inadequate . . .

    (A) Decrease pressure until peaks or baseline break up then increase slightly.

    (B) Increase temperature in 10 °C increments (may require higher pressure).

    (C) Decrease temperature.

(D) Decrease polarity of the modifier (methanol > ethanol > isopropanol > acetonitrile > tetrahydrofuran > methylene chloride > hexane).

(E) Turn off modifier pump (use pure CO_2).

If peaks remain symmetrical but retention remains inadequate . . .

(A) Increase temperature at constant low pressure.

(B) Increase SP polarity (SA > diol > silica > CN > phenyl > C_8 > C_{18}).

If this decreases retention . . .

(A) Use the least polar column.

(B) Increase column surface area (smaller pores, longer column).

Adequate Retention, Adjust Selectivity

If retention is adequate but selectivity poor . . .

(A) Change temperature ± 10 °C. Continue in direction of best a.

(B) Decrease pressure to near two phase region

(C) Consider using longer retention time.

(D) Change SP and start over.

At some point, the speed and resolution goals of the separation should be met. If they are not, there are several further things to try.

Enhancing Resolution

Programming Instrumental Parameters

If some conditions resolve early eluting peaks and others resolve later eluting peaks, attempt linear programs from one set of conditions to the other. Any of the control variables can be programmed without dramatic losses in peak shape, or efficiency. There have been situations where percentage modifier, pressure, temperature, and even flow have all been programmed simultaneously.

Complex Samples

If sample complexity is high, the simplest solution is to increase column length (use higher N). One useful guide to determining efficiency required is:

$$\text{No. of solutes} = (N^{0.5})/4$$

where $N = \text{Length}/2d_p$ (d_p is particle diameter).

For example: 10 000 plates should be used for no more than *ca.* 25 solutes, and even this number is excessive. With 200 000 plates, the upper limit of sample complexity should be in the vicinity of 112 solutes. For more complex samples, it would be desirable to split the sample and do several analyses.

More complex samples can be resolved on relatively low efficiency columns but the effort required is usually extensive. The obvious desire to perform a single analysis must be weighed against the exaggerated time required for method development.

Continuing Selectivity Problems

If unresolved pairs of solutes persist after a number of different SPs have been tried, it is sometimes helpful to combine columns with different SPs in series. This works best when different pairs are only resolved on different SPs. The order in which the columns are connected does sometimes matter. If the order AB doesn't work, the order BA may still work.

Finishing Up

At some point in this procedure the analyst has either given up, or the speed and resolution should meet the analysis goals. It is at this point that the user should worry about detection goals. Trying to optimize the separation and detection at the same time can be confusing and difficult.

8 Alternative Approaches to Method Development

The standard method development scheme outlined above is based on a 'cookbook' approach which has been shown to cut through the confusion surrounding the differences between SFC and the other chromatographic techniques, and produce an optimized separation in minimal time. This 'standard' approach involves sequential changes in instrumental parameters and, when necessary, changes in the phases. In that approach, there is no systematic survey of the effect of each variable. Each parameter adjustment is based on the changes brought about by the previous instrumental change. The emphasis is on first using the variable which most strongly affects retention, moving the peaks around until they are all in a relatively narrow range of retention times. After most of the major adjustment in retention is out of the way, other variables which have a more subtle effect are manipulated, to try to optimize selectivity. This approach does not leave the developer with any strong feeling on the robustness of the method developed. It can also suffer from finding 'local optima', conditions which are superior to any other nearby conditions but are inferior to the true optima which may require substantially different conditions.

There is at least one other way to optimize methods which can give a stronger indication of robustness while also revealing more options for optimization. This method involves systematically studying the effect of each variable, independent of the others. Such an approach is similar to the 'standard' method but more extensive.

Systematic Study of Each Variable

Unlike GC or LC, there are three variables of importance in SFC. This can make it difficult to decide where to start in method development. Fortunately, the high diffusivity and low viscosity of SFs results in rapid equilibration, so the effect of each variable can be systematically studied in modest time. Concentrated injections of standards at three to five modifier concentrations can be accomplished in less than an hour and allow the chromatographer considerable insight into the relative behaviour of each solute. Choosing a modifier concentration then studying another variable provides additional understanding, and so forth.

Initial Conditions

Optimization starts with the selection of initial conditions likely to elute all sample components in minimal time. Choose a high modifier concentration, a relatively high pressure (*e.g.* $P = 2$–$3 \times P_c$), and a relatively low temperature (*e.g.* $T = 1.05$–$1.10 \times T_c$). With MeOH/CO_2: 20–30% MeOH, 150–200 bar, and 40–60 °C. Chromatograms are run, and the retention times of all the solutes are noted. This verifies that all the solutes elute.

If some solutes do not elute, or some elute with poor peak shapes, immediately include an additive in the MP. There are good reasons to try to avoid the inclusion of an additive, but it is also important to push toward some viable solution in minimal time. If an additive improves the peak shapes, it is immediately clear that some SFC approach can be found. Alternatively, if additives do not improve the separation, the likelihood of finding some other viable approach is degraded.

Vary Modifier Concentration First

After the initial experiments, the modifier concentration should be cut in half, and additional chromatograms collected. This process is repeated until retention becomes excessive (*e.g.* retention greater than 20 minutes or $k' > 50$). It is helpful to plot t_R *vs.* modifier concentration as the data are collected.

Choose a modifier concentration at which all solutes have k' between 0.5 and 10 (if possible). Alternatively, choose a modifier concentration where selectivity seems 'best' (minimum overlap among all solutes) but without excessive retention. It is not necessary to baseline resolve all the components changing modifier concentration alone.

Vary Pressure

Holding modifier concentration constant, and temperature at its original value, vary pressure. Plot t_R *vs.* P. The largest changes in α generally occur as P approaches the point where the fluid breaks down into two phases. $P \geqslant 200$ bar seldom results in significant α changes but is worth the little extra time it takes to try.

Vary Temperature Last

If pressure adjustments fail to resolve difficult pairs, increase the pressure modestly $(P \geq P_c + 20\text{–}30 \text{ bar})$, then change the temperature. Generally, collecting two more chromatograms at $+ 10 °C$ and $- 10 °C$ from the original set point gives an indication of the system susceptibility to temperature. If some selectivities change significantly over this 20 °C range, more extensive temperature changes are likely to produce even larger shifts. If modest changes in temperature have little or no effect on α, much larger changes are also unlikely to have much effect.

If temperature alters α significantly, extend the range of temperatures covered to at least 50 and preferably 75 or 100 °C while plotting t_R *vs. T*. Either higher or lower temperature may produce enhanced selectivity. It is unimportant whether the temperature is greater or less than T_c. It may even be desirable to use cryogenic cooling to make the temperature less than ambient.

Excessive temperature tends to degrade peak shapes of solutes that require high modifier concentrations. If the baseline becomes noisy or the peaks start to distort, slightly increase the *pressure* (not temperature) to move away from the point where the MP breaks down. temperatures much greater than 100 °C are seldom effective for such compounds. For many 'low' polarity species, however, much greater temperature is often desirable. Several of the bonded phases appear stable at 150 °C, or even higher, for extended periods. At temperatures much greater than 300 °C, silica starts to dehydrate.

Optimization

After the three plots of t_R *vs.* modifier concentration, pressure, and temperature are collected, it is useful to examine them in detail. In many instances, conditions can be found where all the solutes are resolved in relatively short times without any programming. Such a situation is most likely with simple mixtures, containing fewer than 10–20 components. If early eluting peaks are easily resolved at a particular set of conditions, and later eluting peaks are more easily resolved under a substantially different set of conditions, it may be advantageous to program from one set of conditions to the other. Such an approach becomes more favourable as the range of retention, at any one set of conditions, increases.

Since viscosity is much lower, and optimum flows are much higher, it is substantially easier and more rewarding to use long columns in SFC, compared with LC. If it is difficult to resolve solutes on a 'standard' LC column, it is sometimes more feasible to use either dramatically increased length or to use columns with different stationary phases in series. Either offers an alternative to the traditional LC approach. In the former, either difficult pairs are resolved by brute force (high plates), or more complex samples can be resolved. In the latter, a unique selectivity can be generated.

Additional Approaches to Method Development

There are at least several other approaches to method development. These include: (1) programming a single variable to scout out retention effects, while holding other variables constant, and (2) the use of statistically designed experiments. Since SFC is still a new technique to most chromatographers, and results can sometimes appear confusing, these other methods are probably less desirable than those discussed, at length, above.

Programming a Single Variable

One of the simplest method development strategies involves initially programming a single variable while keeping all others constant. This is similar to both LC and GC, where a single physical parameter is usually varied to optimize a separation. Initial conditions are set so that all the solutes are significantly retained. Conditions are then programmed to elute all the solutes with minimal retention. This works well in GC and LC, because retention is usually a monotonic function of a variable.

In packed column SFC, using pure mobile phases, the primary control variable is mobile phase density. Density is changed through pressure programming. In a typical experiment, the flow and temperature are kept constant while pressure is programmed from a low to high value.

With modified fluids, the modifier concentration is the primary control variable. Pressure and temperature are normally held constant, at least for the initial experiments. After the initial experiment, the primary emphasis is on changing the ramp rate to optimize the separation. If this is inadequate, some other variable is changed. Relative retention is often dependent on both the value and the rate at which variables are changed. This can cause confusion due to peak reversals.

Statistically Designed Experiments

Modern methods for statistically designed experiments are presently popular in both LC and GC. Such methods generally produce optimized methods with the fewest possible experiments and without the need to study the effect of each variable systematically, independent of the others.

Statistical methods have also been shown to be applicable to simple versions of SFC, particularly using pure fluids. However, the number of experiments required to find a performance optimum is a function of the number of variables. Of all the chromatographic techniques, packed column SFC provides the largest number of variables for adjusting retention and selectivity. Since a large number of variables is a feature of packed column SFC, the number of experiments required to find a true optimum is much larger than in LC or GC.

At present, statistical methods are unlikely to produce optimum results in dramatically fewer experiments or shorter time than other methods suggested

Table 7.2   *General attributes of detectors used in SFC*

Name	Type	Response MDQ/pgs^{-1}	Lin range	Select	Fluids
FID	u	5 C	10^7	–	CO_2,CFC,HFC
PID		0.1	10	–	CO_2,
ECD	s	0.02–1000	10^4	1–10^5	CO_2,CH_3OH, others
NPD	s	0.5 N	10^4	10^5	CO_2,CH_3OH
		0.1 P			
SCD	s	1–5 S	10^4	10^4	CO_2,CH_3OH
UV	s	50–1000	10^4	–	many
Fluor	s	1–1000	10^4	–	many

above. In addition they can be less well understood, and less robust. Since the effect of variables on retention is not always monotonic it should be relatively easy to be confused by local performance maxima. With proper experimental design such false leads can be overcome but require extra optimizations. Multiple optimizations require a large number of extra experiments.

Some sort of statistical approach is likely to form the core of future automated methods development software. However, at this time other methods are easier to understand and probably faster.

9 Optimizing the Detection of Real Samples

After the separation is optimized for speed *vs.* resolution, the final step is to optimize detection. Ideally, this would simply involve choosing the right detector so that only the solutes of interest would respond, with adequate sensitivity. Some general attribute of detectors compatible with SFC are listed in Table 7.2.

Choosing a Detector

Detectors are generally considered to be either 'universal' or 'selective', depending on which compounds produce a response. These designations are only approximations. Obviously, even universal detectors are somewhat selective, being capable of ignoring the carrier fluid. Conversely, 'selective' detectors also respond at least slightly to elements or compounds to which they are not 'selective'. Some 'selective' detectors respond well to a wide range of compounds, and are, therefore, only vaguely 'selective'.

A good example of a universal detector is the FID, providing universal response to compounds containing carbon. However, some carbon containing compounds produce little or no response (CO_2, CO, formaldehyde, carbon disulfide, formic acid, *etc.*). In addition, there are a few extreme instances where response factors can be forced to deviate by up to 20% under adverse circumstances.

Not all detectors are compatible with all MPs. For instance, the FID responds to virtually all organic solvents. The addition of even a small concentration of the most desirable (chromatographically) modifiers into the MP produces a large baseline offset, noise and poor sensitivity.

Selective detectors tend only to respond to subsets of all the compounds present. The nitrogen–phosphorus detector primarily responds to nitrogen and phosphorus, but also responds somewhat to carbon (the response factor to nitrogen is generally 2.5×10^4 times higher than the response factor to carbon).

There is also a general correlation between selectivity and sensitivity. The more selective a detector, the fewer the possible interferences, and generally the more sensitive the detection limits.

There are two main reasons to choose a selective detector. With complex matrices, there are often compounds present that elute in the vicinity of the solutes of interest. To detect and quantitate the solutes properly, one could use a more extensive sample preparation to pre-separate the solutes from these interferences, modify the chromatographic conditions to resolve them further, or alternatively, one could use a selective detector that does not respond to the interferences.

Selective detectors are often more sensitive than universal detectors. This enhanced sensitivity can allow direct detection of very dilute solutes, eliminating the need for a solute preconcentration step. In either case, selective detectors are used primarily to simplify sample preparation.

Inadequate detector sensitivity can dictate more extensive sample preparation, such as a major preconcentration of an extremely dilute sample. Alternatively, a derivatization may be required if none of the detectors respond to the solutes of interest.

10 Other Considerations. Injection Volume/Sample Solvent Polarity

There are two aspects of the sample solvent that need to be thought out: the effect of solvent polarity and injection volume on peak shapes and retention.

The sample solvent should be weaker than the mobile phase to avoid potential peak distortions or unstable retention times. Injection solvents can act as 'local' modifiers. If the sample solvent is stronger than the chromatographic mobile phase, the solutes will tend to stay with the plug of solvent as it travels through the column. Larger injections using such solvents will decrease solute retention times, and tend to broaden and distort peaks.

Since SFC is normal phase, it is advisable to try to get the solutes into a low polarity solvent for injection. On the other hand, switching solvents adds to the sample preparation burden and should be avoided whenever possible.

Small injection volumes of polar solvents may not affect retention times or peak shapes. However, as the injection volume is increased (to detect more dilute solutes) greater care must be exercised to avoid problems with the sample solvent.

Guidelines for Sample Solvent and Injection Volume

On 4.6 mm i.d. columns (for other i.d.'s, scale according to cross sectional area):

Water

Peak distortion and the tolerance of water is dependent on the MP and SP. Water is retained by bare silica. With C_{18} (and many other bonded phases), solutes can often be retained whereas water is more or less unretained using pure CO_2 as the MP. After the water passes, the modifier can be turned on. A few μl of water can be tolerated at 40 °C; up to 5 μl at $T > 60$ °C.

For very dilute moderately polar solutes in water, a pre- or guard column can be mounted in place of the loop in a fixed loop injection. Large water samples (*i.e.* 10 ml) can be sucked or pumped through this precolumn. The water is then blown out of the loop and partially dried. The solutes remain on the precolumn. The valve rotated into the inject position.

Methanol

Methanol is generally more polar than the MP. Nevertheless, 5–10 μl injections of MeOH usually cause no problems. With 20 μl of methanol, peaks are measurably broadened ($< 20\%$ loss in N). With 100 μl injections of methanol peaks are severely distorted and retention shifts. Increased temperature seems to make the problem less severe. The ability to tolerate different methanol loadings appears to be somewhat phase dependent.

Ethyl Acetate, Toluene, Hexane

With less polar solvents, 20 μl injections still broaden peaks slightly. Larger injections cause more severe broadening. When there is a severe mis-match between the sample solvent and the MP, the chromatographer should consider why he or she chose such a mismatch.

A few low polarity solvents, like methylene chloride, sometimes cause problems in reproducibility or peak shape. This may be due to disruption of the adsorbed film of mobile phase components.

CHAPTER 8

Pharmaceutical Analysis by Packed Column SFC

1 Introduction

In this chapter, the analysis of pharmaceutically important compounds by packed column SFC is discussed. Chiral analysis is sufficiently important to justify a separate chapter (see Chapter 9).

In recent reviews on pharmaceutical analysis,[1,2] SFC was either not mentioned or covered with a few brief paragraphs. In previous chapters, pharmaceutical analysis was listed as one of the potentially most important application areas for SFC. Although these two statements appear to conflict, one describes the past, the other the likely future.

Pharmaceutical analysis involves qualitative and quantitative analysis on major/minor and trace components with similar and wildly different polarities. The ideal analytical tool would be fast, and efficient, with multiple detection options. A key attribute is the ability to vary selectivity dramatically. Financial conditions generally dictate that long analysis times can only be tolerated when no other option exists and the product under study is valuable enough to support the costs associated with low information density per unit time.

Instrumentation specifically designed for packed column SFC has only existed for a short time. However, it should be clear that it possesses the figures of merit required to compete against LC in pharmaceutical analysis. Compared with reversed phase LC, packed column SFC is faster, more efficient,[3] has a wider range of detector options, has more selectivity options, and generates less toxic waste. Packed column SFC exhibits quantitative reproducibility similar to that of LC. Packed column SFC offers a second selectivity option to complement reversed phase LC.

2 Separations of Some Specific Pharmaceuticals

This section attempts to indicate the breadth of SFC applications. Some compounds have simply been eluted using a supercritical or near critical mobile phase. Others have undergone extensive scrutiny within the matrices in which they are sold. These discussions use an inordinate number of results from the writer's laboratory, simply because that material is the most familiar. The section is not intended to be an exhaustive review of all pharmaceuticals separated by packed column SFC. The intent is to give the viewer a snapshot of the present state of SFC in pharmaceutical analysis.

Miscellaneous

Phenols and hydroxyacids are common in over the counter (OTC) products. Although not specifically aimed at pharmaceutical analysis, several papers deal with the separation of these and related compounds.[6,7] Anilines,[8,9] benzylamines,[9,10] and other aliphatic primary amines[9] and amides[11] are also difficult to elute substances that have been studied by packed column SFC.

Several purines and pyrimidines were eluted from packed columns.[12] They included mercaptopurine, trimethoprim, trifluridine, and zidovudine (AZT), used to treat AIDS. Several other unrelated but important drugs, like triprolidine, pseudoephedrine, and permethrin, were separated in the same reference. In an unpublished study,[13] many of the purines and pyrimidines present in DNA and RNA were studied, as shown in Figure 8.1. Several of the common compounds were not sufficiently soluble in methanol to allow detection. A few percent of trifluoroacetic acid (TFA) improved solubility in methanol but only

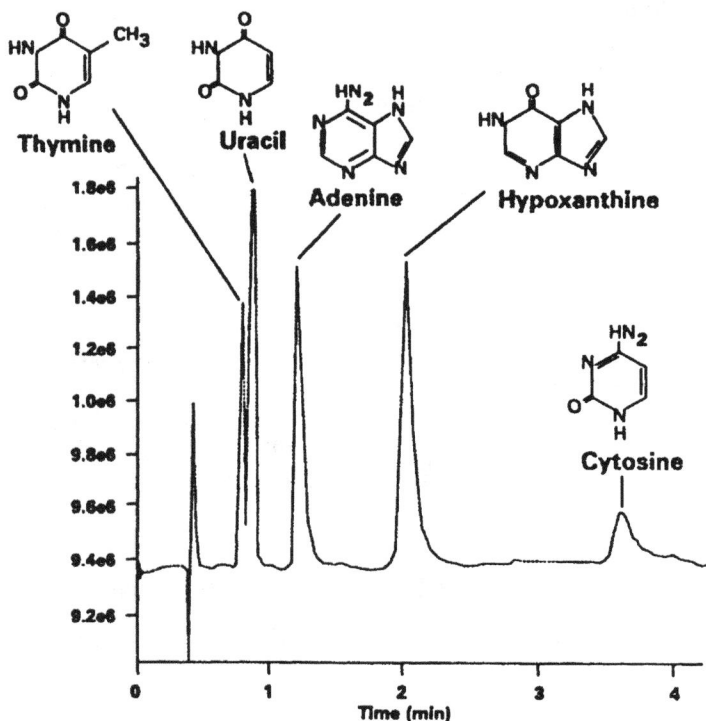

Figure 8.1   *Purines and pyrimidines separated on a short diol column:* 1 ml min^{-1} *of 10% [MeOH + 0.6% IPAm] in* CO_2 *at 50 °C, 200 bar; Column* 2 × 100 mm, 7 μm *Nucleosil Diol*

marginally improved the ability to elute compounds like guanidine. Since conditions in the column provide weaker solvent strength than pure methanol, this result illustrates one of the likely limitations of SFC. Compounds requiring an aqueous environment or a high salt or buffer concentration are unlikely to be separated by SFC. This leads to a rule of thumb: **If a solute will not dissolve in methanol or a less polar solute it is an unlikely candidate for SFC.** The corollary is: **if a solute readily dissolves in methanol (or a less polar solvent) it will likely be separated by SFC.**

Fat soluble vitamins and some water soluble vitamins can be separated, as shown in Figure 8.2. Vitamins A, E, and D_2 and at least eight contaminants were resolved on a 4.6 × 200 mm, 5 μm RP-18 column in 14 minutes using 0.5% MeOH in CO_2 for 7.5 minutes, then programming to 1% at 0.2% min^{-1}.[14] Inlet pressure was 165 bar, and temperature 40 °C. None of the compounds eluted with pure CO_2. Four tocopherols[15] were separated in a few minutes from corn oil using pure CO_2 and a C_{18} column. Similarly, Vitamin A was separated from fortified cod liver oil under similar circumstances. In an unpublished study, Vitamins K_1, E, A, and D_3, with numerous contaminants, were separated in 12 minutes, on a Lichrosphere C_{18} column with 2 ml min^{-1} 4% MeOH in CO_2, at 140 bar, 40 °C.

The PTH derivatives of more than 25 amino acids were separated using tertiary mobile phases[16-18]. Underivatized amino acids have been reported separated[19], as shown in Figure 8.3.

Several reports have appeared about antiparasitic agents. In one,[20] mono- and di-saccharides of 22,23-dihydroavermectin B_{1a}-aglycone were separated on

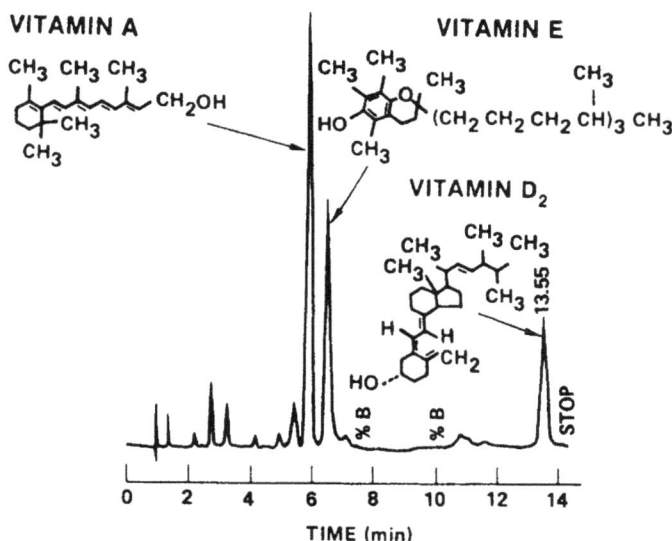

Figure 8.2 *Separation of fat soluble vitamins on a packed column. 0.5% MeOH in CO_2 for 7.5 min then 0.2% min^{-1} to 1.0% at 40 °C, 165 bar column 4.6 × 200 mm, 5 μm RP-18*
(Reproduced by permission from ref. 14)

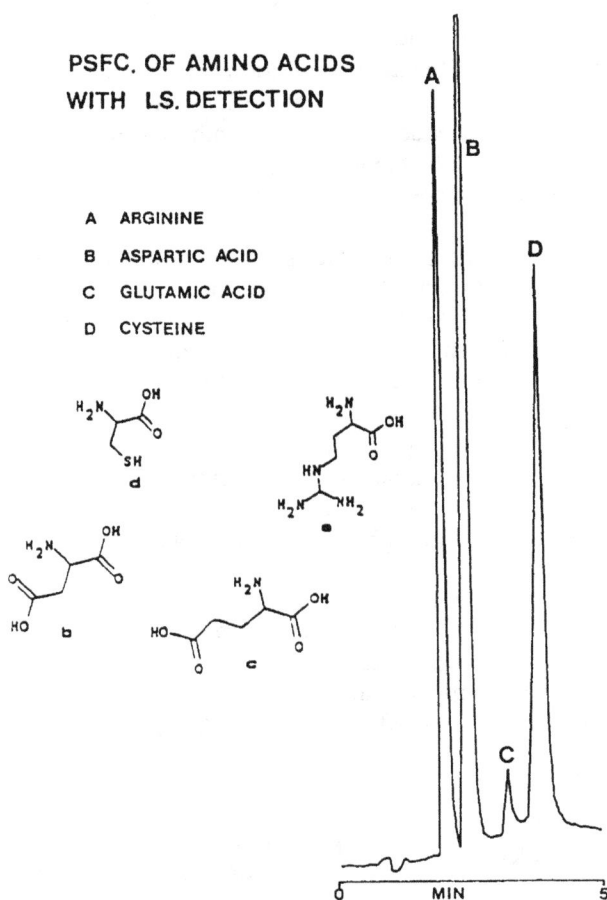

Figure 8.3   *Underivatized amino acids*

silica with 3 ml min^{-1} of 7% methoxyethanol, at 65 °C, 3100 psi. Negative ion CI mass spectral detection with ammonia was performed.

Ibuprofen and noproxen were separated[21] on an APS column with good peak shapes and high capacity (*ca.* 500 μg). The glycine conjugate of salicylic acid was recovered[21] from urine by solid phase extraction and chromatographed. The peak for *o*-hydroxyhippuric acid was manually collected and identified by ^{1}H NMR.

Mitomycin C,[22] cephalosporins,[20] erythromycin A,[23] and mefloquine[24] (with an ECD) have each been eluted in packed column SFC. Games[25,26] has published many SFC papers using MS detection.

Steroids

Many SFC chromatograms of steroids have been published (*e.g.* ref. 27). Steroids with only one or two polar groups can be eluted using pure CO_2 (or

GC). Steroids containing three or more polar groups may require high concentrations of polar modifier to elute.

In Chapter 4 (see also ref. 28), polar steroids were shown to require a moderately polar SP for good peak shapes. Lower polarity phases produced poor peak shapes. A CPS phase provided good peak shapes and even peak spacing for testosterone, estradiol, hydroxycortisone, and estriol. More polar diol and sulfonic acid phases spread out some of the peaks and produced poorer selectivity.

With pure CO_2, many steroids do elute but peak shapes tend to be poor. Even less polar steroids (keto in place of hydroxy) required $MeOH/CO_2$ for both rapid elution and good peak shapes. Ethanol and isopropyl alcohol (IPA) modifiers produced longer retention and slightly degraded peak shapes. Acetonitrile produced less retention than IPA but peaks tailed. Surprisingly, changing from the alcohols to acetonitrile had little effect on selectivity.

Packed column SFC can be a high speed technique, as demonstrated with eight steroids[29] in Figure 8.4. The compounds were resolved in 2.2 minutes with high resolution and symmetric peaks. In an early chromatogram, eight steroids were nearly baseline resolved in 1.5 minutes.

Bile Acids

The bile acids, found in urine, retain the four ring steroid structure with a carboxylic acid on a side chain. Chenodeoxycholic, and ursodeoxycholic

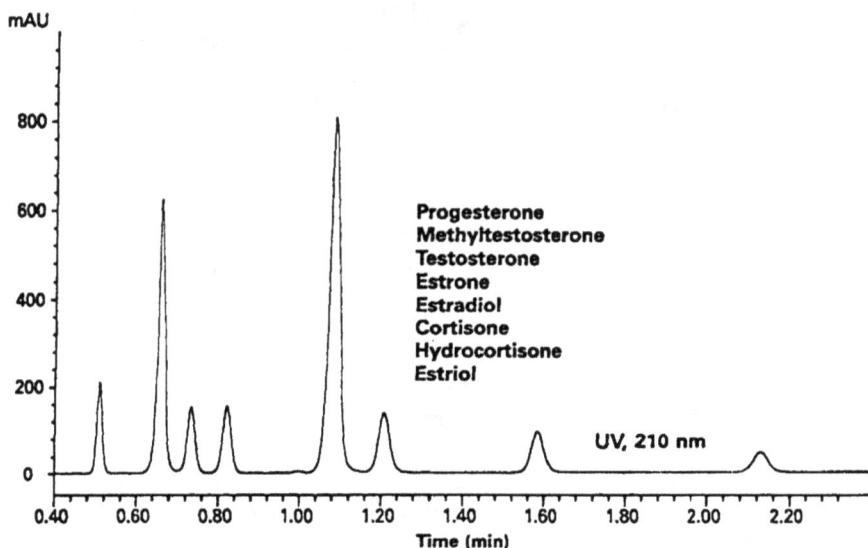

Figure 8.4 *High speed separation of eight steroids by packed column SFC*: 2.5 ml min^{-1} *of 20% MeOH in CO_2 at 70 °C, 200 bar; column 2.1 × 250 mm Lichrosphere Si-60. Progesterone, methyltestosterone, testosterone, estrone, estradiol, cortisone, hydrocortisone, estriol*
(Reproduced by permission from ref. 25)

acids[30] are used in capsules or tablets to dissolve cholesterol gall stones and also have other uses. Ursodeoxycholic acid allows treatment of biliary cirrhosis, primary sclerosing cholangitis, and cystic fibrosis. Some closely related acids present in the animal sources of the therapeutic agents are hepatotoxic, or cause other unwanted reactions. Analytical methods must differentiate all the important agents and potential contaminants.

The SFC separation is normal phase, since solute retention increases[30] with increasing polarity. As with the hydroxysteroids, a C_{18} column and 4% MeOH in CO_2 produced little retention but broad, unseparated peaks. On an APS phase, the compounds were excessively retained and several coeluted even with 25% modifier. Intermediate polarity columns produced acceptable retention and α with the best separation obtained using a phenyl column. Changing the stationary phase produced significant changes in selectivity.

With pure CO_2, the solutes did not elute. The final method used 15% MeOH in CO_2 at 4 ml min^{-1}, at 200 bar, 40 °C. The compounds of interest and likely toxic contaminants were easily resolved in less than 4 minutes. An additive was not required.

Modifier concentration was most effective in changing retention, while temperature and pressure produced modest changes. None of the instrumental parameters appeared to cause significant changes in α. Resolution was improved primarily through increasing retention.

Sample preparation was trivial and consisted of dissolution of the formulations in 50 ml MeOH, 5 minutes of sonication, followed by filtering. An aliquot of the filtrate was then directly injected. Excipients caused no interferences. The acids were detected by UV at 210 nm. Calibration curves were linear with zero intercepts. Recoveries and relative standard deviations for the method were found to be 100.2% and 1.4% for ursodeoxycholic acid and 101.5% and 1.2% for chenodeoxycholic acid, using 10 analyses for each acid. No internal standard was used.

The method was found to be acceptable for routine quality control. The method was eight times faster than HPLC methods reported in the literature, and nearly five times faster than a 'high speed' HPLC method (using a shorter column with a 3 μm packing) developed as part of this work.

Ecdysteroids

Ecdysteroids contain up to seven hydroxyl groups on a four-ring steroid backbone.[31,32] They are polar and thermally unstable. They have been either derivatized to decrease polarity then separated by GC or determined by LC. They could not be eluted from packed columns, and only the less polar members could be eluted from capillaries using pure CO_2. With MeOH/CO_2 all the compounds rapidly eluted with good peak shapes. The elution order was nearly identical to that in normal phase LC, and did not correlate with the number of hydroxyls present. This implies that steric effects are at least as important in retention as the number of polar groups.

Several plant extracts were separated in as little as 1/20th the time by SFC compared with LC, but with somewhat lower resolution. Several instrumental difficulties in optimization reflect the early date of the work and the lack of modern equipment available at that time.

All the compounds were identified by SFC–MS. At low concentrations of MeOH, the results resembled EI spectra. As the MeOH concentration increased, the spectra progressively shifted to resemble CI spectra.

Carbohydrates–Oligosaccharides

Sugars have been separated and detected by packed column SFC with MS detection.[33] Ribose, xylose, fructose, glucose, sucrose, maltose, melezitose, and raffinose were all baseline separated in less than 12 minutes, as shown in Figure 8.5. The mobile phase was 4 ml min^{-1} of 8–18% methanol in carbon dioxide over 4 minutes, at 3400 psi, 40 °C. The column was 4.6×250 mm, 5 μm Hypersil CPS. A β-cyclodextrin phase produced similar retention but required 20–30% methanol programmed over 4 minutes.

Carbohydrates and oligosaccharides modify the structure and properties of glycoproteins and, thus, are important in cell–cell, and cell–molecule recognition. The synthesis of these complex compounds for use as therapeutic agents

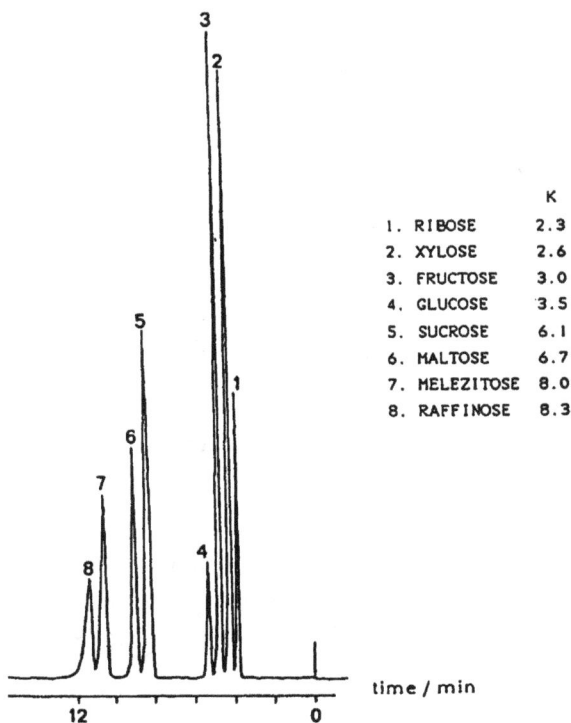

		K
1.	RIBOSE	2.3
2.	XYLOSE	2.6
3.	FRUCTOSE	3.0
4.	GLUCOSE	3.5
5.	SUCROSE	6.1
6.	MALTOSE	6.7
7.	MELEZITOSE	8.0
8.	RAFFINOSE	8.3

Figure 8.5 *Separation of eight saccharides on a packed column with mass spectral detection*
(Reproduced by permission from ref. 34)

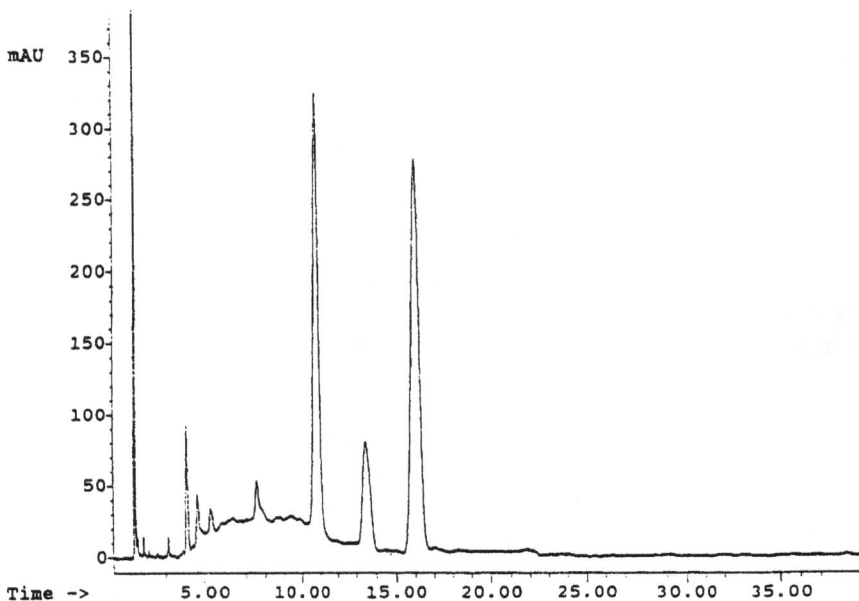

Figure 8.6 *Separation of several pentasaccharides by packed column SFC*
(Reproduced by permission from ref. 34)

has been extremely challenging. Analytical techniques for the synthetic intermediates have been difficult. SFC offers the rapid elution and detection of
many of the important compounds.[34] Important factors included elution at low
temperatures and the low molar absorptivity of $MeOH/CO_2$ mixtures at low
wavelengths compared with ethyl acetate and chloroform used in LC.

The compounds were acetylated to suppress polarity. Mono-, di-, and penta-
saccharides were eluted with good peak shapes and high efficiency in relatively
short times (< 15 minutes). Diastereomers of mono- and penta-saccharides
were separated. For example peracetylated galactose and glucose were resolved
in just over three minutes with a resolution > 1.5. Various other derivatives
were also eluted. Some pentasaccharides are separated in Figure 8.6.

Barbiturates

Phenobarbitone was chromatographed by capillary SFC using pure carbon
dioxide.[36] Many barbiturates were separated with poor peak shapes from a
polystyrene–divinylbenzene (PS–DVB) column with pure CO_2, but would not
elute from a C_{18} silica based column.[37] With $MeOH/CO_2$, many barbiturates
eluted in a short time with good peak shapes[37] from either PS–DVB or C_{18} on
silica. Some have also been separated on a CPS column and simultaneously
detected using several UV wavelengths plus the NPD using a $MeOH/CO_2$[38]
mobile phase.

Opium Alkaloids

Several opium alkaloids have been separated and quantitatively determined.[39–41] In poppy straw extracts,[41] narcotine, papaverine, thebaine, ethylmorphine, codeine, cryptopine, and morphine were found. These compounds are strongly retained on bare silica or APS using pure CO_2. MeOH/CO_2 mixtures produced inadequate α and peak shapes. The best peak shapes were obtained using CO_2/MeOH/strong amine/H_2O mixtures. The addition of amines caused a general decrease in retention.

Imidazole Derivatives

Many imidazoles are used as building blocks in the synthesis of therapeutic agents. In LC these compounds require derivatization for detection. In GC they produce tailing peaks. Initial attempts[42] using pure fluids failed to elute anything from either capillary or packed columns. The authors stated that the compounds appeared to react with CO_2 and plug the system. They could not be resolved with nitrous oxide. SF_6–NH_3 produced severe plugging problems and Vespel damage. It was concluded that these compounds could not be separated with fluids compatible with the FID. The compounds could all be rapidly eluted with good peak shapes using an APS column and CO_2/MeOH/amine (*e.g.* methylamine).

Crotamitron in Creams and Lotions

Crotamitron[43] is a scabicide applied in a cream or lotion, often in combination with hydrocortisone. Propylparabene or methylparabene is used as a preservative in creams. In the lotion, 2-phenylethanol or scorbic acid is used as a preservative. There are often additional related by-products present in commercial products.

The typical analytical procedures for quality control involve up to four different LC analyses with up to four different sample preparations. In LC, the excipients must be removed before the analysis of the active ingredients to avoid precipitation and plugging of the column.

In SFC,[43] the creams were dissolved in THF, filtered and injected. Simultaneous (or serial) detection by UV for the active ingredients and light scattering for the excipients provided complete quality control information in a single run with minimal sample preparation. The addition of both a strong acid and a strong base in the mobile phase allowed the elution of scorbic acid and a primary aliphatic amine contaminant.

SFC provided an overall time saving of 3–4 times through faster analysis and reduced sample preparation. Fewer analyses were required to characterize the commercial products completely. This analysis represented one of the first uses of SFC in routine quality control.

Taxol

Taxol is a promising drug for the treatment of ovarian cancer. The parent compound consists of a number of fused and attached hydrocarbon rings with numerous attached polar functional groups. The polar groups include three hydroxyls, a secondary amine bridge between two benzyl groups, two keto groups, and several acetates and esters.

The primary source of the compound is a rare, slow growing, and now threatened tree. Several alternative sources are under development but production is hindered by complex matrices. Taxol can be produced by cell culture or from a fungus. Analytical methods must separate taxol from a large number of potential interferences with wide variations in polarity.

Useful chromatograms were obtained using moderately polar to polar SPs on 5 μm particles in 4.6 × 250 mm columns. The most useful bonded phase was diol, with 2–2.5 ml min^{-1} MeOH/CO_2, 150 or 250 bar, and 30 °C. No additive was required in the MP. The authors characterized the separation as normal phase since retention increased with solute polarity.

Taxol could be rapidly separated from the related compounds baccatin, and cephalomannine with high resolution in 8 minutes.[44] A packed column SFC method was developed[44] which can separate and detect taxol in the presence of more than 60 contaminants present in a taxus bark extract in less than 35 minutes, as shown in Figure 8.7a. The same method can also be shown to isolate taxol from an even more complex matrix, a cell culture extract, as shown in Figure 8.7b. The method employs gradient elution with several ramp rates between 5 and 35% methanol to separate and elute both low and high polarity contaminants. A calibration curve covering 25–1000 ng injected was linear. Relative standard deviations on area using concentrated samples were better than ± 0.2%. Overall, the SFC method offered an alternative selectivity to reversed phase LC and cut analysis time by more than 50%.

Caffeine, Theophyline, Theobromine

Caffeine, theophyline, and theobromine[14,39,45–46] are present in many OTC medications. All three compounds are easily separated in a few minutes using 8% MeOH in CO_2 and a 4.6 × 250 mm, 5 μm CPS phase. Changing the SP produced dramatic changes in α, including peak reversals.

All three compounds contain nitrogen and can be detected using either a UV or an NPD,[46] as shown in Figure 8.8. Ten injections produced relative standard deviations of better than ± 0.25% in retention times, and ± 0.75 to ± 1.9% in peak areas. After the column, a 'tee' split a small fraction of the total flow to the NPD. Both chromatograms were, thus, obtained from the same injection. All three compounds could be easily detected well below 1 p.p.m. using either detector.

Several OTC medications were analyzed for caffeine, as shown in Figure 8.9. The tablets were crushed and dissolved in methanol, filtered, and injected. The broad peaks are due to excipients present as binders and diluents.

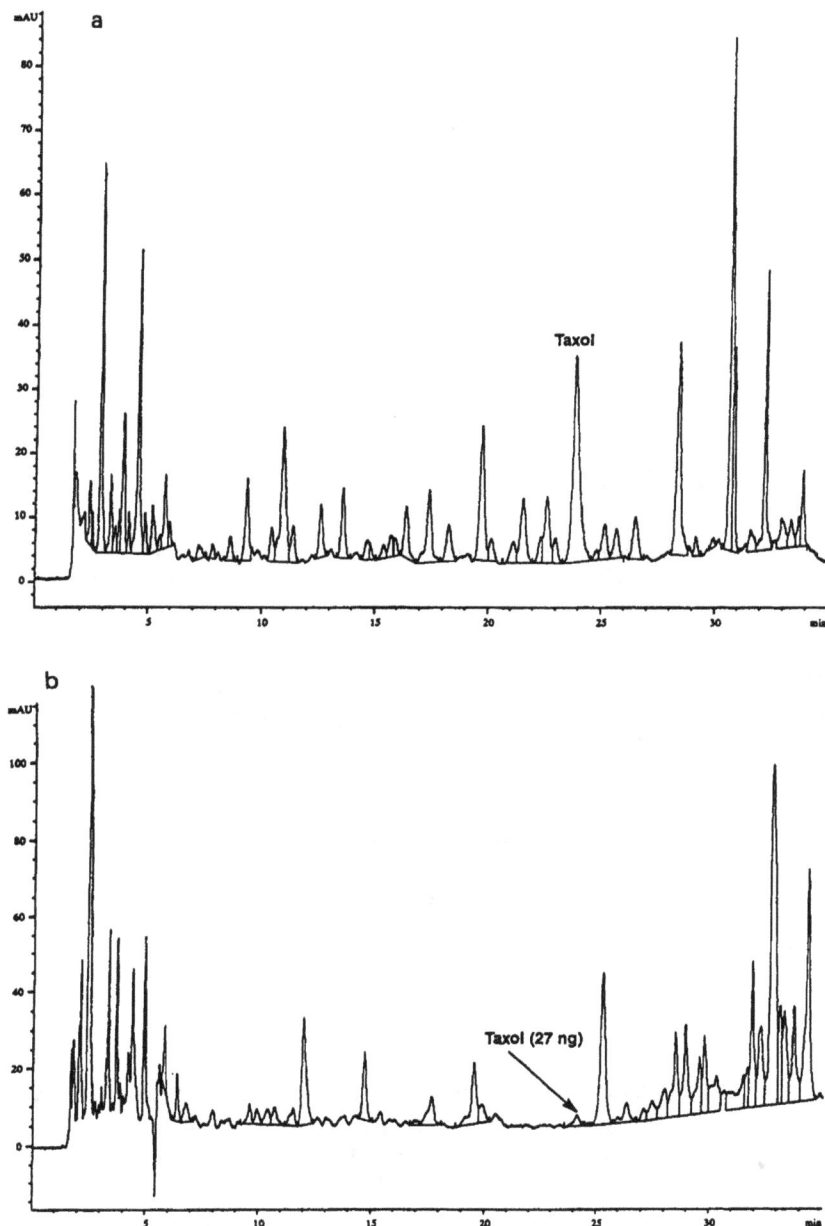

Figure 8.7 *The separation and determination of taxol from complex matrices*: (a) *taxus bark extract*; (b) *cell culture extract. Conditions*: 2.0 ml min^{-1} *of MeOH/CO$_2$, 8% to 18% at 0.4% min^{-1}, then 18% to 35% at 3% min^{-1}, hold 4 min, 30 °C, 150 bar; column 4.6 × 250 mm, 5 μm Lichrosphere diol* (Reproduced by permission from ref. 44)

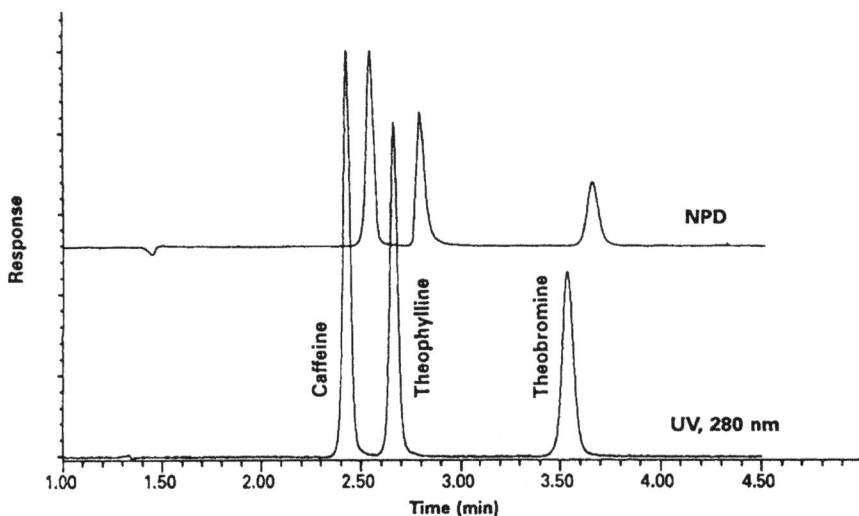

Figure 8.8 *Separation of caffeine analogues with NPD and UV detection.* 2.5 ml min^{-1} *of 8% MeOH/CO$_2$ at 80 °C, 250 bar; column 4.6 × 250 mm, 5 μm cyanopropyl* (Reproduced by permission from ref. 46)

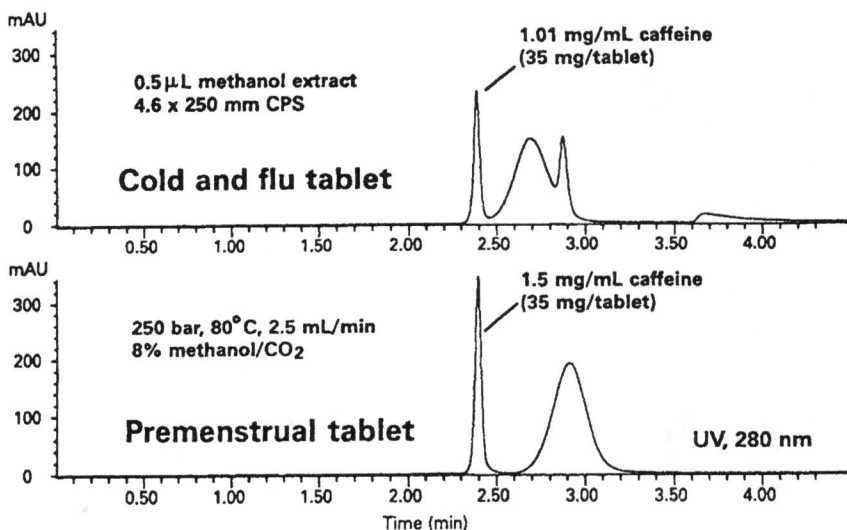

Figure 8.9 *Caffeine in several OTC medications. Tablets crushed, dissolved in 50 ml methanol, filtered, and 0.5 μl injected directly. Cold and flu tablet, Premenstrual tablet.* 2.5 ml min^{-1} *of 8% MeOH/CO$_2$ at 80 °C, 250 bar; column 4.6 × 250 mm, 5 μm cyanopropyl* (Reproduced by permission from ref. 46)

Benzodiazepines

Benzodiazepines were extensively studied[47] by packed column SFC on poly-styrene–divinylbenzene (PS–DVB) and both ODS and CPS silica columns. With pure CO_2, none eluted from the silica based columns and only a few eluted as broad, long retained peaks from the PS–DVB column. With either methanol or acetonitrile modified CO_2, all the benzodiazepines eluted rapidly with good peak shapes.

Metabolites of Benzodiazepines

In a recent study,[48] the *in vitro* and *in vivo* metabolites of several compounds were studied. For example, eight metabolites of camazepam were resolved from the parent in less than 10 minutes, as shown in Figure 8.10. Many benzodiaze-pines and their metabolites contain a chiral center. The simultaneous chiral separation of the parent compounds and the metabolites is the subject of a separate section in Chapter 9.

Achiral separations were mostly performed on a 4.6 × 250 mm, 5 μm Lichro-sphere APS column, with ethanol/CO_2 at 2.5 ml min^{-1}, 200 bar, 30 °C. Many of the separations employed an initial 4 minute hold at 13% ethanol/CO_2, then programming to 30% modifier, to ensure that polar contaminants were washed from the column.

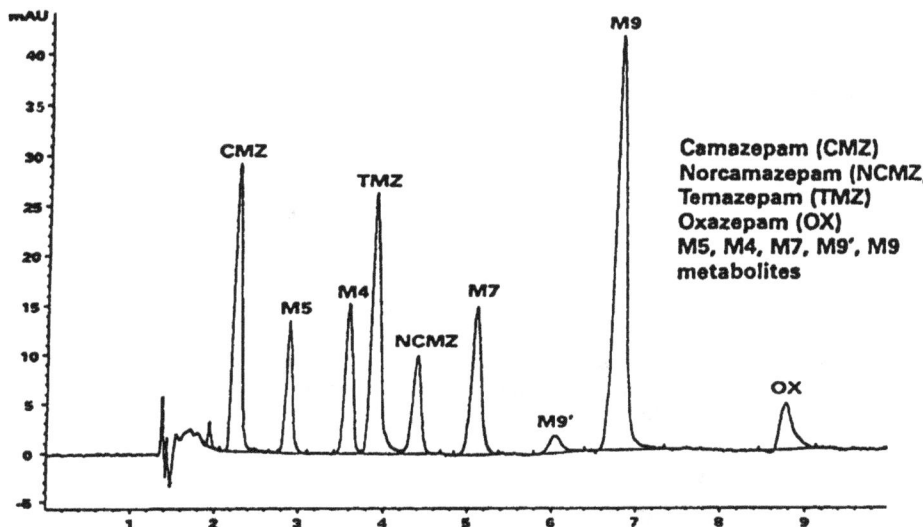

Figure 8.10 *The separation of camazepam and eight metabolites. Conditions: 2.5 ml min EtOH/CO_2; 13% EtOH for 4 min, then 3% min^{-1} to 30%; 30 °C, 200 bar; column 4.6 × 250 mm, 5 μm Lichrosphere NH_2. Key: CMZ is camazepam; NCMZ is norcamazepam; TMZ is temazepam; Ox is oxazepam; M5, M4, M7, M9', and M9 are unnamed metabolites*
(Reproduced by permission from ref. 48)

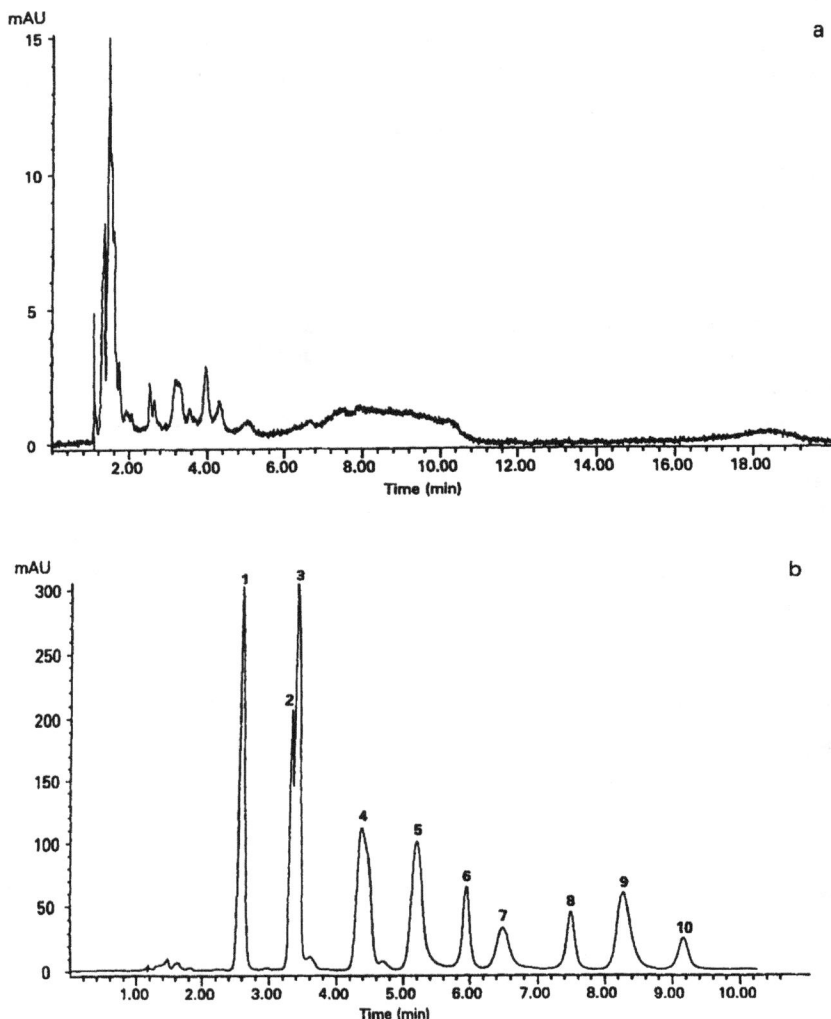

Figure 8.11   *The effect of additive on the elution and peak shape of phenothiazine anticonvulsants:* (a) *elution without additive* (b) *with 0.5% IPAm in MeOH* (Reproduced by permission from ref. 50)

Compared to HPLC, SFC was more than twice as fast with both higher efficiency and much higher resolution. Full UV spectra allowed some degree of *comparison* with standards.

Phenothiazine Anticonvulsants

The phenothiazine anticonvulsants[49,50] are moderately polar drugs containing hindered amines. Many contain a linear three ring structure with various side chains generally attached to a nitrogen in the central ring. They can be eluted in either GC or LC although they often yield tailing peaks in GC.

Because of their intermediate polarity, a CPS column with MeOH/CO$_2$ was chosen. No additive was initially included because the aliphatic nitrogens, although strong bases, appeared to be hindered. The first few injections of a mixture of 10 compounds rapidly confirmed the need for an additive in the MP. None of the 10 compounds injected eluted in 20 minutes, as shown in Figure 8.11a. With 0.5% IPAm in the MeOH, all the solutes eluted with good peak shapes in less than 10 minutes, as shown in Figure 8.11b.

Doubling the modifier concentration approximately halved retention, as shown in Figure 8.12. Selectivity was not strongly affected by modifier concentration. Pressure changed retention less dramatically than modifier concentration, as shown in Figure 8.13. Selectivity was moderately affected with a single peak reversal.

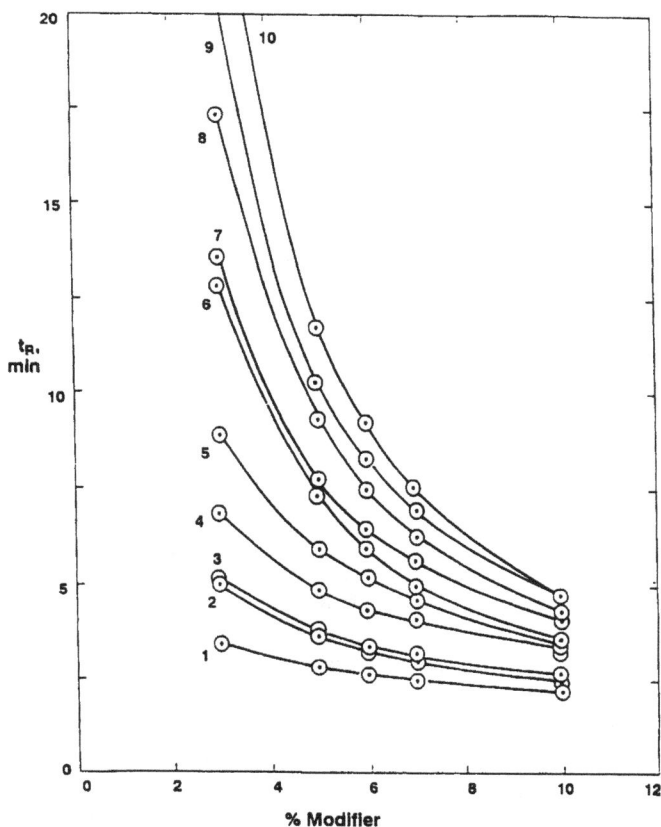

Figure 8.12 *The effect of modifier concentration on the retention of phenothiazine anticonvulsants, indicating that modifier concentration has the largest impact on retention but little impact on selectivity*
(Reproduced by permission from ref. 50)

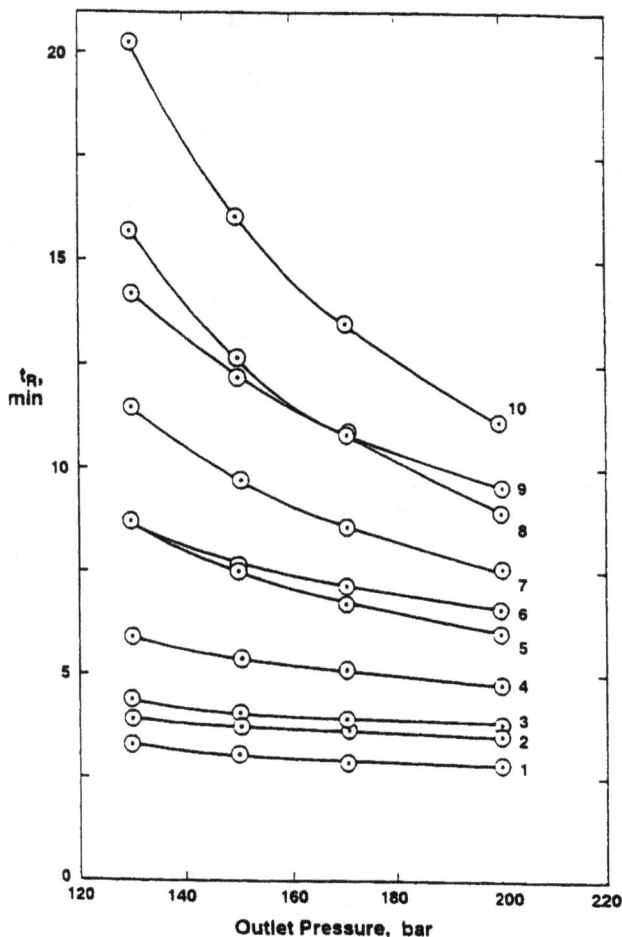

Figure 8.13   *The effect of pressure on retention of the anticonvulsants in Figure 8.12. The
results indicate that pressure is a secondary control variable compared with
modifier concentration. For these solutes, pressure has only a modest impact
on either retention or selectivity*
(Reproduced by permission from ref. 50)

Temperature changes resulted in dramatic α shifts, as shown in Figure 8.14.
The retention of some solutes increased, that of others decreased, and still
others showed first a decrease then an increase from $30 \leq T \leq 65$ °C.

None of the steady state conditions resolved all the solutes in this arbitrary
mix. Programming temperature (!) produced the best resolution of all the
programs tried. Analysis time was short and resolution was adequate. There
was no need to change either SP or MP, once it was clear that an additive was
required. The peaks were reasonably spread out and did not bunch into
clusters. This indicates that the phases were appropriate for these solutes. The
physical parameters provided a wide range of both retention and α adjustment.

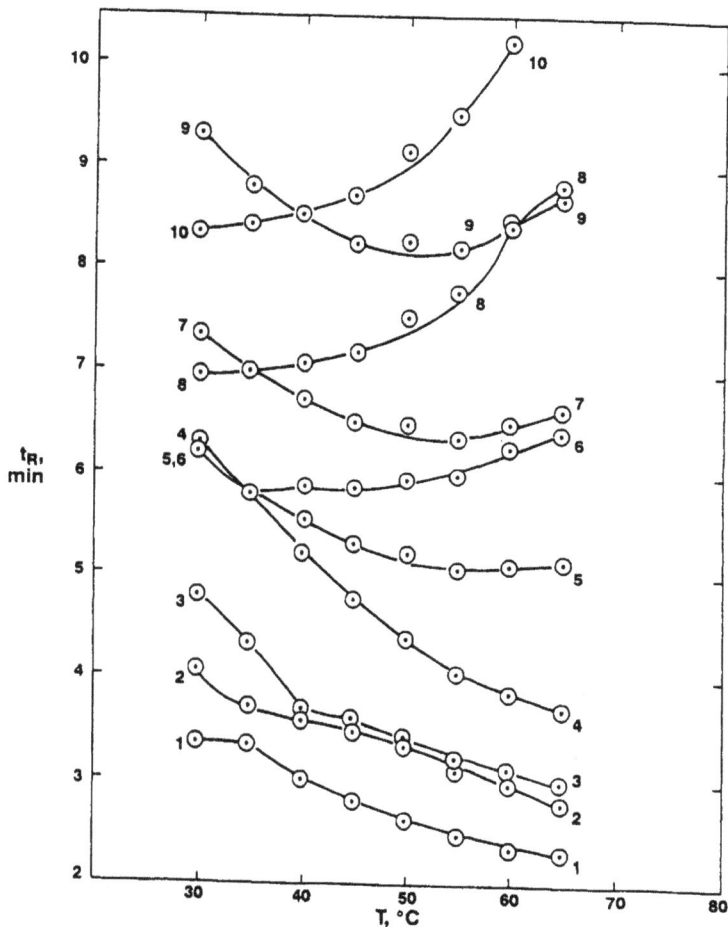

Figure 8.14 *The effect of temperature on retention of the anticonvulsants in Figure 8.12. Crossing lines indicate peak reversals. For these solutes, temperature has only modest impact on retention but dramatic impact on selectivity* (Reproduced by permission from ref. 50)

Tricyclic Antidepressants

'Tricyclic' antidepressants are actually compounds with a range of structures. Some contain the 'classic' linear three-ring structure, or the central ring can be seven membered. Such molecules are similar to the anticonvulsants, so similar phases were chosen for preliminary SFC experiments.[51] A CPS column with CO_2/[MeOH/0.5% IPAm] eluted all the compounds with good peak shapes. Five tricyclic compounds could be resolved in less than 3 minutes, as shown in Figure 8.15. Modifier concentration was again the most effective factor in adjusting retention, as shown in Figure 8.16. Pressure had only modest effect on retention and almost no effect on a. Temperature had the biggest impact

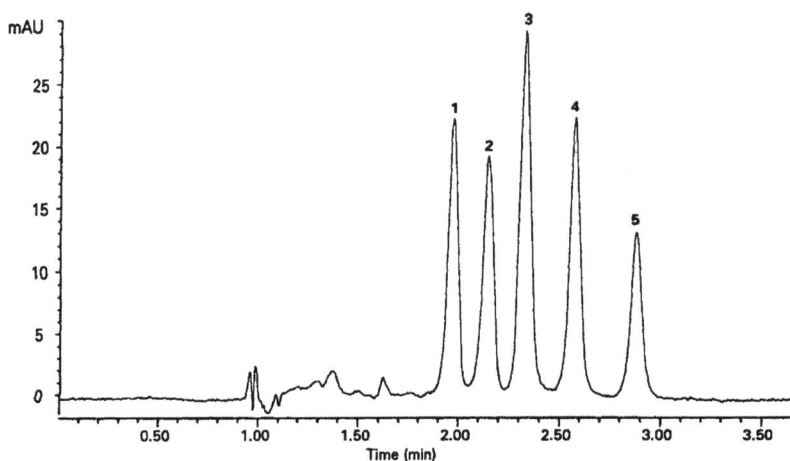

Figure 8.15 *High speed separation of five tricyclic antidepressants by packed column SFC*
(Reproduced by permission from ref. 51)

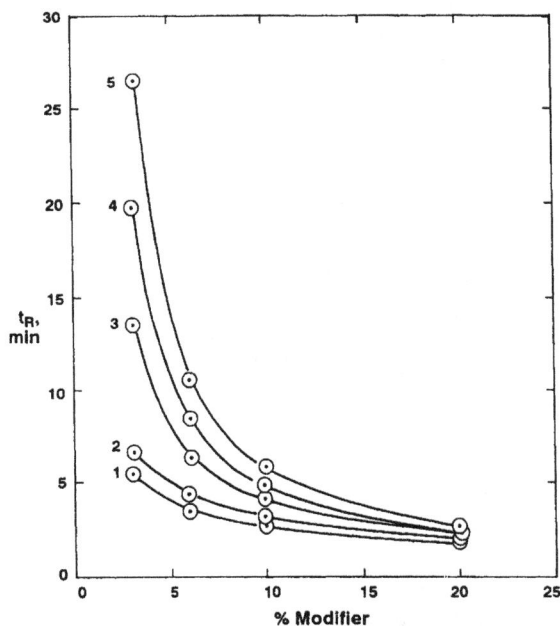

Figure 8.16 *The effect of modifier concentration on the retention of tricyclic antidepressants. Modifier concentration is again shown to be the most significant control
variable for changing retention*
(Reproduced by permission from ref. 51)

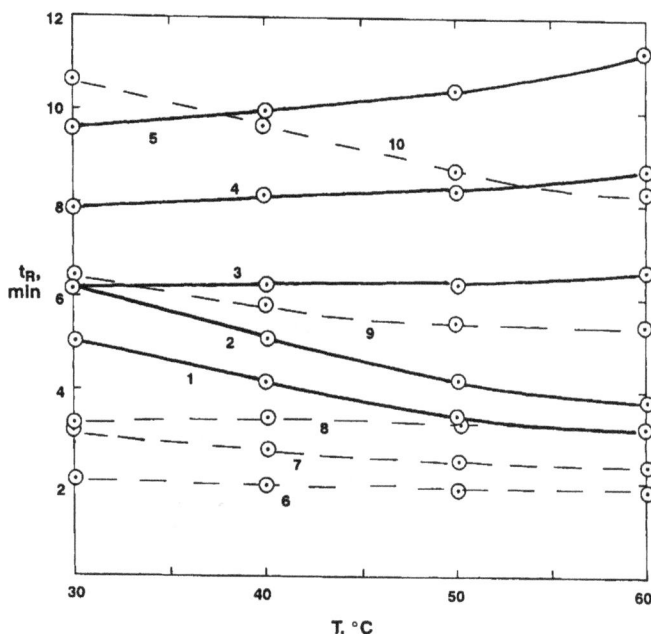

Figure 8.17 *The effect of temperature on retention in the separation of tricyclic anti-depressants and various contaminants. Again, retention only changed modestly but selectivity was dramatically affected* (Reproduced by permission from ref. 51)

on α, as shown in Figure 8.17. Changes in retention were not as dramatic as with the anticonvulsants drugs. However, the retention of some compounds increased, some decreased, and others were unaffected from $30 \le T \le 60$ °C. This behaviour made it simple to resolve arbitrarily chosen mixtures.

All the solutes tested eluted with good peak shapes in short times. There was no need to change either the SP or MP after the initial experiments. The control variables provided adequate adjustments for both retention and α to optimize this separation.

Stimulants

Many drugs of abuse, including cocaine and amphetamines ('speed'), are stimulants. Some can be eluted in GC since, although polar (including aliphatic primary amines), they are small, volatile molecules. Low polarity, FMOC derivatives of four amphetamines have been separated[52] by packed column SFC. However, the parent underivatized compounds can also be separated.

Many stimulants are primary or secondary aliphatic amines with multiple polar functional groups. Nevertheless, the same phases were chosen for the initial SFC experiments.[53] A CPS bonded phase and [0.5% IPAm + MeOH]/CO_2 were used.

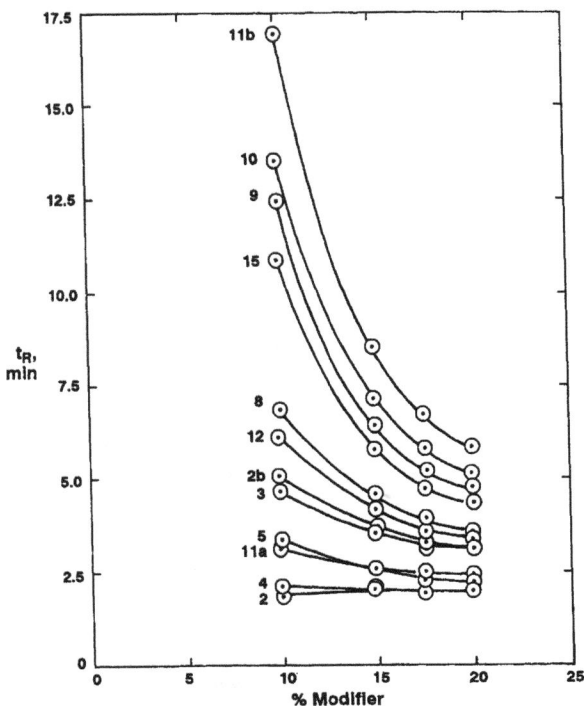

Figure 8.18   *The effect of modifier concentration on the retention of stimulants: 2 amphetamine sulfate; 3 methamphetamine; 4 benzphetamine; 5 phenmetrazine; 8
ephedrine; 9 phenylephrine; 10 hydroxyamphetamine; 11 nylidrine; 12 phenylpropanolamine; 15 baraphazoline. Conditions*: 2 ml min^{-1} [0.5% IPAm
+ MeOH] in CO_2, 40 °C, 200 bar; *column* 4.6 × 250 mm, 5 μm *Lichrosphere Cyanopropyl*
(Reproduced by permission from ref. 53)

Many stimulants were strongly retained, and produced multiple peaks. The
instability appeared to have occurred over time (not on the column) since
different sets of conditions produced the same number of peaks, with the same
area ratios.

Doubling the modifier concentration roughly halved retention, as shown in
Figure 8.18. Several minor peak reversals accompanied changes in modifier
concentration.

Pressure changes produced minor changes in retention and minimal changes
in α. Retention changed less than 20% over the region where density changes
the most.

Temperature produced the greatest changes in α but only minor changes in
retention over the range 28–90 °C, as shown in Figure 8.19. In a few cases,
retention changed by as much as 50%. Again, the retention of some solutes
increased, some decreased, and some remained unchanged when the temperature was increased.

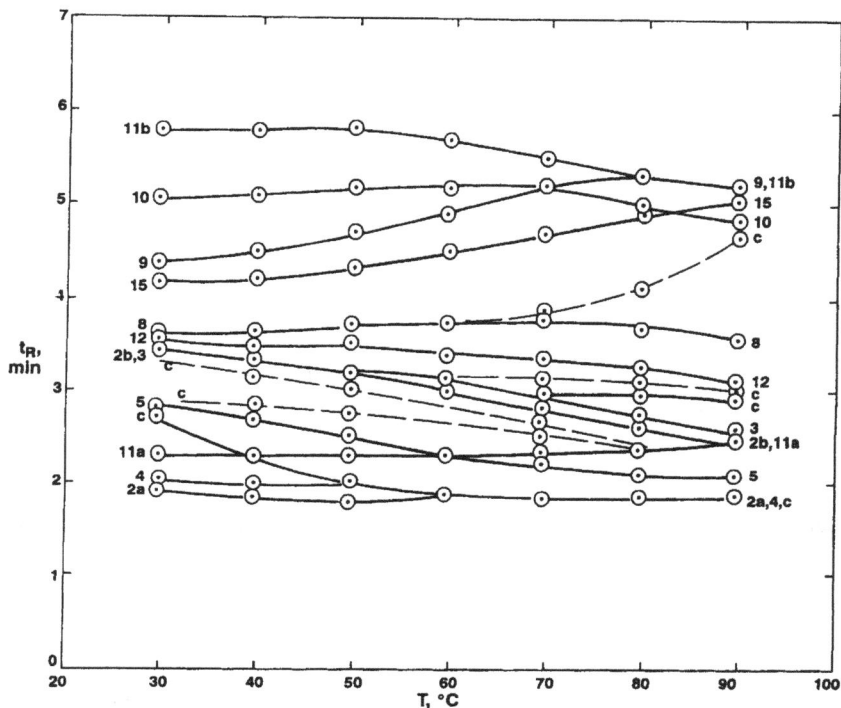

Figure 8.19 *The effect of temperature on the retention of some of the stimulants listed in Figure 8.18 plus some contaminants. The main point of this Figure is the peak reversals evident from the various lines crossing each other. Conditions the same as Figure 8.18, except 10% modifier, varied temperature* (Reproduced by permission from ref. 53)

Programming produced better separations than steady state conditions. The main difficulty was resolution of the least retained solutes while retaining good peak shapes for the later eluting solutes.

All the solutes eluted with good peak shapes even though many were primary aliphatic amines, including several containing hydroxyl groups. All the solutes could be resolved by changing physical parameters to adjust retention and α.

Tetrahydrozoline is used in over the counter eye drops to decrease redness in the eye. The drops are a boric acid aqueous buffer with nominally 0.05% tetrahydrozoline. The commercial product was diluted 4 to 1 with MeOH and injected without further preparation. A chromatogram is presented in Figure 8.20, indicating good peak shapes. A number of injections were made without any apparent problems. No attempt was made to determine the long term viability of this sort of sample preparation and injection.

Sulfonamides

The sulfonamides were one of the first families of modern antibiotics but continue to have many uses. They generally contain an aniline ring bonded at

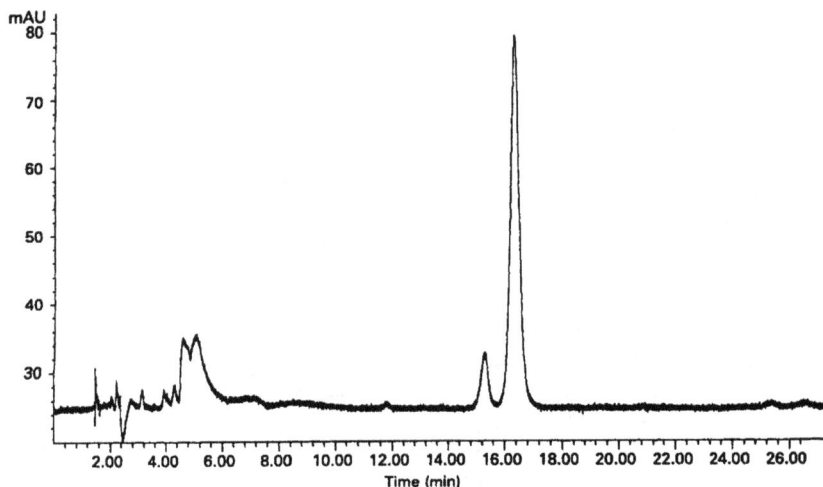

Figure 8.20   *Determination of tetrahydrozaline in a commercial eye drop. The sample was a borate buffer, diluted 4 to 1 with MeOH and injected*
(Reproduced by permission from ref. 53)

its *para* position to the SO_2NH group. On the 'back' of the sulfonamide group, a wide range of structures and functional groups are substituted. The simplest involves a single hydrogen. Others include five, and six membered rings, and multiple rings. The rings sometimes contain one, two, or three nitrogens, oxygen(s), and various side groups. With such a range of structures one should expect a range of retention.

Previous experience showed that anilines[8] and benzamides easily elute with good peak shapes[9] from packed columns using binary fluids. The amides were found to be more retained than anilines but less retained than primary aliphatic amines. It was also shown that the presence of both an aromatic amine and an amide functional group in the same molecule dramatically increased retention.

All the sulfonamides tested,[39,54–56], including triazines and those with multiple polar functional groups, could be eluted with $MeOH/CO_2$ mixtures from a wide range of stationary phases. Surprisingly, most exhibited similar retention despite major differences in the structures of the 'back' side of the molecules.

Many SPs were investigated, including several CPS phases from different manufacturers. A phenyl column produced little retention, and allowed the partial resolution of a number of peaks in only a few minutes, using 3% MeOH in CO_2. However, the peaks tended to tail. This is consistent with previous observations that lower polarity stationary phases are inadequate for the separation of polar solutes. Some of the most polar sulfonamides tested are shown separated in Figure 8.21.

Modifier concentration was again most powerful for adjusting retention. Both pressure and temperature also affected retention. On a CPS column, the retention order of sulfamethoxypyridazine, and sulfisomidine reversed[56] between 28 and 60 °C. On a diol column, retention increased and temperature

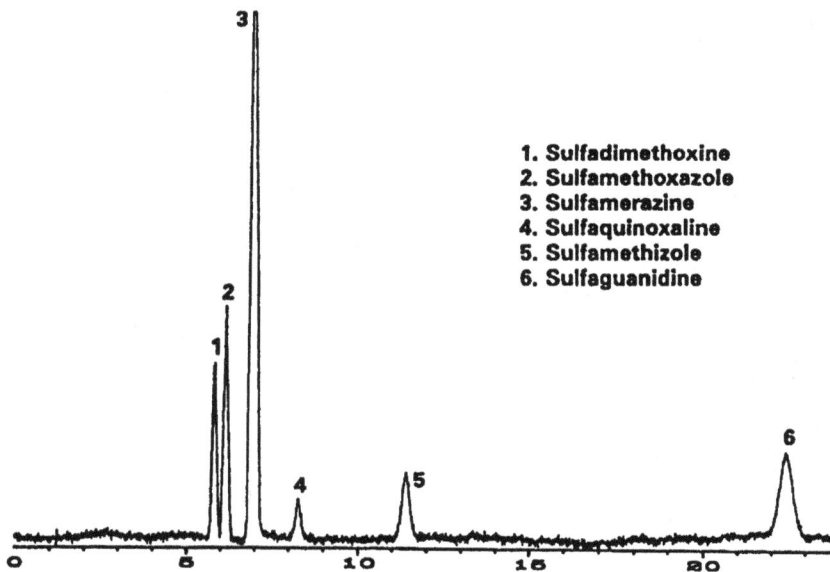

Figure 8.21 *Separation of some of the most difficult to elute sulfonamides by packed column SFC*

had less impact on α. The SP identity had a significant impact on elution order.

Diuretics

Diuretics are also widely used in OTC and prescription drugs. The use of such compounds in sports is banned, and they are routinely screened for at international sporting events.

These compounds can be quite polar. Some require additives in the MP. In, so far, unpublished work,[57] eleven diuretics, banned from sports, were studied. Six were found to elute rapidly and efficiently using $MeOH/CO_2$ mixtures. Three more produced good peak shapes when an acidic additive was included in the MP. All those compounds plus two more eluted when a basic additive was used instead of the acidic additive. However, the next day the peak shapes of a few of the compounds began to degrade. The compounds which had produced good peak shapes with an acidic additive showed degraded peak shapes on the second day that basic additive was used. This implies that the SP or more probably the silica support was attacked by the strong base additive. Since most of the compounds continued to show good peak shapes even after the peaks for a few of the others began to degrade, it is unclear whether this is a general phenomenon or specific to certain classes of compounds. Other families of solutes have not shown such degradation of performance over relatively long periods of nearly continuous use of strong bases as additives with silica based columns. Several columns have been used for many thousands of injections without apparent problems.

3 References

1 R.P. Gilpin and L.A. Pachla, *Anal. Chem.*, 1991, **63**, 130R.
2. T.L. Chester, J.D. Pinkston, and D.E. Raynie, *Anal. Chem.*, 1992, **64**, 153R.
3. T.A. Berger and W.H. Wilson, *Anal. Chem.*, 1993, **65**, 1451.
4. J.F. Deye, T.A. Berger, and A.G. Anderson, *Anal. Chem.*, 1990, **62**, 615.
5. T.A. Berger and J.F. Deye, *Anal. Chem.*, 1990, **62**, 1181.
6. T.A. Berger and J.F. Deye, *J. Chromatogr. Sci.*, 1991, **29**, 54.
7. T.A. Berger and J.F. Deye, *J. Chromatogr. Sci.*, 1991, **29**, 26.
8. T.A. Berger and J.F. Deye, *J. Chromatogr. Sci.*, 1991, **29**, 390.
9. T.A. Berger and W.H. Wilson, *J. Chromatogr. Sci.*, 1993, **31**, 127.
10. T.A. Berger and J.F. Deye, *J. Chromatogr. Sci.*, 1991, **29**, 310.
11. T.A. Berger, unpublished work.
12. L.J. Mulcahey and L.T. Taylor, *J. High Resolut. Chromatogr.*, 1990, **13**, 393.
13. T.A. Berger, unpublished work.
14. D.R. Gere, R. Bored, and D. McManigill, Presented at 1982 Pittsburg Conference on Analytical Chemistry, and Hewlett-Packard Application Note, Publication Number 43–5953–1647, 1982.
15. T.A. Berger, unpublished work.
16. D.E. Games, *Lab. Practice*, 1987, Feb., 45.
17. M. Ashraf-Korassani, M.G. Fessahaie, L.T. Taylor, T.A. Berger, and J.F.Deye, *J. High Resolut. Chromatogr.*, 1988, **11**, 352.
18. T.A. Berger, J.F. Deye, M.Ashraf-Khorassani, and L.T. Taylor, *J. Chromatogr. Sci.*, 1989, **27**, 105.
19. D.E. Games, unpublished work.
20. S.J. Lane, in 'Supercritical Fluid Chromatography', ed. R.M. Smith, The Royal Society of Chemistry, London, 1988, pp. 175.
21. D.W. Roberts and I.D. Wilson, in 'Analysis of Drugs and Metabolites', E. Reid, and I.D. Wilson, The Royal Society of Chemistry, Cambridge, 1990, pp. 257.
22. W.M.A. Niessen, P.J.M. Bergers, U.R. Tjaden, and J. VanDerGreef, *J. Chromatogr.*, 1988, **454**, 243.
23. W.M.A. Niessen, U.R. Tjaden and J. VanDerGreef, *J. Chromatogr.*, 1989, **492**, 167.
24. D.L. Mount, L.C. Patchen, and F.C. Churchill, *J. Chromatogr.*, 1990, **527**, 51.
25. E.D. Ramsey, J.R. Perkins, D.E. Games, and J.R. Startin, *J. Chromatogr.*, 1989, **464**, 353.
26. J.R. Perkins, D.E. Games, J.R. Startin and J. Gilbert, *J. Chromatogr.*, 1991 **540**, 257.
27. T.A. Berger and J.F. Deye, in Atlas of Chromatograms, *J. Chromatogr. Sci.*, 1988, **26**, 131.
28. T.A. Berger and J.F. Deye, *J. Chromatogr. Sci.*, 1991, **29**, 280.
29. Steroid chromatogram from Hewlett-Packard SFC Slide Set.
30. S. Scalia and D.E. Games, *J. Pharm. Sci.*, 1993, **82**, 44.
31. E.D. Morgan, S.J. Murphy, D.E. Games, and I.C. Mychreest, *J. Chromatogr.*, 1988, **441**, 165.
32. M.W. Raynor, J.P. Kitinji, K.D. Bartle, D.E. Games, I.C. Mylchreest, R. Lafont, E.D. Morgan and I.D. Wilson, *J. Chromatogr.*, 1989, **467**, 292.
33. D.E. Games, unpublished work.
34. Jose Baptista, unpublished work.
35. D.W. Later, B.E. Richter, D.E. Knowles, and M.R. Andersen, *J. Chromatogr. Sci.*, 1986, **24**, 249.
36. R.M. Smith and M.M. Sanagi, *J. Pharm. Biomed. Anal.*, 1987, **6**, 837.
37. R.M. Smith and M.M. Sanagi, *J. Chromatogr.*, 1989, **481**, 63.
38. T. A. Berger, unpublished work.
39. A.J. Berry, D.E. Games, and J.R. Perkins, *J. Chromatogr.*, 1986, **363**, 147.
40. J.R. Perkins, D.E. Games, J.R. Startin, and J. Gilbert, *J. Chromatogr.*, 1991, **540** 239.
41. J.L. Janicot, M. Caude, and R. Rosset, *J. Chromatogr.*, 1988, **437**, 351.
42. D. Parlier, D. Thiebaut, M. Caude, and R. Rosset, *Chromatographia*, 1990, **31**, 293.
43. K.Anton, M. Bach, and A. Geiser, *J. Chromatogr.*, 1991, **553**, 71.
44. M.S. Klee and M.Z. Wang, Poster presented at the 5th International Symposium on Supercritical Fluid Chromatography and Extraction, Baltimore, MD, January 1994.
45. D.R. Gere, *Science*, 1983, **222**, 253.
46. M.Z Wang and M.S. Klee, Hewlett-Packard Application Note, 228.
47. R.M. Smith and M.M. Sanagi, *J. Chromatogr.*, 1989, **483**,51.
48. M.Z. Wang, M.S. Klee, and S.K. Yang, Poster presented at the 5th International Symposium on Supercritical Fluid Chromatography and Extraction, Baltimore, MD, January 1994.

49. T.A. Berger and J.F. Deye, Atlas of Chromatograms, *J. Chromatogr. Sci.*, 1988, **26**, 131.
50. T.A. Berger and W.H. Wilson, *J. Pharm. Sci.*, 1994, **83**, 281.
51. T.A. Berger and W.H. Wilson, *J. Pharm. Sci.*, 1994, **83**, 287.
52. J.L. Veuthey and W. Haerdi, *J. Chromatogr.*, 1990, **515**, 385.
53. T.A. Berger and W.H. Wilson, *J. Pharm. Sci.*, 1995, **84**, 489.
54. Hewlett Packard SFC Slide Set, 1992.
55. T.A. Berger and E.A. Berger, unpublished work.
56. T.A. Berger, *J. High Resolut. Chromatogr.*, 1991, **14**, 312.
57. Personal communications with M.Z. Wang and M.S. Klee, Hewlett-Packard.

CHAPTER 9

Chiral Analysis of Drugs

1 Introduction

The separation of enantiomers is well established[1] but is still one of the fastest growing fields in chromatography. The development of chiral HPLC[2-5] and, to a lesser extent, GC[6] separations has aided in the rapid commercialization of enantiomerically pure pharmaceuticals, by providing simple, fast methods for determining the purity of both finished products and intermediates. Many thousands of HPLC chiral separations have been developed.

Caude and Rosset have been the leading academic proponents of packed column SFC for chiral separations[7-15] starting in 1985.[7] They were quickly joined by others in 1986.[16,17] The lack of commercial equipment for packed column SFC has been a major handicap. However, three different commercial instruments, two within the last two years, are now available, and this handicap is being overcome. Ciba Geigy[18] has been one of the longest supporters of SFC and has started using it in routine analysis. ICI[19] has concluded that packed column SFC should be the method of choice for chiral separations. DuPont and now DuPont Merck has invested in chiral separations by SFC for many years and is now beginning to publish[20-24] some of their results. In a recent paper[21] they listed the reasons for using SFC instead of LC. They include: (1) higher efficiency, (2) faster analysis, (3) faster column equilibration, (4) faster method development, (5) reduced generation of hazardous waste, (6) less pressure drop across the column. They also discussed the use of GC detectors with LC-like separations.

Other aspects include the ease of solvent removal, and sample collection, and use of only biocompatible solvents (*e.g.* CO_2 and ethanol).

Professor Pirkle, one of the best known inventors of chiral LC stationary phases, has only briefly used SFC but has, in a few months, demonstrated a wide range of rapid separations.[23-25] He has indicated that SFC appears generally superior to LC in terms of speed, efficiency and ease of use.

SFC is not a panacea. As with achiral separations, solutes amenable to SFC include most compounds that will dissolve in MeOH or a less polar solvent. Compounds requiring an aqueous or ionic environment are unlikely candidates for SFC.

As the means to monitor purity has expanded, regulators have shown an increased interest in the production of enantiomerically pure drugs.[26-28] In the United States, the Food and Drug Administration (FDA) published guidelines[29] which requires substantial extra testing during development of all new drugs with chiral centres. One intent is to determine the degree of chiral conversion from one form into the other. The enantiomeric composition of all bulk and formulated drug materials must be known. Pharmacokinetic and toxicological studies must be undertaken at the enantiomeric level until the situation is completely understood. Some countries prefer that only pure enantiomers be sold as drugs. Alternatively, each enantiomer should be individually tested to assure its safety, since the body often converts one enantiomer into the other. Any contaminant over some limit, usually 0.1% of the main component, is tested as rigorously as the main component.

To manufacture an enantiomerically pure drug, each production step must be monitored for purity and interconversion. Sample types include chiral and achiral raw materials, bulk drug, through finished drug potency in the presence of excipients and coatings, and the analysis of both chiral and achiral contaminants in various stages of manufacture. Many different analytical methods may be required for the manufacture of one drug. Fortunately, most samples from production steps contain only a few components.

Another aspect of testing involves clinical samples of biological matrices, such as urine, tissue, plasma, or whole blood. These matrices generally require extensive sample clean-up. Such testing is also likely to require additional methods for the separation of chiral or achiral metabolites.

2 Characteristics of Chiral Separations

Dalgliesh[1] proposed a three point interaction as the basis of all chiral separations. At least one of the three must depend on the stereochemical structure of the solute.

A nomenclature has been developed[2] describing types of interactions between solutes and chiral stationary phases (CSPs). Five types of CSP are recognized.

In Type I, the solutes form complexes with the SP through attractive forces such as hydrogen bonding, dipole dipole interactions, stacking, *etc.* Many such SPs are based on derivatives of amino acids.

Type III involves the formation of an inclusion complex by the solutes within a cavity in the SP. Type III phases include cyclodextrins, and derivatized cyclodextrins (Cyclobond). Cyclodextrins consist of seven or eight sugar molecules connected end to end in a circle which creates a cup shape with a well defined size. Hydroxyl groups on the sugar molecules can be derivatized to change *α*.

Another Type III SP consists of synthetic polymers (mostly methacrylates) with a helical structure which turns only one way. The shape of some enantiomers allows one to interact intensely with the helix while the other has less intense interactions, allowing differential elution (some of the Chiralpak family).

Type II SPs employ a combination of both attractive interactions and inclusion complexes. Many are based on derivatized cellulose or amylose, including the Chiracel family.

Type IV involves formation of a diastereomeric complex between the solutes and a metal containing ligand.

Type V involves the use of a protein as the SP, forming complexes with the solutes based on polar and hydrophobic interactions.

Another major approach, besides using CSPs, involves preferential interaction of a MP additive with one enantiomer. Like SPs, the additive may be either chiral or achiral. Cyclodextrins are among the MP additives of choice since they are only moderately expensive. However, molecules much larger or bulkier than a benzene ring will not fit into such cups. There is also some polarity on the outside of the rings, creating non-stereospecific interactions which add retention without enhancing α.

3 Developing a Chiral Method

Developing a method for chiral analysis is similar to that for achiral applications, with a few differences. Most important is the choice of the SP. For chiral separations there are hundreds or even thousands of SPs. Fortunately, the vast experience already accumulated in LC can be applied directly to SFC. In most cases, a column that will separate a chiral pair in LC will separate the same pair in SFC.

In chiral separations, as in achiral work, a few MPs cover almost the entire useful range of SFC. If CO_2 can solubilize the molecules, they must be of low to moderate polarity. If pure CO_2 cannot elute the molecules, $MeOH/CO_2$ mixtures are likely to. If $MeOH/CO_2$ mixtures do not elute the solutes, or produce poor peak shapes, then an additive may be needed. If such tertiary mixtures cannot elute the solutes, SFC will probably not work for the application.

MeOH can be replaced with EtOH with only minor changes in α. However, EtOH is less polar than MeOH and a higher concentration is sometimes required to get the same retention. On the other hand, MeOH and EtOH produced virtually the same separation[11] of 3–hydroxymethyldihydrobenzofuran, whereas isopropyl alcohol produced longer t_R.

The Effect of Physical Parameters on Chiral Separations

The composition of the MP continues to have the largest effect on retention. Many chiral phases are more retentive than achiral phases, requiring a higher modifier concentration. Retention appears to be a steeper function of concentration than with achiral columns, more than doubling with a two fold decrease. At low modifier concentration, solutes often exhibit long t_R and poor peak shapes. Peak shapes on CSPs tend to degrade more rapidly at low modifier concentration than those in achiral separations. This can cause unexpected complications for little retained compounds.

Temperature appears to be the second most important parameter for adjusting α. The best resolution is usually (but not always) found at lower temperatures. The temperatures producing the best α are often actually subcritical. By definition, such operation is defined as LC. However, the viscosity of CO_2 is much lower, and diffusivity much higher than in water at room temperature. A backpressure regulator is still required to prevent the fluid from expanding into a gas. The transition from supercritical to subcritical temperatures is a non-event, provided $P > P_c$. Several groups[30–32] have reported subambient operation ('cryo' oven) which produced very rapid analyses.

Pressure tends to do little to change either retention or a. At low pressure, peak shapes tend to degrade.

Since chiral columns tend to be many times more expensive than standard LC columns, greater care is usually exercised in their use. SFC places far less stress on columns owing to the much lower viscosities, and subsequent lower pressure drops, typical when using these fluids.

Compared with LC, SFC Van Deemter curves appear quite flat, owing largely to higher values for μ_{opt} in SFC. This means that $2 \times \mu_{opt}$ in SFC may be 6–10 times μ_{opt} in LC, yet produce modest pressure drops.

Initial Conditions

The separation of the enantiomers of propranolol,[33] a β-blocker, will be used as an example of chiral method development. A Chiracel OD column separates the enantiomers of propranolol in approximately 17–25 minutes by LC,[34] so that column was chosen for trial in SFC. Being aminoalcohols, the solutes are polar and will probably require a polar modifier. MeOH/CO_2 mixtures were chosen. It was uncertain, *a priori*, whether an additive was required. Additives were initially avoided.

After the column and MP were chosen, the physical parameters were set to try to elute the constituents rapidly with minimal retention. At this point, no effort was wasted in trying to guess at the likely optimum conditions. In general it is useful to choose for initial conditions: $30 < T < 40\,^\circ\text{C}$, $150 < P < 200$ bar, and 30% MeOH in CO_2 at a flow rate of 2–2.5 ml min^{-1} for a 4 or 4.6 mm i.d. column.

The two enantiomers of propranolol eluted within 6 minutes and were partially resolved. However, the peak shapes were less than ideal, as shown[33] in Figure 9.1a. Isopropylamine (IPAm) (0.5%) was added to the MeOH. The resulting tertiary mobile phase dramatically improved peak shapes as shown[33] in Figure 9.1b.

Optimization after the Initial Experiment

Optimization of a chiral separation is similar to that for any achiral separation. If the initial conditions produce little retention or R_s, decrease the percentage modifier by a factor of two, approximately doubling retention. If the enantiomers are still inadequately resolved, cut the percentage modifier in half, again. Repeat until they resolve, $t_R > 30$–40 minutes, or until the peak shapes degrade.

If the enantiomers show no sign of separation up to this point, changing temperature, or pressure is unlikely to provide dramatic improvements. However, before changing the column or the MP, make sure this is true. Set the

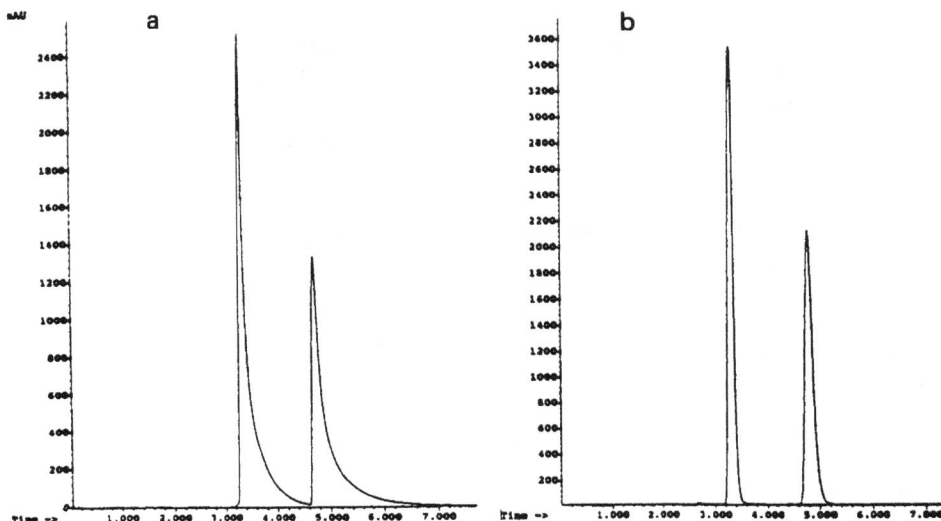

Figure 9.1 (a) *Resolution of propranolol enantiomers on a Chirocel OD column using a binary mobile phase*: (b) *using the same conditions except that 0.5% of isopropylamine (a strong base) was added to the methanol. Column* 4.6 × 50 mm, 10 μm, *30% methanol in carbon dioxide at* 200 bar *outlet pressure,* 30 °C
(Reproduced by permission from ref. 33)

temperature to the lowest value obtainable (usually a few degrees above ambient) or, if a cryo option is installed in the oven, set the temperature to 20 or 30 °C below the initial temperature. Alternatively, increase the temperature to 20 or 30 °C above the initial temperature. If either produces a noticeable improvement in resolution, continue in the direction showing improvement.

If the enantiomers do not show a significant tendency to separate between ± 30 °C of the initial temperature, the most fruitful next step is to try a different SP. If, on the other hand, some temperature produces partial resolution, changing pressure may further improve the separation.

The analyst needs always to be aware of the pressure below which the fluid breaks down into two phases. This is not difficult. With downstream control the user always knows p_o and it is fairly easy to see when the baseline breaks up. The breakup can be subtle since the detector may be at a lower temperature than the column. The fluid can fall apart in the column and go back to a single phase in the detector. However, peaks broaden and loose their symmetry when two phases form. Approaching the pressure where two phases form usually results in the biggest changes in the fluid characteristics due to density effects. The modifier tends to form bigger clusters in the MP, and adsorption tends to change on the SP. This is the pressure region most likely to provide a change in R_s compared with the initial pressure. Much higher pressure is unlikely to have much, if any effect.

The modifier identity provides the least understood way to change α.

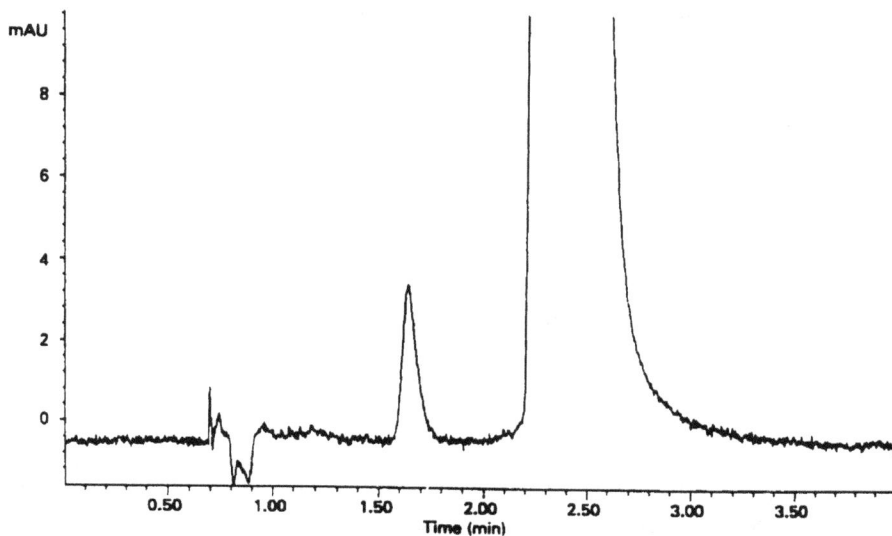

Figure 9.2   *Trace determination of 0.1% R-propranolol in the presence of 99.9% S-propranolol. Conditions as in Figure 9.1b*
(Reproduced by permission fromm ref. 33)

However, it is easier to change the modifier than it is to change the column. The analyst might consider trying a modifier from a different solvent family before changing the column. However, if a solute is strongly retained using high MeOH concentration, it is likely to be even more retained with a less polar solvent. Some solvents are incompatible with some SPs. Since chiral columns are extremely expensive, great care should be exercised in choosing a modifier.

Other Figures of Merit of Chiral Analysis

Other important figures of merit include: linearity, linear dynamic range, reproducibility, and injection size. A UV detector can provide a linear dynamic range of at least 2500 at most wavelengths. This allows both one isomer present at 0.1% and the other present at 99.9% to be quantified in a single run, as shown[33] in Figure 9.2. Calibration curves were linear over a wide range.

Reproducibility of packed column SFC can be extremely good, as indicated[33] in Table 9.1. Higher concentrations produced area RSDs of less than ± 0.1%. The RSDs increased as signal to noise (S/N) degraded, as expected. The RSD of retention times was also consistently small, degrading with S/N.

Packed column SFC allows a wide range of injection volumes to be used. Fixed loops as small as 0.06 μl and as large as 30 μl can be used without significant problem, provided the sample solvent is not a stronger solvent than the MP. This *ca.* 500 fold variation allows the analyst to keep samples of various concentrations within the linear dynamic range of the detector, while maintaining high separation speed and good reproducibility.

Table 9.1 *Reproducibility and linearity of propranolol*[33]

Conc./p.p.m.	R-Propranolol T_R RSD	Area RSD	S-Propranolol T_R RSD	Area RSD
0.25	0.3%	9.1%		
2.5	0.1%	3.4%	0.2%	5.9%
25	0.2%	0.9%	0.2%	1.1%
250	0.1%	0.2%	0.2%	0.2%
2500			0.1%	0.02%
Linear regression	Coefficient 0.99999		0.99998	

Packed column SFC also allows numerous detector options. Variable wavelength, multiwavelength, and full spectra UV detectors are readily available. For compounds with good chromophores detection limits in the middle part per (American) billion (p.p.b.) ($1/10^9$) range (*e.g.* 20–50 p.p.b.) are common. Fluorescence detectors can also be used. In addition both the nitrogen-phosphorus (NPD), and electron capture (ECD) detectors can be used with most mobile phases.

4 Some Examples

Isolation and Detection of Ibuprofen from a Biological Fluid

Racemic ibuprofen (Motrin) has been sold as an OTC pain killer for years. Ibuprofen is slow acting, compared with acetaminophen or aspirin. In the competitive market, this slowness is a serious disadvantage. Tests indicate that one enantiomer is considerably faster than the mixture.[35] The ineffective isomer is converted into the more active form over *ca.* 90 minutes.

If a racemic mixture is used, the ratio of isomers changes over time, with the less active component slowly disappearing. This can be demonstrated by observing the change in the enantiomeric ratio in urine samples. Acidified urine is sucked through a C_{18} solid phase extraction cartridge, which was then washed with distilled water. The drug was eluted with a low polarity solvent.

A chromatographic system was developed,[36] employing an automated six-port valve and two columns. Flow proceeded from the pump to the six-port valve, the chiral column, the injection valve, the achiral column, the UV detector, back through a loop in the 6-port valve, to the back pressure regulator. On injection (**between** the columns), the sample passed through the achiral column which separated ibuprofen from contaminants. When the detector indicated that the ibuprofen peak was in the 100 μl loop, the valve was thrown. This injected only the ibuprofen onto the Chiracel OD column, where the enantiomers were resolved. After the chiral column, the enantiomers made a second pass through the achiral column, then through the detector again. The contaminants never entered the expensive chiral column.

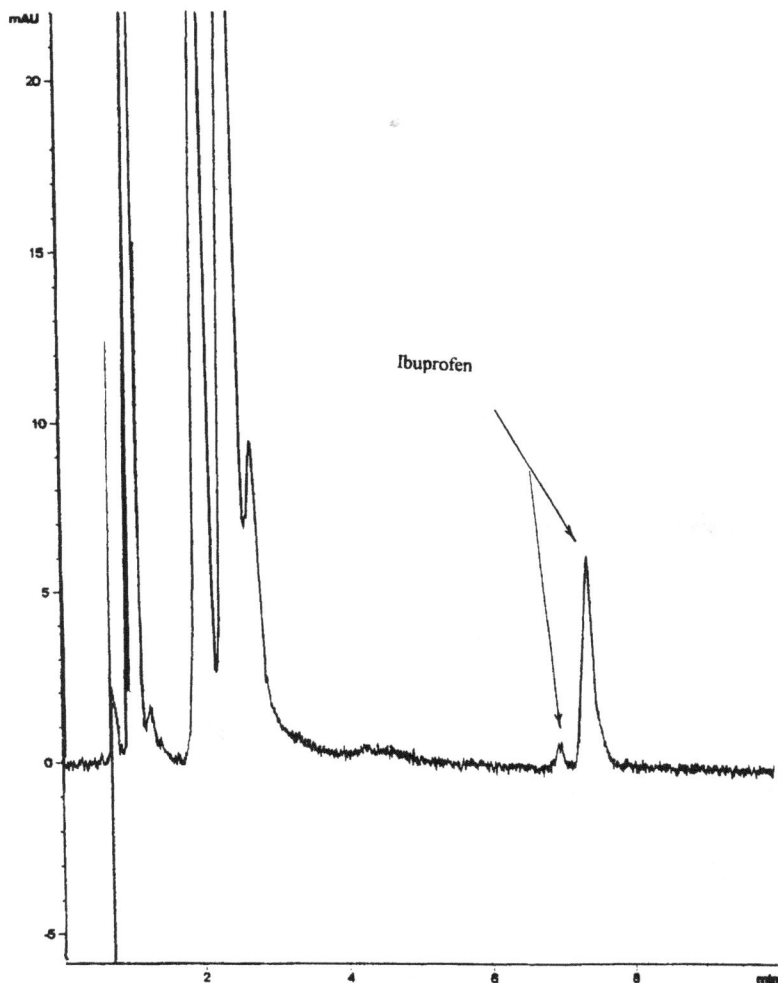

Figure 9.3 *Enantiomers of ibuprofen resolved in a urine extract. The administered pain killer was a racemic mixture. The body converts the inactive form into the active form resulting in the unequal concentrations shown*

A chromatogram of unresolved ibuprofen, just after the solvent front, contaminant peaks, then resolved enantiomers (near 8 minutes) is shown in Figure 9.3. A system to eliminate the second pass through the achiral column was developed but required an additional six-port valve. A similar scheme with a switching valve could be applied to diverting pure enantiomers for collection out of a complex matrix.

Mixed 'Universal' Phases?

Column selection is probably the most difficult task in chiral chromatography, in part because there are more than 100 diffierent commercial CSPs. Sandra[37,38]

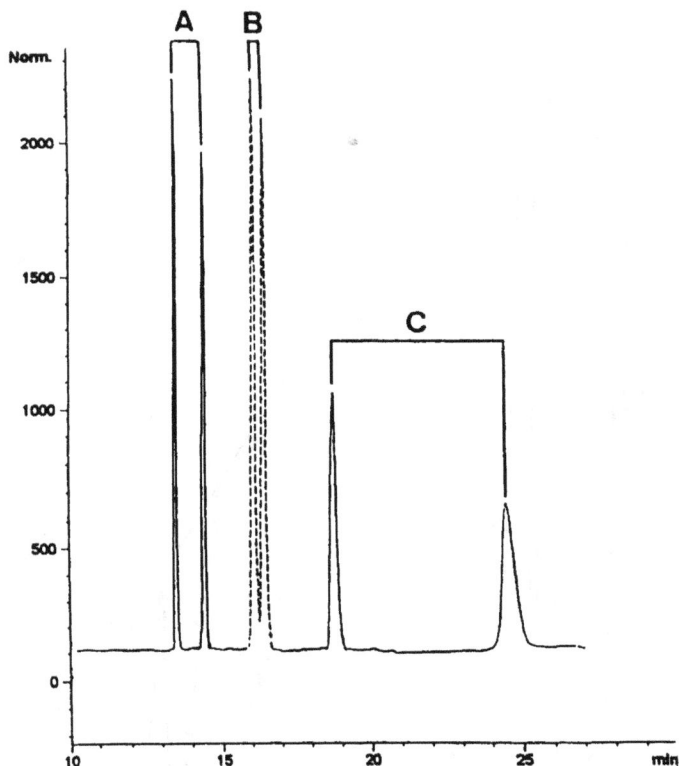

Figure 9.4 *An attempt at creating a 'Universal' CPS. Chiracel OD, AD, and Chirex 3022 columns connected in series. Solutes are: A metoprolol, B tiaprofenic acid, and C lormetazepam, representing three completely different families of chiral compounds. In later work, now in press, the same authors included an additional column in series to resolve barbiturates also*

connected Chiracel OD, AD, and Chirex 3022 columns in series and resolved, for example; metoprolol, tiaprofenic acid, and lormetazepam, in a single run, as shown in Figure 9.4. The addition of a fourth column (brush type, with π-acceptor characteristics) further allowed the simultaneous resolution of barbiturates. These compounds are from completely different families. Their resolution in a single run implies that such an approach should see more widespread use, at least in screening, in the future.

Benzodiazepams and Metabolites

Many metabolites of chiral drugs are also chiral. It would be cost effective to separate all the enantiomers of parents and metabolites from the matrix in a single chromatogram.

In Chapter 8, the separation of various benzodiazepams and their metabolites was discussed. In many cases, both the parent and metabolites were chiral.

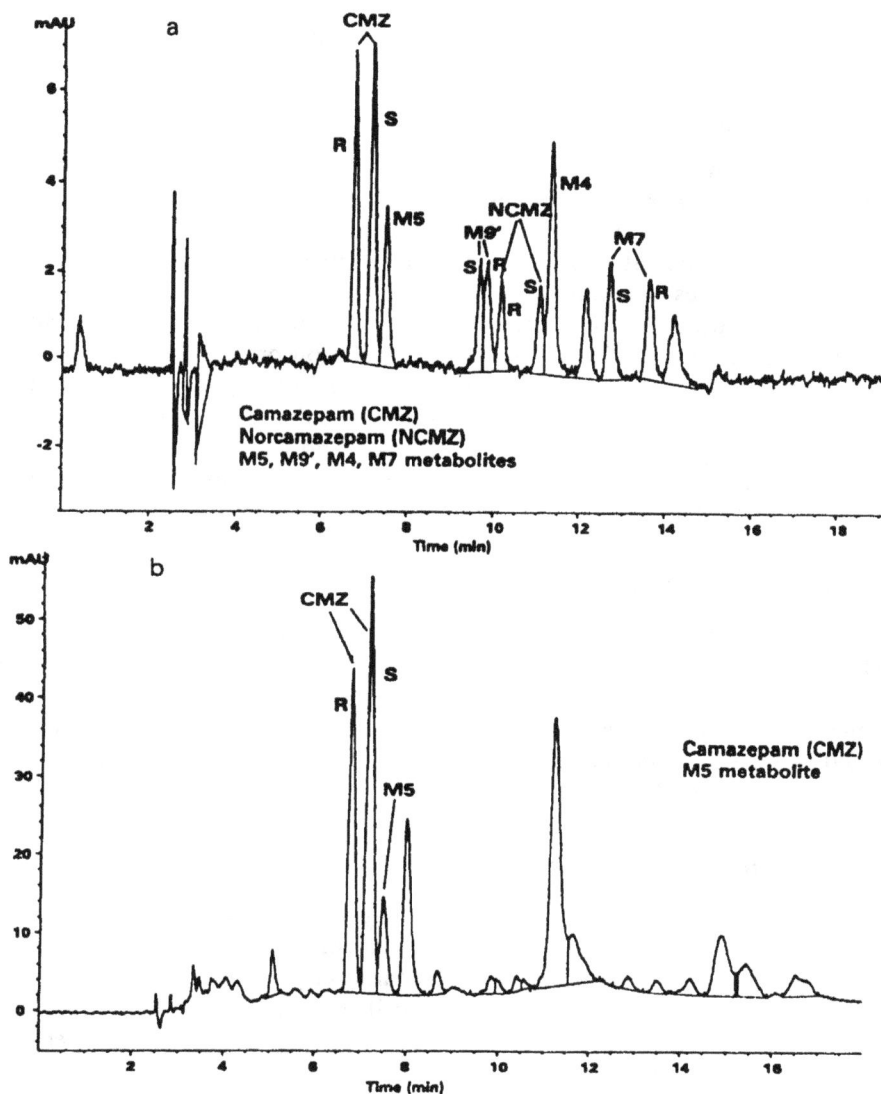

Figure 9.5 (a) *Chiral separation of the enantiomers of camazepam and five metabolites using a Lichroshpere CN column in front of a Chiracel ODH column. Without the CN column several metabolites completely overlapped;* (b) *similar but is from metabolized camazepam*

The metabolites of camazepam could not be resolved[39] from each other on a Chiracel ODH column. Placing an achiral cyanopropyl column in front of the chiral column changed the selectivity enough so that the each had a unique retention time, although a few were still partially overlapped, as shown with standards in Figure 9.5a. This seperation represents a significant improvement in both time and resolution over the best achieved in LC. A chromatogram of metabolized camazepam is presented in Figure 9.5b.

Long Column as Easiest Way to Increase Resolution

At very low modifier concentration, peak shapes can degrade. Despite a tendency for greater retention on chiral columns, some solutes are little retained, even at relatively low modifier concentration. If a pair of enantiomers is partially resolved at 2% modifier, it is often true that a lower modifier concentration does not improve R_s, because peak shapes degrade.

One could attempt to find either a different modifier or a different SP to resolve the peaks fully. However, partial resolution on the first column indicates that a brute force approach should provide an adequate separation with the least development time. Combining identical columns in series increases R_s by $L^{0.5}$. Thus, two columns provide 1.41 times, and four columns provide two times the R_s of one column. Since t_R increases with L such an approach is expensive but rapid, requiring the least risky development.

LC Worked, SFC Didn't

The reader should not interpret the results discussed above as indicating that SFC is always superior to LC. The Chiracel OD column can separate many different but related enantiomeric pairs. However, CGP 49823 [(2R,4S)-2-Benzyl-1-(3,5-dimethylbenzoyl)-N-(4-quinolinemethyl)-4-piperidineamine] could be separated on this column in LC but was not separated[11] on the same column using a SF MP phase.

Other Chiral Separations

Chiral separations tend to appear quite boring with two resolved peaks on a flat baseline. The only major change is usually in the time base. Columns seldom partially work. Either a specific column separates a specific pair of enantiomers or it does not. Because of this characteristic, it is not particularly informative to show many chromatograms of two peaks. Instead, Table 9.2 was prepared which lists many of the compounds reported separated by packed column SFC up to the early part of 1994. Other papers[41–47] also cover chiral separations but are not included in the Table.

Table 9.2 *Enantiomers separated by 'SFC'*

Compound/family	Column	Mobile phase
Ref. 7		
Phosphine oxides		
1-(4-methylnaphthyl)	(*R*)-*N*-(3,5-Dinitrobenzoyl	
	phenylglycine	CO_2 + MeOH
1-(2-methylnaphthyl)		or EtOH
9-phenanthryl		or isopropyl alcohol,
		260 bar, 20 °C
1-(2-methylnaphthyl)		
Ref. 11		
Methyl *p*-methylphenyl	(DACH-DNB(5))	5% ethanol/CO_2 sulfoxide
150 bar, 25 °C		
Oxazepam	(*S*)-thio-DNB Tyr-A	8% EtOH/CO_2, 200 bar
		25 °C
α-Methylene-*γ*-lactone	Cyclobond I	1% methanol/CO_2,
		200 bar, 25 °C
α-Methylene-*σ*-lactone	ChiraCel OB	8% isopropanol/CO_2,
		140 bar, 25 °C
α-Methylene-*γ*-Lactam		
o-Anisyl methylcyclohexyl	Cyclobond I	1% methanol/CO_2,
		phosphine oxide
150 bar, 25 °C		
Various amides	Chiracel OB	7% 2–propanol/CO_2
Albendazole sulfoxide	ChiraCel OB	
Bi-*β*-Naphthol	ChiraCel OB	other alcohols
Ref. 12		
α-Amino esters	(*S*)-DNB-tyrosine	CCl_4 or CH_2Cl_2/CO_2,
α-Aminoamides	(*R*)-DNB-phenylglycine	200 bar, 25 °C
α-Amino alcohols	(*R*)-DNB-p-OH-	
	phenylglycine	
Ref. 14		
β-Blockers	(*S*)-ChyRoSine-A	tertiary
Acebutolol	Chiraline	mobile phases
Alprenolol	(*S*)-DNBLeu	CO_2 + modifier
Atenolol	(*R*)-DNBPG	+ base
Betaxolol	Nucleosil Chiral-2	200 bar (av.), 25 °C
Metoprolol	DACH-DNB	
Oxprenolol		
Propranolol		
β-Propranolol		
Pindolol + derivatives		
Ref. 16		
Amino acid derivatives		
N-Acetylamino acid	(*N*-formyl-L-valylamino)	200 bar, 60 °C, 2 l min[-1]
	propyl	CO_2
t-Butyl esters		
Leu Ile Val	silica gel	(expanded), 0.5 ml min[-1]
Abu Phe Pro		methanol

Table 9.2   (*cont.*)

Compound/family	Column	Mobile phase
α-Phenyl-Gly		
S-Benzyl-Cys		
N-t-Butyl-Trp		
O-Benzyl-Tyr		
O-t-Butyl-Ser		
O-t-Butyl-Glu		
Ref. 18		
Benzoin	ChiraCel OBH	20% MeOH, 150 bar, 35 °C
2,2,2-Trifluoro-1-(9-anthryl)-EtOH	ChiraCel OD	?EtOH
CGP 39540A	Chiracel OD	4.8% MeOH + NH₄CH₃COOH, 225 bar, RT
Phenylalaninol	Chiracel OD	6.2% IPA, 225 bar, RT
CGP 47900	Chiracel OD	3.5% EtOH, 275 bar, RT
4-Phenyl-1,3-dioxane	ChiraCel OD	4.2% EtOH, 225 bar, RT
Hydroxyzine	Chiracel OD	20% MeOH + IPAm, 200 bar, 35 °C
3-OH-Methyldihydro-benzofuran	Chiracel OD	2.5% MeOH, EtOH, IPA 225 bar, RT
Ref. 30		
Sulfoxides	(R,R)-DACH-DNB-Nucleosil	10% MeOH, 100 bar, 25 °C
β-Blocker-oxazolidine-2–one derivative	(R,R)-DACH-DNB LiChrosorb	25% MeOH, 2 ml min⁻¹, 140 bar, 60 °C
Aryloxypropionic-COOCH₃	(R,R)-DACH-DNB LiChrosorb	5% MeOH/CO₂, 100 bar, 25 °C
Binaphthyl derivatives	(R,R)-DACH-ACRYL	40% MeOH, 80 bar, 25 °C
Ref. 33		
Propranolol	Chiracel OD	30% MeOH + IPAm 200b, 30 °C
Atenolol		
Metoprolol		
Ref. 34		
1,2–Aminoalcohols	cyano	N-benzoxycarbonylglycyl-L-Proline + triethylamine, + acetonitrile in CO₂
Pindolol		
Metoprolol		
Oxprenolol		
Propranolol		
DPI 201–1061		
Ref. 36		
Oxyphenylcyclimine	Chiracel OBH	20% MeOH, 150 bar, 35 °C
Oxprenolol	Chiracel OD	30% MeOH, 200 bar, 35 °C
Pindolol	ChiraCel OBH	10% MeOH, 150 bar, 35 °C
Hydroxyzine	ChiraCel ODH	20% MeOH + IPAm, 200 bar 35 °C
Naphthylethylamine	ChiraCel OD	20% MeOH, 150 bar, 35 °C

Table 9.2   *(cont.)*

Compound/family	Column	Mobile phase
Propafenone	ChiraCel OD	10% MeOH, 200 bar, 35 °C
Flurbiprofen	ChiraCel OBH	10% MeOH, 150 bar, 30 °C
Ketoprofen	ChiraCel OJ	5% MeOH,150 bar, 30 °C
Ibuprofen	ChiraCel OBH	2% MeOH, 150 bar, 35 °C
Ref. 38		
Camazepam	Chiracel ODH	8% EtOH, 200 bar, 30 °C
Norcarbazepam		25%
Oxazepam		8%
Temazepam		15%
CMZ-M4		6%
CMZ-M5		2.2%
CMZ-M7		12%
CMZ-M9		15%
CMZ-M9′		8%
3-*O*-Methyloxazepam		25% EtOH, 200 bar, 30 °C
Oxazepam 3-acetate		
3-*O*-Ethyloxazepam		15%
Temazepam		8%
3-*O*-Methyltemazepam		20%
3-*O*-Ethyltemazepam		3%
Lorazepam		20%
Loretazepam		15%
Ref. 40		
β-Blockers	numerous columns	
Nadolol (4 peaks)	ChiraCel OD	20% MeOH
Betaxolol		in CO_2, 200 bar, 35 °C,
Pindolol		
Metoprolol		
Cicloprolol HCl		
Ref. 41		
Ibuprofen	Chirapak AD	5% MeOH, 100 bar, 35 °C
Flurbiprofen	Chirapak AD	20% MeOH, 150 bar, 35 °C
Ethosuximide	Chirapak AD	15% MeOH, 150 bar, 35 °C
Metobarbital	Chiropak AD	10% MeOH, 150 bar, 35 °C
Warfarin	Chiropak AD	20% MeOH, 150 bar, 35 °C
Suxazole	Chiropak AD	20% MeOH, 150 bar, 35 °C
Bendroflumethazide	Chiracel AD	15% MeOH, 150 bar, 35 °C
Sulconazole	Chirapak AD	20% MeOH, 150 bar, 35 °C
Hydroxyzine	Chiracel OD	10% MeOH + IPAm, 200 bar, 35 °C
Buclizine	Chiracel OD	10% MeOH + IPAm, 200 bar, 35 °C
Indapamide	ChiraCel OD	30% MeOH + 0.5% IPAm, 200 bar, 35 °C
Propranolol		30% MeOH + 0.5%IPAm, 250 bar, 40 °C

5 References

1. C.E. Dalgliesh, *J. Chem. Soc.*, 1952, 3940.
2. I.W. Wainer, in 'Drug Stereochemistry, Analytical Methods and Pharmacology', 2nd Ed., ed. I.W. Wainer, Marcel Dekker, Inc. New York, 1993, Chapter II.6.
3. A.M. Krstulovic, 'Chiral Separations by HPLC: Applications to Pharmaceutical Compounds', Wiley, New York, 1989.
4. W.J. Lough, 'Chiral Liquid Chromatography', Blackie, Glasgow, 1989.
5. D. Stevenson and I.D. Wilson, 'Recent Advances in Chiral Separations', Plenum Press, New York, 1990.
6. W.A. Konig, in 'Drug Stereochemistry, Analytical Methods and Pharmacology', 2nd Ed., ed. I.W. Wainer, Marcel Dekker, Inc. New York, 1993, Chapter II.5.
7. P.A. Mourier, E. Eliot, M.H. Caude, and R.H. Rosset, *Anal. Chem.*, 1985, **57**, 2819.
8. P. Macaudiere, A. Tambute, M. Caude, R.H. Rosset, M.A. Alembik, and I.W. Wainer, *J. Chromatogr.*, 1986, **371**, 159.
9. P. Macaudiere, M. Caude, R. Rosset, and A. Tambute, *J. Chromatogr.*, 1987, **405**, 135.
10. P. Macaudiere, M. Caude, R. Rosset, and A. Tambute, *J. Chromatogr. Sci.*, 1989, **27**, 383.
11. P. Macaudiere, M. Caude, R. Rosset, and A. Tambute, *J. Chromatogr. Sci.*, 1989, **27**, 583.
12. P. Macaudiere, M. Lienne, M. Caude, R. Rosset, and A. Tambute, *J. Chromatogr.*, 1989, **467**, 357.
13. R. Rosset, M. Caude, and A. Jardy, in 'Chromatographies en Phase Liquide et Supercritique', Publ. Masson, Paris, pp. 555, 1991.
14. L. Siret, N. Bargmann, A. Tambute, and M. Caude, *Chirality*, 1992, **4**, 252.
15. N. Bargmann-Leyder, J.C. Truffert, A. Tambute, and M. Caude, *J. Chromatogr.*, 1994, **666**, 27.
16. S. Hara, A. Dobashi, K. Kinoshita, T. Hondo, M. Saito, and M. Senda, *J. Chromatogr.*, 1986, **371**, 153.
17. S. Hara, A. Dobashi, T. Hondo, M. Saito, and M. Senda, *J. High Resolut. Chromatogr.*, 1986, **9**, 249.
18. K.Anton, J. Eppinger, L. Fredriksen, E. Francotte, T.A. Berger, and W.H. Wilson, *J. Chromatogr.*, 1994, **666**, 395.
19. *Lab. News*, December 1992.
20. K.G. Lynam and E.C. Nicholas, *J. Pharm. Biom. Anal.*, 1993, **11**, 1197.
21. A.M. Blum, K.C. Lynam, and E.G. Nicolas, *Chirality*, 1994, **6**, 302.
22. R.W. Stringham, K.G. Lynam, and C.C. Grasso, *Anal. Chem.*, 1994, **66**, 1949.
23. G.J. Terfloth, W.H. Pirkle, K.G. Lynam, and E.C. Nicolas, 'Broadly applicable polysiloxane-based stationary phases for HPLC and SFC', in press.
24. G.J. Terfloth, W.H. Pirkle, K.G. Lynam, and E.C. Nicolas, 'Use of polysiloxane-based chiral stationary phases derived from (*S*)-naproxen diallyl amide with supercritical carbon dioxide', in press.
25. W.H. Pirkle, personal communication.
26. W.H. Decamp, *Chirality*, 1989, **1**, 2.
27. H. Shindo and J. Caldwell, *Chirality*, 1991, **3**, 91.
28. A.J. Hutt, *Chirality*, 1991, **3**, 161.
29. 'Stereoisomeric Drugs', US Food and Drug Administration, Washington DC, 1992.
30. F. Gasparrini, D. Misiti, and C. Villani, *Trends Anal. Chem.*, 1993, **12**, 137.
31. F. Gasparrini D. Misiti, and C. Villani, *J. High Resolut. Chromatogr.*, 1990, **13**, 182.
32. W.H. Wilson, in preparation.
33. P. Biermanns, C. Miller, V. Lyon, and W.H. Wilson, *LC–GC*, 1993, **10**, 744.
34. W. Steuer, M. Schindler, G. Schill, and F. Erni, *J. Chromatogr.*, 1988, **447**, 287.
35. S.C. Stinson, *Chem. Eng. News*, 1993, Sept. 27, 0.38.
36. W.H. Wilson, *Chirality*, 1994, **6**, 216.
37. A. Kot, P. Sandra, and A. Venema, *J. Chromatogr. Sci.*, 1994, in press.
38. P. Sandra, A. Kot, and F. David, presented at, 17th International Symposium on Capillary Chromatography, Riva del Garda, Italy, Sept. 1994.
39. M.Z. Wang, M. Klee, and S.K. Yang, Poster presented at 5th International Symposium on SFC and SFE, Baltimore, Md. January 1994.
40. C.R. Lee, J.-P. Porziemsky, M.-C. Aubert, and A.M. Krstulovic, *J. Chromatogr.*, 1991, **539**, 55.
41. W.H. Wilson, Hewlett-Packard Application Note 228–275, May 1994.
42. I.W. Wainer, *Trends Anal. Chem.*, 1987, **6**, 125.
43. I.W. Wainer, M.C. Alembik, and E. Smith, *J. Chromatogr.*, 1987, **388**, 65.

44. F.J. Ruffing, J.A. Lux, and G. Schomburg, *Chromatographia*, 1988, **26**, 19.
45. A. Dobashi, Y. Dobashi, T. Ono, S. Hara, M. Saito, S. Higashidate, and Y. Yamauchi, *J. Chromatogr.*, 1989, **461**, 121.
46. T. Nitta, Y. Yakushizin, T. Kametani, and T. Katayama, *Bull. Chem. Soc. Jpn.*, 1990, **63**, 1365.
47. R. Brugger, P. Krahenbuhl, A. Marti, R. Straub, and H. Arm, *J. Chromatogr.*, 1991, **557**, 163.

Separation of Agricultural Chemicals by Packed Column SFC

1 Introduction

There are more than 20 000 pesticide formulations registered in the USA. Fortunately, a few hundred compounds make up the vast majority of commercial use. New Zealand, may represent an opposite since it is well isolated with a small population. Despite this, foods there still need to be screened for > 130 different pesticides. Even pesticides banned more than 20 years ago, like DDT, are still sometimes found in products. Such a large number of target compounds mean that only parent compounds, but few if any metabolites, are monitored.

Older pesticides were highly effective, **stable**, and were used in relatively concentrated forms. Most are moderately polar organic compounds containing nitrogen, phosphorus, or chlorine. Specific, sensitive GC detectors exist for each of these elements. These characteristics made them easy to analyse by GC. Organochlorine pesticides are almost universally analysed by GC,[1] often using the electron capture detector (ECD) or the electrolytic conductivity detector (ELCD). Organophosphorous pesticides are another, older group of pesticides, most of which can be separated with care by GC and detected[2] with the flame photometric detector (FPD) or the nitrogen–phosphorus detector (NPD).

More modern pesticides are designed to break down rapidly in the environment into less toxic products. In addition, newer products tend to be more potent so they can be applied in lower doses. Less stability and lower concentrations make these newer pesticides more difficult to analyse by GC owing to limited thermal stability. Contaminants accumulating on the head of the column can cause troubles with peak shapes. Metabolites are often too polar to elute and may degrade the peak shapes of analytes. There has been a trend to switch some methods to LC since it requires less sample preparation.

LC is slower, less efficient, and has fewer detection options than GC, but offers a means of analysis. LC lacks the specific and sensitive detectors to detect trace levels of pesticides directly. Consequently, pesticide analysis by LC often involves derivatization for detection or preconcentration.

Phenylurea herbicides[3] can be separated by cool on-column GC but today they tend to be analysed by LC. The sulfonylureas are not stable enough to be analysed by GC. The standard method uses LC. Even triazines are often determined by LC. Methyl carbamates and some other compounds are separated by LC followed by two postcolumn reactions to create an *o*-phthaldehyde (OPA) fluorescence derivative.[4,5]

Whenever LC shows improvement over GC, SFC is likely to do an even better job. Capillary SFC can be used for pesticide analysis with GC detectors, such as the NPD,[6] FPD,[7] and ECD[8,9] or just the FID.[10-14] One paper compared LC, capillary SFC, and packed column SFC[15] in pesticide analysis. Several have dealt with general aspects of packed column SFC.[16,17] Finally, others[18-26] have shown the analysis of specific families of pesticides.

2 Trace Contaminants/Quality Control

There has been no discussion of quality control product monitoring of agricultural chemicals using SFC and selective detectors in the open literature. In addition, trace compounds which might be present in products are seldom monitored after application of the product. These facts are brought up simply to point out that SFC should offer enhanced capabilities for such analysis. Chromatograms of two organophosphorus pesticides are presented in Figure 10.1. Both a UV and a NPD trace from each injection are presented. Each set of chromatograms shows many very small peaks in addition to those of the parent(s). The NPD tends to show many peaks not observed in the UV chromatogram. Other detectors (ECD, SCD) could be used instead of, or with, these two detectors. Most pesticides contain nitrogen, phosphorus, chlorine, bromine, fluorine, and/or sulfur.

These compounds are probably poor examples in that they can mostly be analysed by GC, which can also use selective detectors. It is unknown whether the breakdown products are stable enough or not too polar to be analysed by GC. With GC, the analyst may have difficulty deciding whether the small peaks are generated in the hot injection port or during the separation. SFC allows monitoring of thermally labile compounds using low temperatures with a different separation mechanism. Most if not all of the pesticides presently analysed using LC are amenable to SFC analysis with enhanced detection capabilities.

3 Residue Analysis

Carbamate Pesticides

US EPA Method 531.1[4] covers analysis of eleven compounds: seven carbamates and four metabolites in ground water. Metabolites include aldicarb sulfoxide, aldicarb sulfone, 3–hydroxycarbofuran, and 1–naphthol. The Association of Official Analytical Chemists (AOAC) has a similar method[5] for ten compounds (see Figure 1.2), excluding the naphthol.

The AOAC and EPA methods involve gradient elution LC, followed by two postcolumn derivatizations. The mobile phase is changed from 25 to 75%

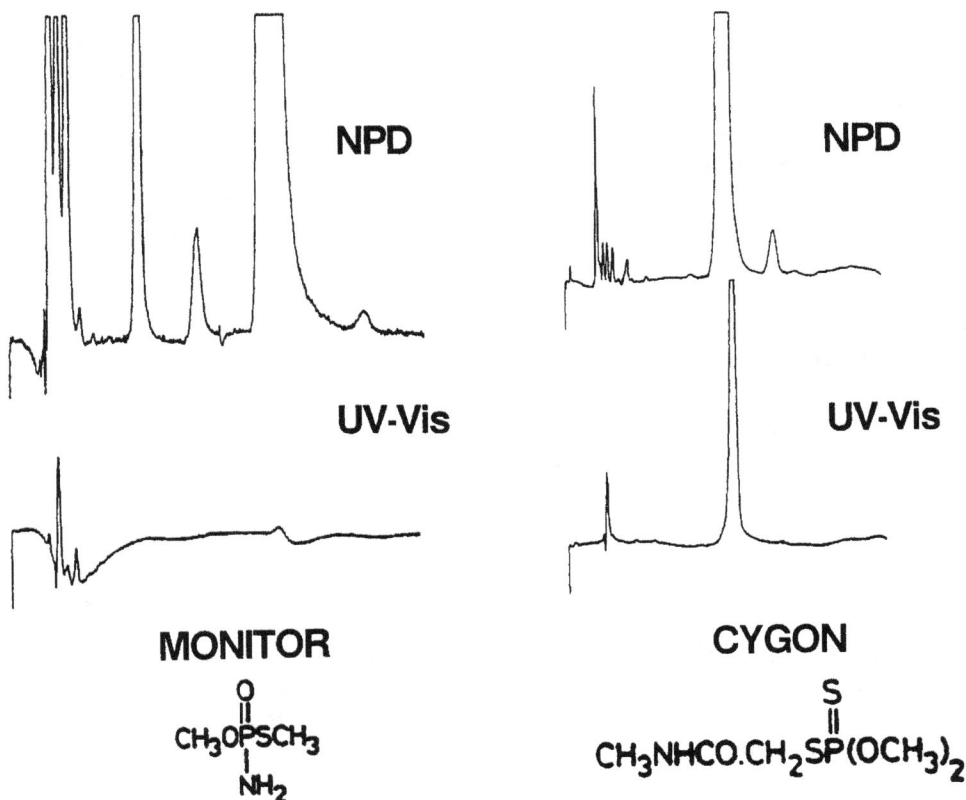

Figure 10.1 *Chromatograms of high concentrations* (mg ml^{-1}) *of two organophosphorus pesticides using UV and NPD detection. Note the small contaminant peaks, expecially in the NPD trace. Left, Monitor; Right, Cygon. They were collected using a 2 × 100 mm column with 7 μm diol, 0.5 ml min^{-1} of 2% [MeOH + 0.1% TFA] in CO$_2$, at 40 °C, 130 bar outlet*

MeOH in H$_2$O over *ca.* 24 minutes. The cycle time is *ca.* 1 hour. The column effluent is reacted with NaOH, to produce amines. A third high pressure pump delivers *o*-phthaldehyde (OPA) which reacts with the amines to form a fluorescent product.

EPA Method 531.1 is a waste water method. To minimize sample preparation in LC, water samples are filtered and 400 μl is injected. Since the postcolumn reactions dramatically degrade efficiency, and the gradient tends to refocus the peaks, band broadening from such large injections is not a problem.

There are several problems with this standard method. The gradient requires a significant equilibration time between runs. The postcolumn reactions dilute the solutes and broaden the peaks. In the past, back flow of the NaOH often destroyed the silica column. Recent advances have replaced the first postcolumn pump with an ion exchange column loaded with base. However, the addition of reagents still results in a time delay, dilution, and peak broadening.

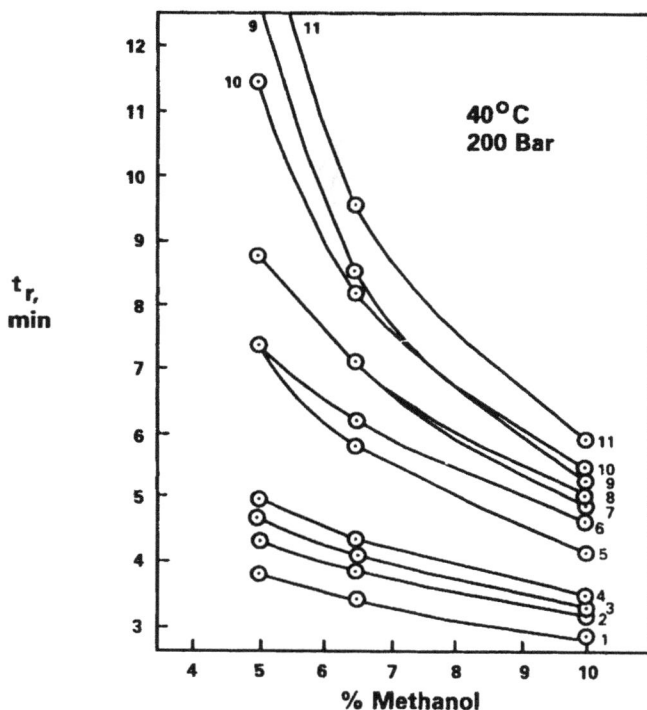

Figure 10.2 *The effect of MeOH concentration on the t_R of the carbamate pesticides of EPA method 531.1. Column: 4.6 × 250 mm, 5μm Lichrosphere diol, 2.0 ml min^{-1} MeOH/CO$_2$ at 40 °C 200 bar, 1. Aldicarb; 2 Baygon; 3 Carbofuran; 4 Methiocarb; 5 Aldicarb sulfone; 6 Carbaryl; 7 Methomyl; 8 1-Naphthol; 9 3–Hydroxycarbofuran; 10 Aldicarb sulfoxide; 11 Oxamyl* (Reproduced by permission from ref. 24)

SFC

SFC offers improved[24] analysis for the carbamates. Initial experiments showed symmetrical peaks from a cyanopropyl column with MeOH/CO$_2$ mixtures. However, α was poor. Lower modifier concentration produced adequate retention but groups of poorly resolved peaks separated by wide empty spaces on the baseline. Changing pressure or temperature only partially improved the separation. Hypersil silica also produced poor selectivity. A 4 × 250 mm, 5 μm Lichrosphere diol column produced the best selectivity among the columns tried, and resolved nearly all of the peaks in the first few experiments.[24]

Using the diol column, the 11 carbamates of EPA Method 531.1 were baseline resolved in less than 10 minutes without a gradient[24] (see Figure 1.1). An alternative, but less robust, set of analysis conditions allowed separation in less than 6 minutes. Even the longer SFC analysis allows at least 6 times higher throughput than the standard method.

Many peak reversals occurred when conditions were changed only modestly.

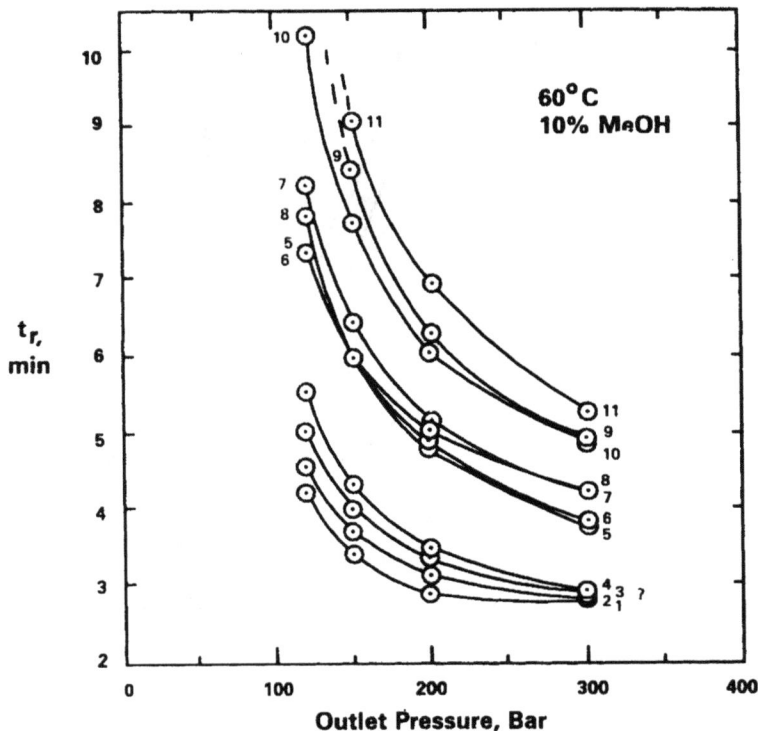

Figure 10.3   *The effect of outlet pressure on the t_R and α of the carbamate pesticides. Column as in Figure 10.1; 10% MeOH in CO_2, 60 °C*
(Reproduced by permission from ref. 24)

In such situations, simplex optimization strategies can lead to local optima but not the best separation. Programming a single parameter can also result in confusion since peak reversals can occur at intermediate conditions, and can also depend on the rate of programming. For complex separations with many peak reversals, it is often easiest to generate plots like Figures 10.2, 10.3, and 10.4. Because equilibration is so rapid in SFC, one can often generate such curves in between half and one day.

Changing modifier concentration, pressure, or temperature takes no more than a few minutes and usually can be seen by an offset in the baseline. Other methods often take substantially longer and leave the user with less understanding of the trade-offs available.

The modifier concentration changed retention but had only minor effect on α, as shown[24] in Figure 10.2. The outlet pressure, p_o, also changed retention modestly, with many minor peak reversals, as shown[24] in Figure 10.3. However, α changes were not extensive enough to resolve all the solutes, at any fixed temperature and modifier concentration.

Temperature adjustments produced the smallest changes in t_R, but significantly changed α, as shown[24] in Figure 10.4. Without a plot of k' or t_R *vs.* T, like Figure 10.4, it would be difficult to find the optimum.

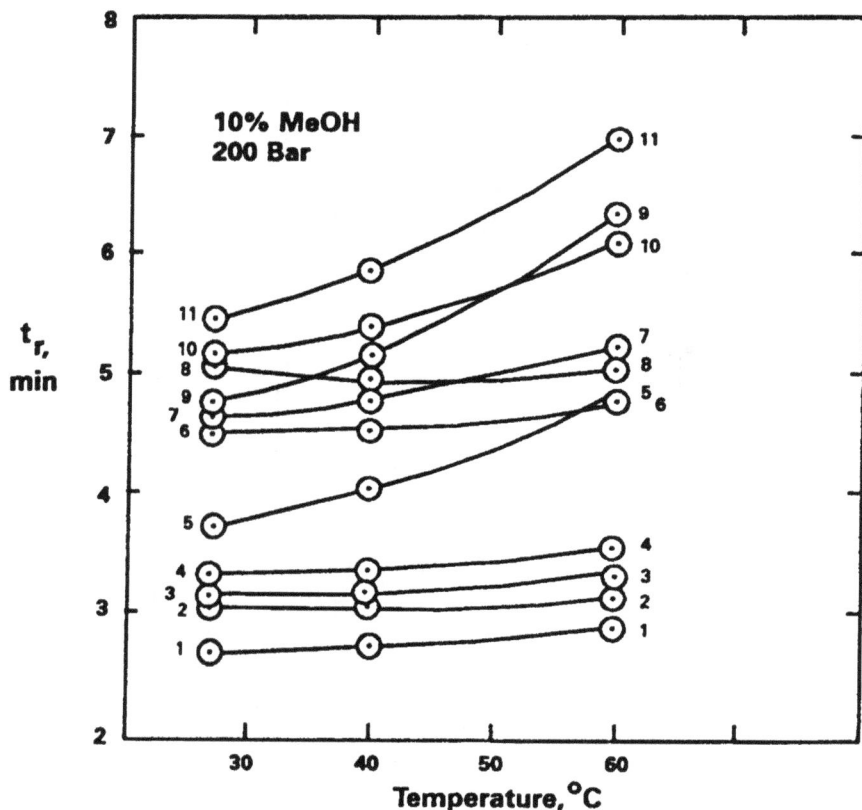

Figure 10.4 *The effect of temperature on the t_R and α of the carbamate pesticides. Column as in Figure 10.1; 2 ml min^{-1}, 10% MeOH in CO_2, 300 bar (Reproduced by permission from ref. 24)*

Peaks 5, 9, and 11 are all breakdown products with functional groups more polar than the parent compounds. Note that the slopes of the curves (in Figure 10.4) for these three compounds are similar, and differ from the rest of the solutes. Peak 8 (the naphthol) is a different kind of breakdown product, and it, too, exhibits a very different slope from the majority of the solutes, as well as from peaks 5, 9, and 11.

Detection

The compounds were simultaneously detected using the UV and NPD detectors.[24] Flow into the NPD was limited to a few ml min^{-1}, expanded, so the split ratio was of the order of 1000:1 (ng s^{-1} on-column, pg s^{-1} into the detector). The ratio of response factors of the two detectors provides additional confirmation of identity. The UV detector can also take spectra, which should provide additional confirmation.

The NPD offers significant selectivity toward all the solutes except naphthol,

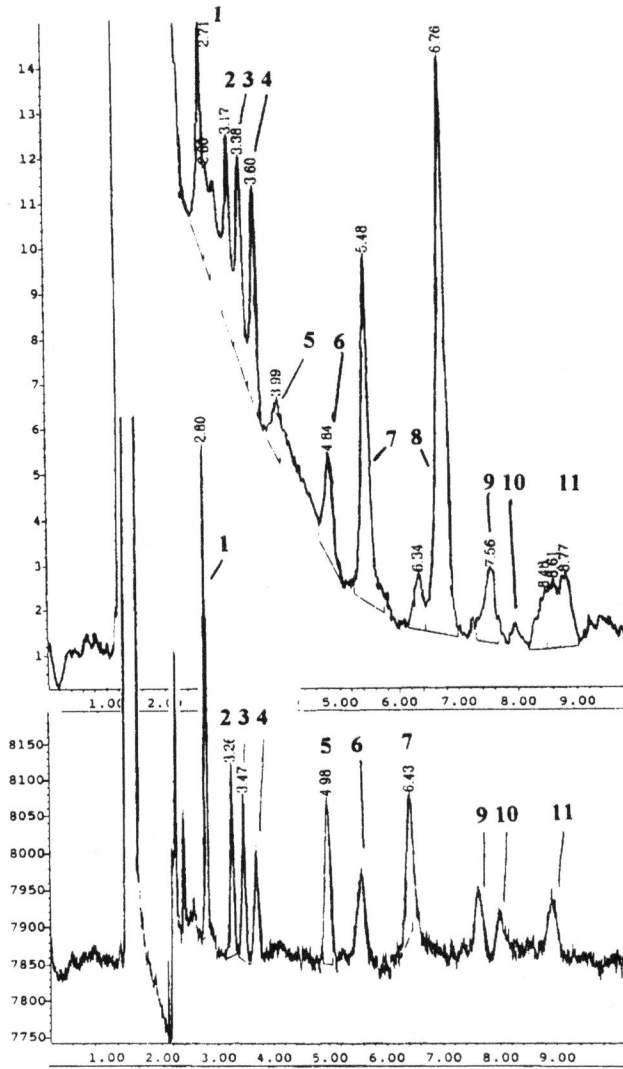

Figure 10.5 *Chromatograms of an on-line SPE extract of 5 ml water from a duck pond spiked with 10 p.p.b. each of 11 carbamates. Top UV; bottom: NPD* (Reproduced by permission from ref. 24)

which contains no nitrogen. On the other hand, it possesses an extremely good chromophore and is easily observed with the UV detector.

Injecting Water

SFC using CO_2 based MPs are incompatible with more than *ca.* 5 μl of water. However, 400 μl and larger water samples can be conveniently injected by replacing the fixed loop of the injection valve with a cartridge precolumn. The aqueous sample is passed through the cartridge, trapping the organics. The system is washed with distilled water, then dried with nitrogen. An organic solvent is injected, it is allowed to stand a few seconds, and the valve is thrown. This process is similar to solid phase extraction except that all the sample is 'injected' onto the column.

A chromatogram is presented in Figure 10.5, representing 5 ml of a water sample taken from a heavily polluted duck pond spiked at 10 p.p.b.. Detection limits were as low as < 100 p.p.t. The cartridge used was C_{18}. Subsequent work suggests that a mixed phase (C_{18} + silica) allows better recovery of the more polar solutes.

Summary

The SFC separation was superior in all aspects compared with the standard method. It was at least six times faster, without a gradient, and without the two postcolumn reactions. The NPD provided the selectivity normally obtained with the fluorescence detector.

Phenylurea Herbicides

Phenylurea herbicides are also moderately polar. Hypersil silica and MeOH/CO_2 mixtures were tried[23] with a set of eight randomly chosen solutes, including Linuron, and Diuron. All the solutes readily eluted,[23] producing symmetrical peaks without additive. However, α was less than ideal, since peaks bunched. Changing the SP to Lichrosphere diol produced more evenly spaced peaks and an easier separation.

The effects of percentage modifier, temperature, and pressure were determined.[23] The numerous peak reversals seen with the carbamates were not evident. The percentage modifier dramatically changed retention but not selectivity. Pressure had only a small impact on either retention or α. Retention uniformly increased with increasing temperature, with no reversals. Efficiency was high, sometimes > 25 000 plates.

At 120 °C, the peaks degenerated into broad humps, as shown in Figure 10.6, probably indicating thermal breakdown. A plot of efficiency *vs.* T indicates degraded peak width even as low as 60 °C. Consequently, temperature was kept low.

The phenylurea pesticides could be rapidly and efficiently separated using

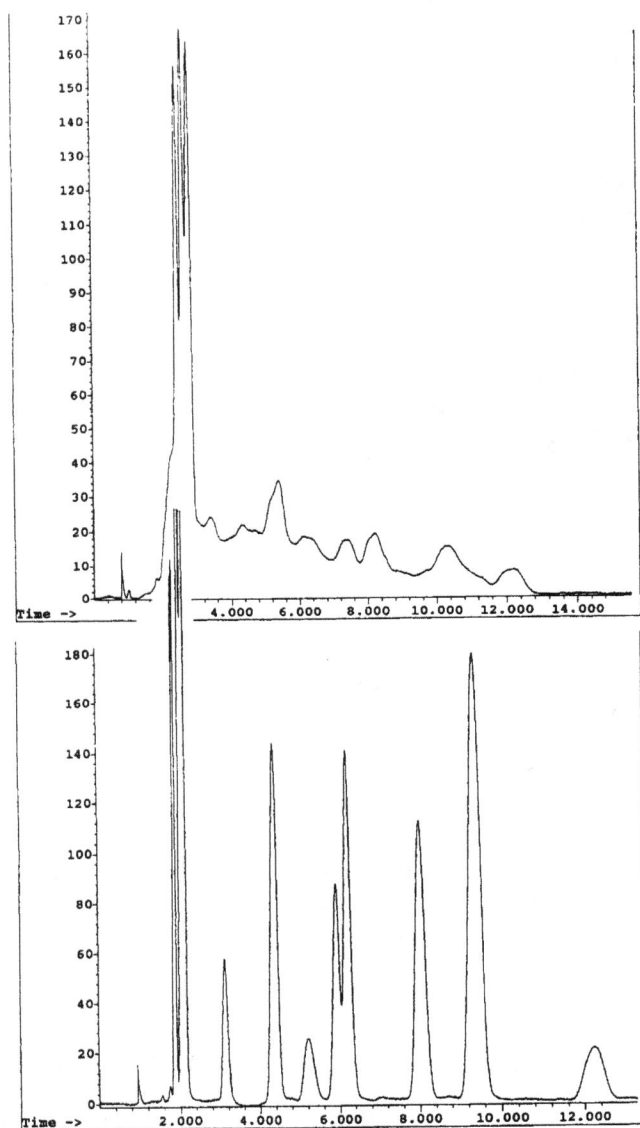

Figure 10.6. *Thermal breakdown of phenylurea herbicides on a silica column. Top: 80°C,*
bottom: 120°C. Other measurements indicated that at 80°C the peak width
had already degraded slightly, implying thermal breakdown. 2 ml min⁻¹,
6% MeOH in CO₂, 200 bar
(Reproduced by permission from ref. 23)

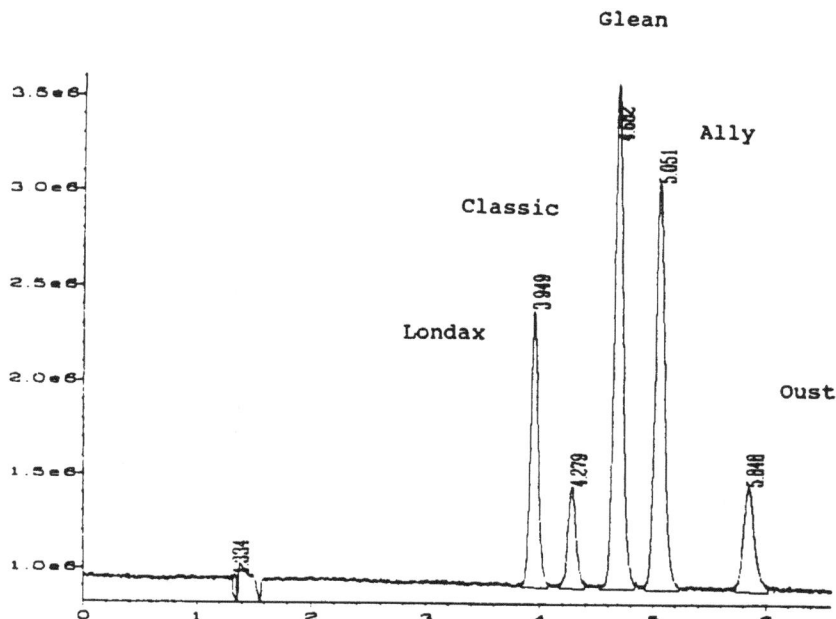

Figure 10.7   *Separation of five sulfonylurea herbicides on a diol column. Column:*
4.6 × 250 mm, 5 μm Lichrosphere diol, 2 ml min⁻¹ 10% MeOH in CO₂,
40 °C, 200 bar

supercritical fluids with standard LC columns. They are all candidates for the
NPD and many respond to the ECD.

Sulfonylureas

The sulfonylureas have been separated by SFC.[18] A MP of 10% MeOH in CO_2,
at 2 ml min⁻¹, 200 bar, 40 °C eluted all the solutes[19] in less than 6 minutes from
a 4 × 250 mm, 5 μm Lichrosphere diol column. The peaks were nearly ideally
spaced, as shown in Figure 10.7. Changing the percentage modifier caused all
the peaks to shift similarly with no peak reversals.

Increasing the temperature resulted in an increase in retention. Over the
ragne $30 \leq T \leq 70$ °C, retention increased by < 50%. As with the phenylureas,
there was little change in α when the temperature was increased.

Efficiency was very high. At 50 °C, 200 bar, and 2 ml min⁻¹ 10% MeOH in
CO_2, all but one of the peaks produced reduced plate heights (h_r) < 2, indicating that the columns were producing full theoretical efficiency. In LC, these
columns seldom produce more than 50% of theoretical efficiency at an
optimum flow rate of *ca.* 0.7 ml min⁻¹, which is less than one-third of the
optimum in SFC.

Breakdown products of the sulfonylureas can also be chromatographed by
SFC. In Table 10.1 and Figure 10.8 the retention times of both the parent and
breakdown product of five sulfonylureas are listed. In addition the presence or

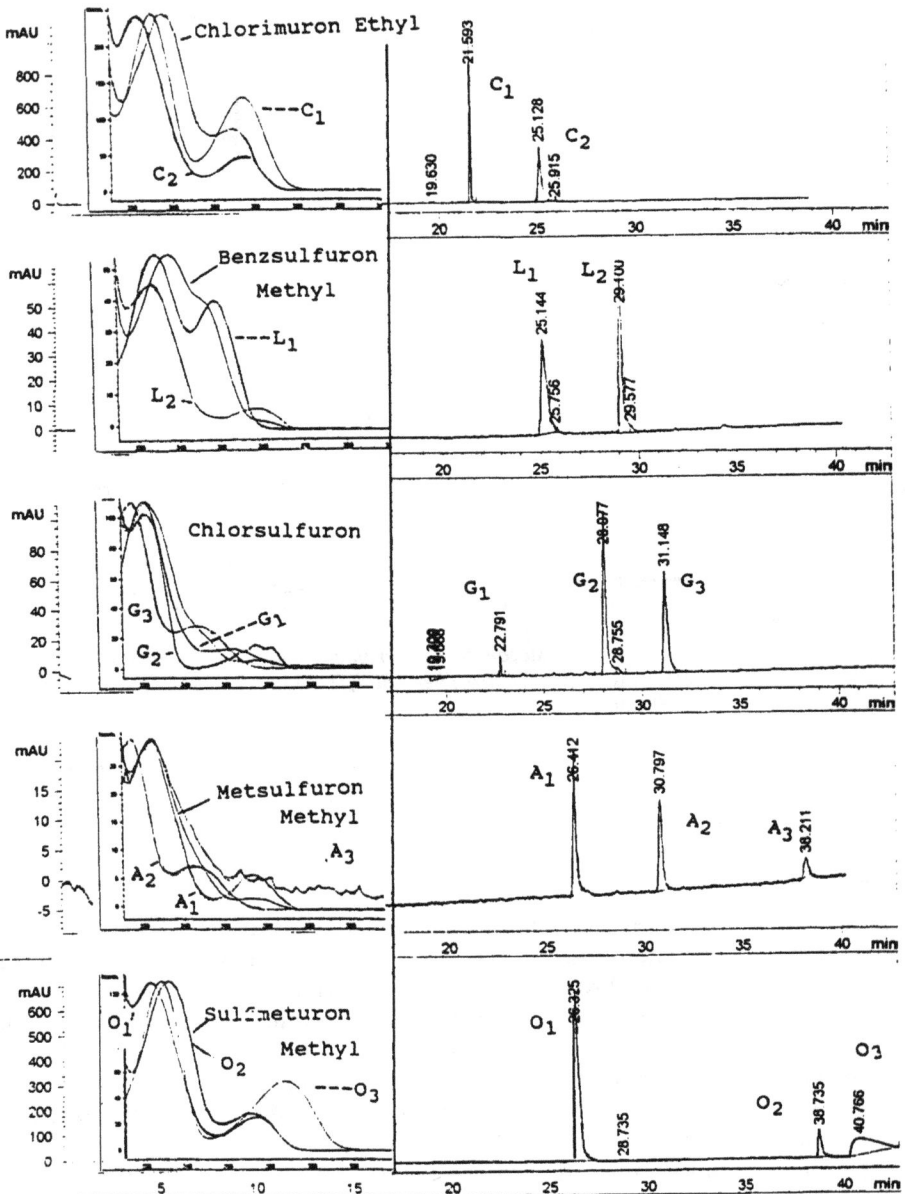

Figure 10.8 *Chromatograms of the breakdown products products of five sulfonylurea herbicides dissolved in methanol. Few of the peaks represent the parent compounds. Spectra of parent and breakdown peaks are also presented*

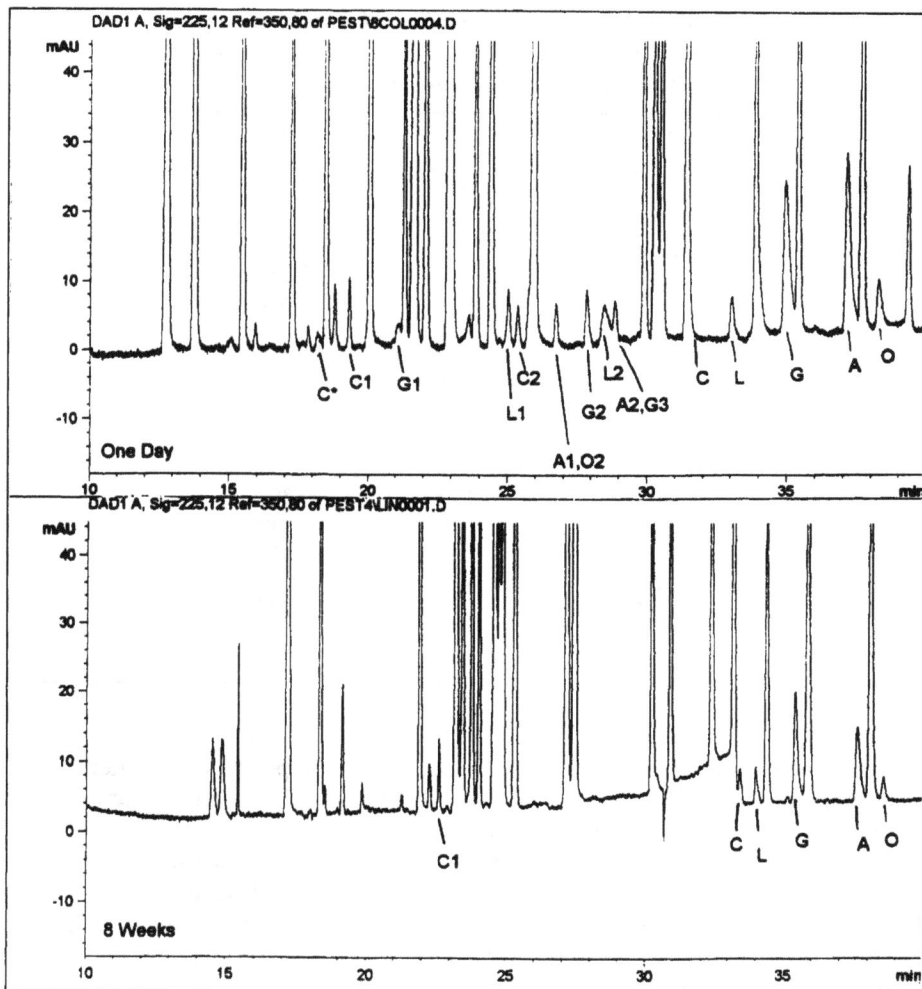

Figure 10.9 *Chromatograms of extracts frofm water samples spiked with 31 pesticides, including five sulfonylurea herbicides. In the top chromatogram, traces of both the parents and metabolities are indicated from spiking of the parent at the p.p.b. level. After eight weeks most of the breakdown products had disappeared*

Table 10.1   *Retention times of five sulfonylurea herbicides and their breakdown products. The presence/absence of peaks in the simultaneous NPD and ECD chromatograms is also listed*

	Classic	*C**	*C1*	*C2*
225 nm	32.9 min	19.6 min	21.6 min	25.1 min
NPD	yes	no	yes	no
ECD	yes	yes	yes	no
	Londax		*L1*	*L2*
225 nm	34.6		25.1	29.1
NPD	yes		yes	no
ECD	no		no	no
	Glean	*G1*	*G2*	*G3*
225 nm	36.1	22.8	28.1	31.1
NPD	yes	no	no	yes
ECD	yes	yes	yes	no
	Ally	*A1*	*A2*	*(parent?)*
225 nm	38.3	26.1	30.8	38.3
NPD	yes	no	yes	yes
ECD	no	no	no	no
	Oust	*O1*	*(parent?)*	*O3*
225 nm	38.7	26.3	38.7	40.7
NPD	?	?	?	?
ECD	no	no	no	no

absence of nitrogen and chlorine in each of the compounds (obtained from simultaneous NPD, ECD, and UV chromatograms) is noted in Table 10.1. Further, spectra of each of the compounds are presented.

The upper trace in Figure 10.9 shows the presence of trace levels of many sulfonylurea breakdown products extracted from a 10 ml water sample a day after it was spiked with the parent compounds plus 26 other pesticides. In the bottom trace, collected 8 weeks later, the breakdown products are shown to have largely disappeared.

Triazines

Atrizine is probably the most extensively used agricultural chemical. Consequently, there are many analytical methods for atrizine, including various kinds of test kits. However, other triazines used for similar reasons may not respond to the kits. Standard methods still rely on laboratory chromatography with confirmation based on selective detectors and t_R windows. Triazines are moderately difficult to separate by GC owing to the multiple nitrogens in the ring. More importantly, the build-up of dirt and contaminants on the column from real samples tends to degrade GC analysis. There is a trend toward the use of LC, since the MP phase can wash away contaminants.

Triazines and triazoles were first separated by packed column SFC several years ago.[20] The SPE pre-concentration of triazines in groundwater followed

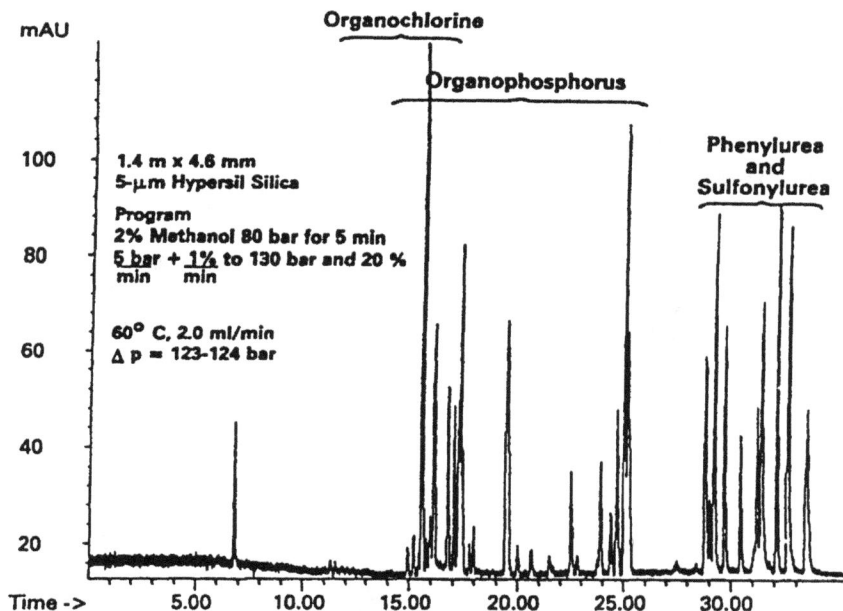

Figure 10.10 *Separation of 44 pesticides, including organochlorine, organophosphorus, phenylurea, and sulfonylurea pesticides, from a 4.6 mm, 1.4 m column packed with 5 μm Hypersil silica. Programs: 2 ml min^{-1}, MeOH in CO$_2$, 2%, 80 bar for 5 min, 5 bar min^{-1} and 1% min^{-1}, to 130 bar, 20%, 60°C*

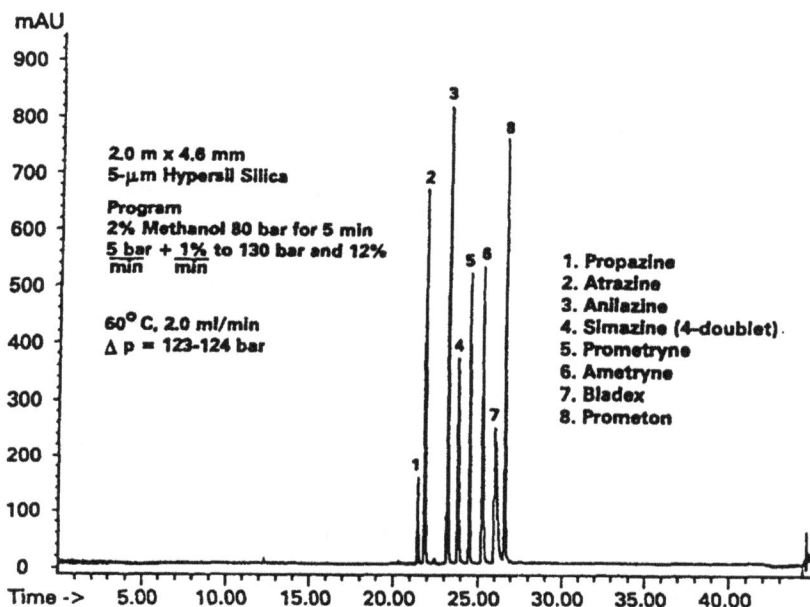

Figure 10.11 *Separation of triazine herbicides on a 2.0 m × 4.6 mm Hypersil silica column, 5 μm d$_p$, 2% MeOH in CO$_2$ at 80 bar for 5 min, 5 bar min^{-1} and 1% min^{-1} to 130 bar and 12% MeOH, 60°C*

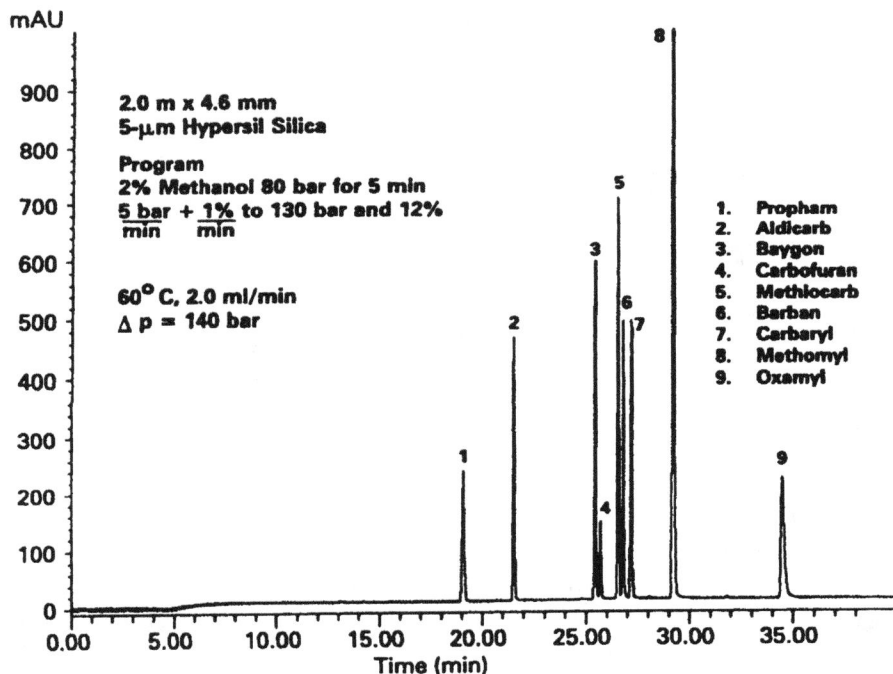

Figure 10.12 *Separation of some carbamates using the same conditions as in Figure 10.11*

by packed column SFC was also recently reported.[25] Highly reproducible
separations produced detection limits ≪ p.p.b., using the NPD.

3 Screening Methods for Multiple Pesticide Residues

Pesticide screening is a major business. The vast majority of screening is done
using GC methods, since GC is fast, efficient, sensitive and relatively inexpensive. In the past, there was a standard method for each family of pesticides. The
more recent trend has been toward the use of GC–MS or GC with a selective
detector, such as the NPD, to screen for larger numbers of pesticides, from
many different families, in a single run.

On the other hand, many pesticides can not be analysed by GC. The most
modern pesticides are becoming much more potent (*e.g.* a few grams, instead of
kilograms, per acre) and much more unstable. GC is also susceptible to
problems where involatile components or breakdown products of the sample
build up on the head of the column, changing the chromatography. GC tends to
require extensive sample clean up to minimize these problems, negating some of
its cost advantage. In addition, many metabolites are hydrolysed, to more polar
forms that are often not volatile or are thermally labile.

Because of these problems there has been a second trend toward the use of
LC. The LC mobile phase is a solvent which can elute much more polar
solutes, breakdown products, and polar contaminants and, thus, requires

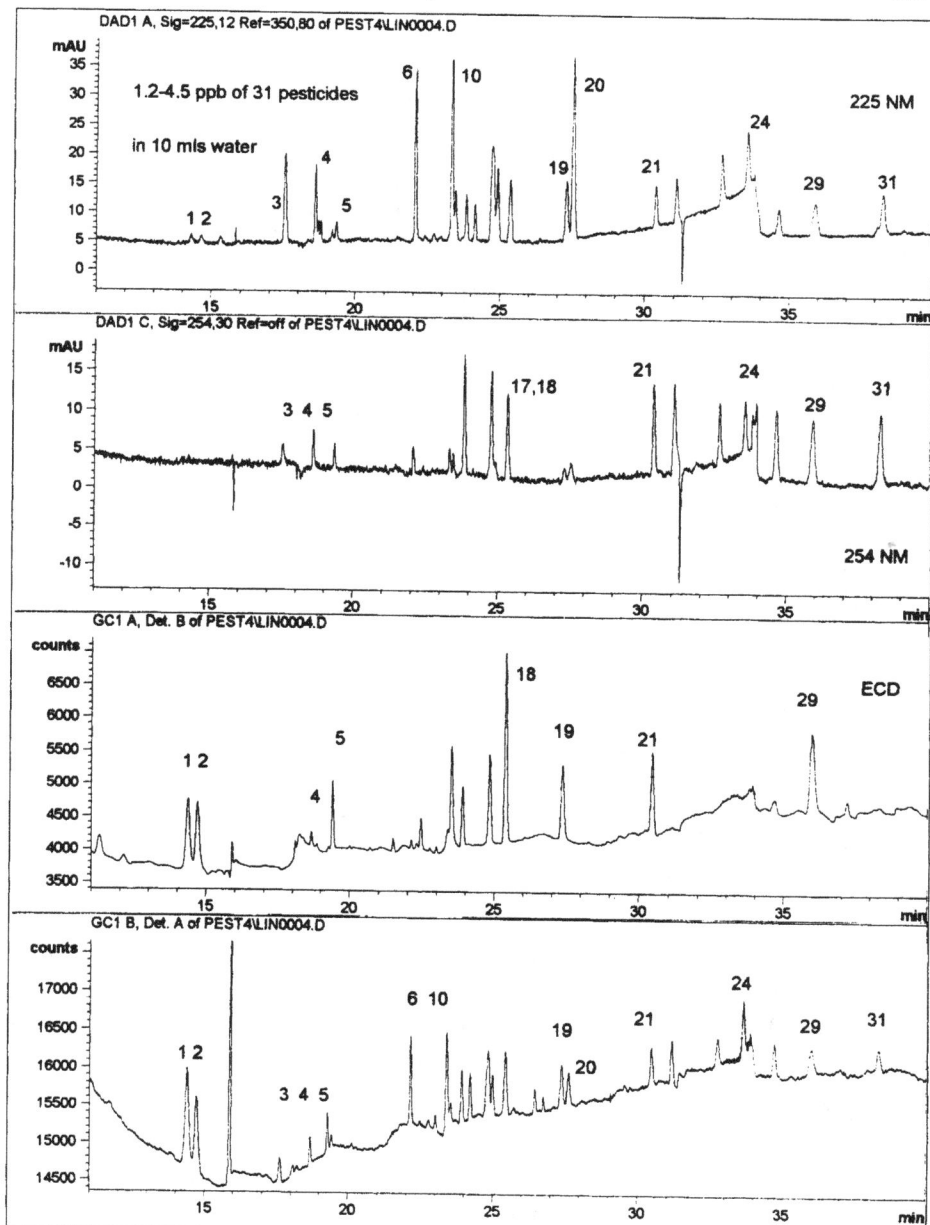

Figure 10.13 *Chromatograms collected simultaneously using the NPD, an ECD, and two UV wavelengths from 1–4 p.p.b. of 31 pesticides spiked into 10 ml water. The pesticides were extracted using an in-line SPE apparatus, then separated on a 1.6 m long, high efficiency column*

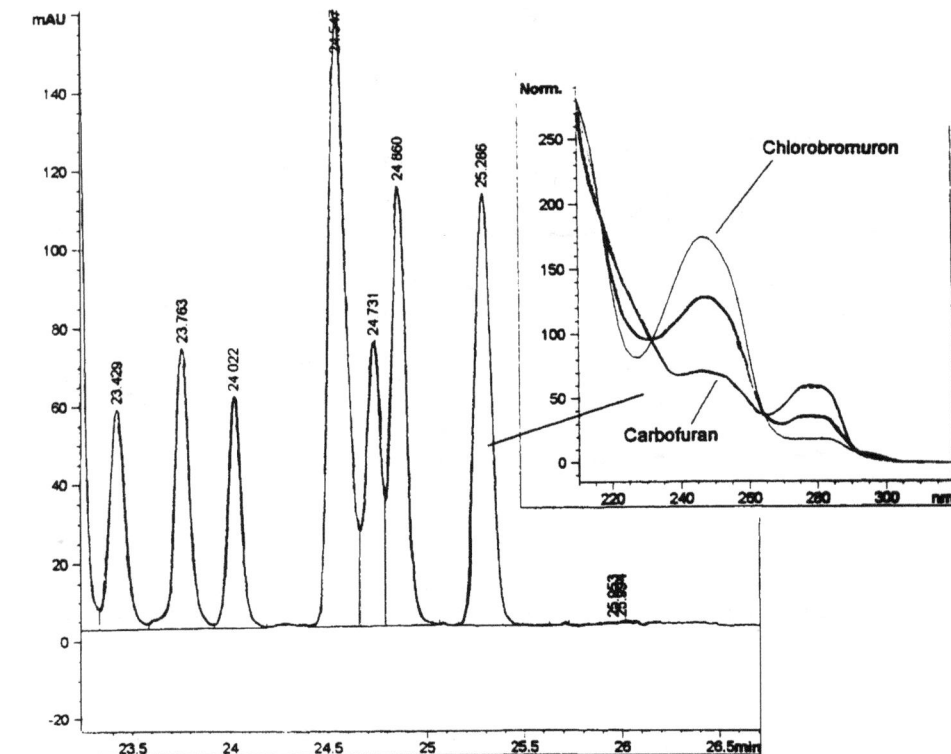

Figure 10.14 *A chromatogram showing an apparently symmetrical peak which actually consists of co-eluting phenylurea and carbamate pesticides. Spectra collected on the leading and trailing edges are also presented along with the spectrum at the apex. The detector allows automatic peak purity measurements and quantitation of the individual components*

somewhat less sample clean-up. In addition, LC elution is primarily a function of solvation not volatility, so labile solutes are not destroyed. Unfortunately, LC is a relatively low efficiency technique that generates much less information per unit time, has few detector options, and is more expensive than GC. High efficiency LC requires excessive[28] analysis times and makes programming difficult.

Packed column SFC exhibits high efficiency[27] with reasonable speed, and multiple selective detectors. Such a combination makes pesticide screening feasible. Note that it is more useful to minimize the total number of analyses than to differentiate between GC-able and non GC-able solutes.

Organochlorine, organophosphorus, phenylurea, and sulfonylurea pesticides were separated on a 1.4 m long column, with a pressure and modifier concentration program. Apparent efficiencies were very high, producing a peak capacity, $n > 150$. In previously unpublished work from 1992, a chromatogram of 44 organophosphorus, organochlorine, phenylurea, and sulfonylurea pesticides is presented in Figure 10.10. Two other chromatograms from the same

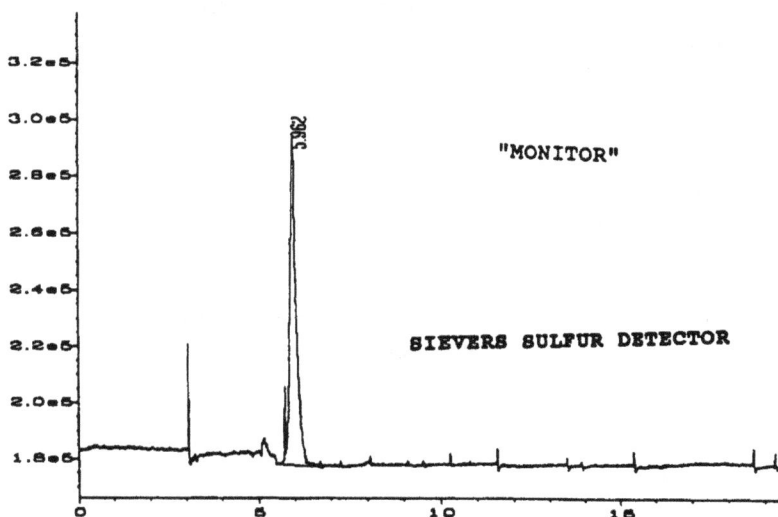

Figure 10.15 *Chromatogram of Monitor using the sulfur chemilumenesence detector. 4.6 × 200 mm, 5 μm diol, 3 ml min⁻¹ of 5% [MeOH + 0.1% TFA] in CO_2, 40 °C, 134 bar*

period of triazine and carbamate pesticides using ten columns in series (2 m total) are shown in Figures 10.11 and 10.12.

Programming Selectivity and Retention

With long columns, the user retains the ability to adjust both t_R and α through programming. Reproducibility in t_R and peak areas was determined for ten consecutive chromatograms of phenylurea pesticides separated using both a pressure and composition program. The column was 2.0 m long packed with 5 μm Hypersil silica. Detection was by UV at 254 nm. The peaks produced a range of signal to noise ratios from as little as 15–20 to > 100. Retention time reproducibility was ± 0.2–0.3% RSD, while area reproducibility ranged from < 1% to > 6.5% depending on S/N ratio.

Multiple Detectors

Direct detection is inadequate to measure the extremely low concentrations of pesticides in real samples. Some form of preconcentration is required. For water samples, up to 10 ml was passed through precolumns mounted in place of the loop in a six-port injection valve. After drying, the loop was switched into the flowing stream. An example, with 2.4–9 p.p.b. of 30 solutes, including phenylurea, sulfonylurea, and carbamates, was presented in Figure 1.3.

An NPD, an ECD, and multiple UV wavelengths can all monitor the effluent from a single injection. Chromatograms from an on-line solid phase extract from a 10 ml water sample is presented in Figure 10.13. Detection limits of many of the pesticides were well below 1 p.p.b.

Spectra can provide peak purity calculations. In Figure 10.14, a single apparently symmetrical peak actually contained a phenylurea and a carbamate co-eluting.

Other detectors, such as a sulfur chemiluminescence detector, can be connected in a similar fashion. An example of the detection of 'Monitor' an old organophosphorus pesticide is presented in Figure 10.15. Other compounds such as thiocarbamates and sulfonylureas should be amenable to such analysis.

The GC detectors are connected to the column effluent using a 'tee' with a fused silica fixed restrictor. Only a small fraction of the total flow passes into the GC detector. Flow continues to be controlled by the pumps, while pressure continues to be controlled by the electronic BPR. It should be recognized that the flow through the fixed restrictor does change when conditions, particularly pressure, are changed.

5 Summary

Packed column SFC is capable of generating the high efficiencies necessary to separate complex mixtures. Multiple detectors can be used simultaneously to yield the selectivity and sensitivity necessary for pesticide residue analysis. Multiple parameters can be programmed yet yield reproducible retention times. In short, the technique appears to provide all the necessary characteristics for screening a broad range of pesticides normally requiring many different GC and LC methods.

6 References

1. For example, EPA Method 608.
2. EPA Method 1618.
3. J. Tekel, K. Schultzova, P. Farkas, and J. Kovacicova, *J. High Resolut. Chromatogr.*, 1991, **14**, 423.
4. EPA method 531.1, US EPA, Washington, DC, 1988.
5. First Action Method, *J. Assoc. Off. Anal. Chem.*, 1985, **68**, No. 2.
6. P.A. David and M. Novotny, *Anal. Chem.*, 1989, **61**, 2082.
7. K.E. Markides, E.D. Lee, R. Bolick and M.L. Lee, *Anal. Chem.*, 1986, **58**, 740.
8. S. Kennedy and R.J. Wall, *LC–GC*, 1988, **6**, 930.
9. H.-C.K. Chang and L.T.Taylor, *J. Chromatogr. Sci.*, 1990, **28**, 29.
10. B.W. Wright and R.D. Smith, *J. High Resolut. Chromatogr.*, 1985, **8**, 8.
11. B.W. Wright and R.D. Smith, *J. High Resolut. Chromatogr.*, 1986, **9**, 73.
12. H.T. Kalinoski and R.D. Smith, *Anal. Chem.*, 1988, **60**, 529.
13. S. Ashraf, K.D. Bartle, A.A. Clifford, I.L. Davies, and R. Moulder, *Chromatographia*, 1990, **30**, 618.
14. S. Ashraf, K.D. Bartle, A.A. Clifford, and R. Moulder, *J. High Resolut. Chromatogr.*, 1991, **14**, 29.
15. J.R. Wheeler and M.E. McNally, *J. Chromatogr.*, 1987, **410**, 343.
16. M.E. McNally and J.R. Wheeler, *LC/GC*, 1988, **6**, 816.
17. A.J. Berry, D.E. Games, I.C. Mylcreest, J.R. Perkins, S. Pleasance, *J. High Resolut. Chromatogr.*, 1988, **11**, 61.
18. M.E. McNally and J.R. Wheeler, *J. Chromatogr.*, 1988, **435**, 63.
19. T.A. Berger, and W.H. Wilson, accepted by *Chromatographia*, 1995.

20. S. Shah and L.T.Taylor, *J. Chromatogr.*, 1990, **505**, 293.
21. J.G.J. Mol, B.N. Zegers, H. Lindeman, U.A. Th. Brinkman, *Chromatographia*, 1991, **32**, 203.
22. B. N. Zegers, A.C. Hogenboom, S.E.G. Dekkers, H. Lingeman, and U.A.Th. Brinkman, *J. Microcol. Sep.*, 1994, **6**, 55.
23. T.A. Berger, *J. Chromatogr. Sci.*, 1994, **32**, 25.
24. T.A. Berger, W.H. Wilson, and J.F. Deye, *J. Chromatogr. Sci.*, 1994, **32**, 179.
25. W.H. Wilson, unpublished results.
26. Submitted to *Chromatographia*, 1995.
27. T.A. Berger and W.H. Wilson, *Anal. Chem.*, 1993, **65**, 1451.
28. K.-E. Karlsson and M. Novotny, *Anal. Chem.*, 1988, **60**, 1622.

CHAPTER 11

SFC and the Petroleum Industry

1 Group Separations

The first standard method of any kind to use SFC was ASTM Method D 5186–91 for determining the percentage of aromatics in diesel fuel. With the SFC method, aromatic content in diesel fuels can be quantified below 1%, in as little as two minutes. At the likely new legal maximum concentration of aromatics of 10%, the SFC method can provide a reproducibility better than ± 1%. The Western States Petroleum Association (WSPA) Round Robin diesel samples tested in one laboratory[1] produced the results in Table 11.1.

The analysis is robust and can be repeated on different instruments. For example, the aromatic content of two jet fuels was determined on four different instruments and showed excellent agreement, as shown in Table 11.2.

The SFC method uses a silica column, pure CO_2 at constant pressure, and a FID as the detector. A sampling valve with a fixed loop is used to introduce the sample.

The ASTM method uses C_{22} and toluene as probes of resolution. The specifications called for a resolution > 4 between these solutes. The standard

Table 11.1 *Aromatic content found in round robin samples*

	WX-1	WX-2	WX-3	WX-4	WX-5	WX-6	WX-7
Wt. %	24.52	1.04	18.19	6.34	12.24	23.06	18.46
SD	0.035	0.036	0.076	0.070	0.015	0.076	0.010
Rel. SD %	0.1	3.5	0.4	1.1	0.1	0.3	0.1

Analysis time 1.5 min, $n = 3$

Table 11.2 *Instrument to instrument reproducibility*

Instrument No.	Jet Fuel No. 1	Jet Fuel No. 2
1	16.78	23.52
2	16.83	23.74
3	16.72	23.58
4	16.71	23.44

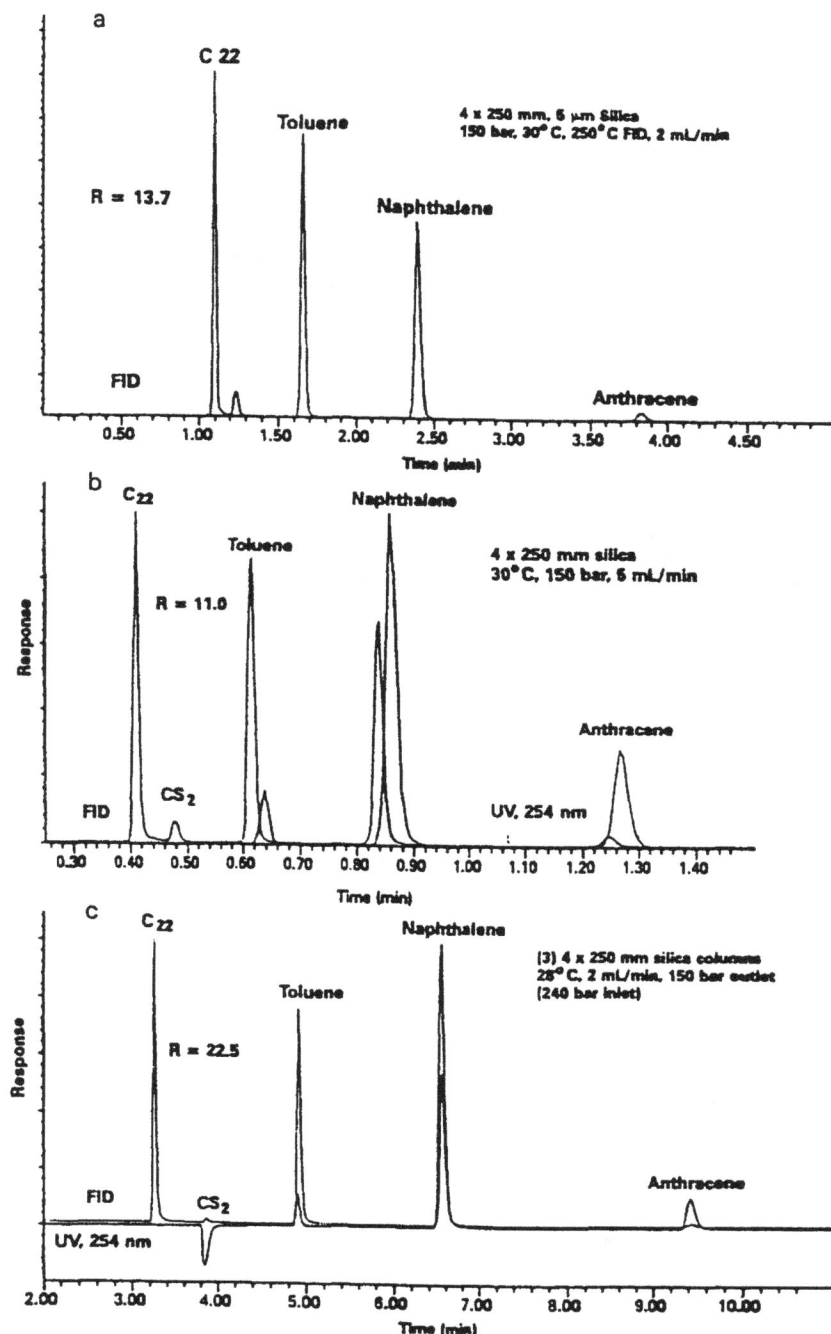

Figure 11.1 *SFC chromatograms of a resolution standard under three different sets of conditions.* (a) *Hydrocarbon Group Separation column* 4 × 250 mm, 5 μm, 2 ml min^{-1} CO_2 at 30 °C, 150 bar; (b) *Higher speed: conditions same as in* (a) *except* 5 ml min^{-1}; (c) *Higher resolution using a* 75 cm *long column*, 2 ml min^{-1} CO_2, 28 °C, 150 bar.

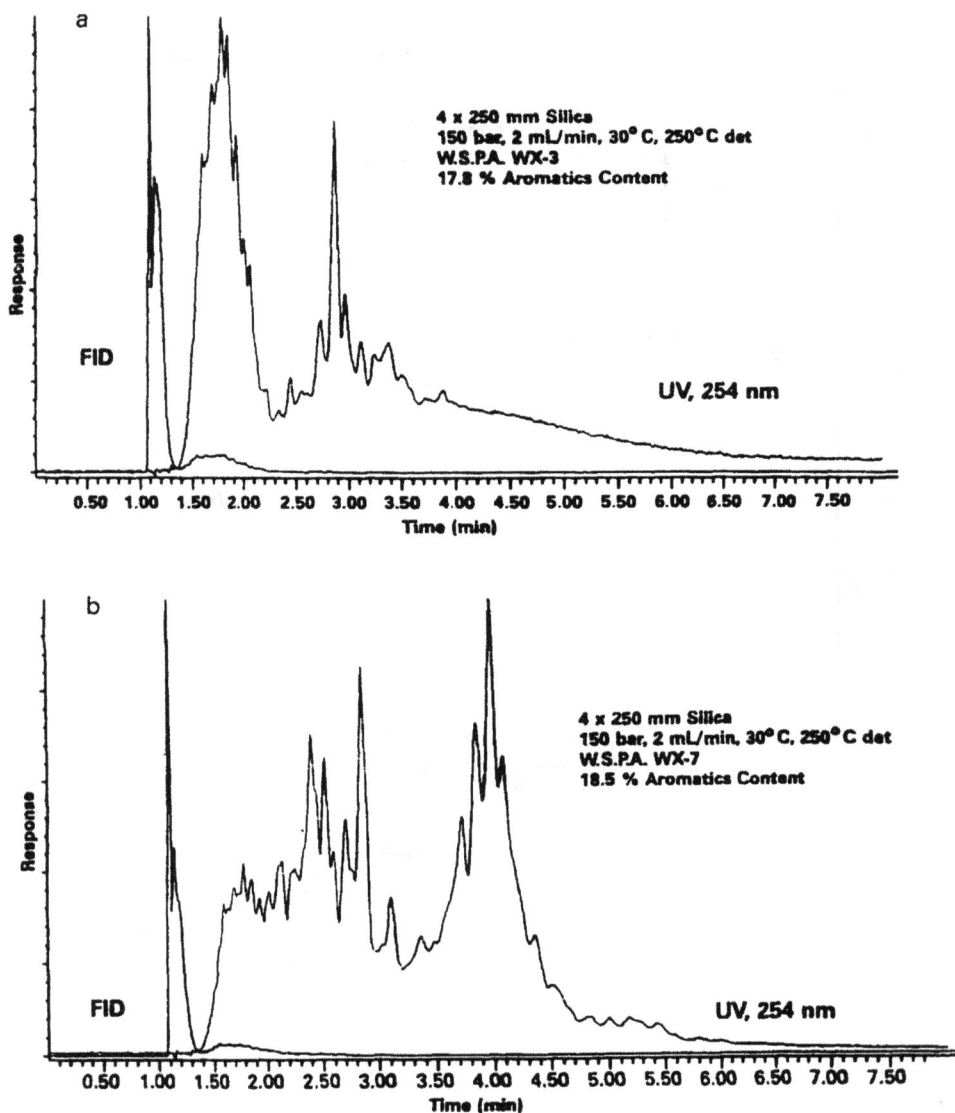

a

4 x 250 mm Silica
150 bar, 2 mL/min, 30° C, 250°C det
W.S.P.A. WX-3
17.8 % Aromatics Content

Response

FID

UV, 254 nm

0.50 1.00 1.50 2.00 2.50 3.00 3.50 4.00 4.50 5.00 5.50 6.00 6.50 7.00 7.50

Time (min)

b

4 x 250 mm Silica
150 bar, 2 mL/min, 30° C, 250°C det
W.S.P.A. WX-7
18.5 % Aromatics Content

Response

FID

UV, 254 nm

0.50 1.00 1.50 2.00 2.50 3.00 3.50 4.00 4.50 5.00 5.50 6.00 6.50 7.00 7.50

Time (min)

Figure 11.2 *Two chromatograms of actual diesel fuels using both the FID and the UV
detector. The sharper peak on the left is the FID trace of the saturates and
olefins. The FID traces of the aromatics appear as small bumps between ca.
1.25 and 2.25 min. The aromatics content of the two fuels is nearly equal
(17.8% at top, 18.5% at bottom). Column: 4×250 mm, 5 μm Silica.
Conditions: 2 ml min^{-1} CO_2, 30 °C, 150 bar, FID at 250 °C, UV detection at
254 nm*

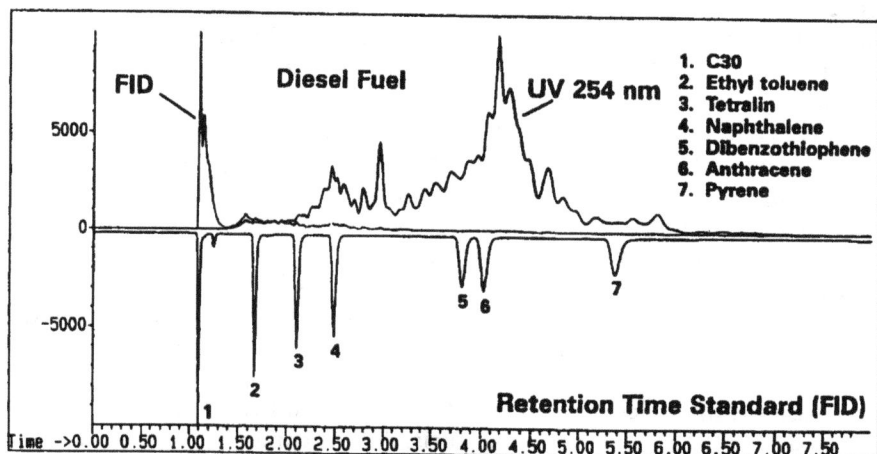

Figure 11.3 *Canadian General Standards Board proposed method for determining the quantity of larger, multi-ring aromatics in diesel fuels. The method is similar to the ASTM method except that it is standardized to show the elution times of multi-ring aromatics. The inverted trace indicates the elution order and time of six aromatics. The upper trace shows both UV and FID detection of a diesel fuel*

resolution using a 25 cm long column can be *ca.* 13.5, as shown in Figure 11.1a. By using higher flow rates the analysis time can be significantly shortened (less than two minutes) with minimal loss in resolution ($R_s = 11$), as shown in Figure 11.1b. Combining three columns in series produced a resolution > 22 with $t_R < 10$ minutes, as shown in Figure 11.1c. This enhanced resolution can be used to resolve aliphatics from olefins, or provide more speciation within the groups.

The ASTM method calls for quantification using the FID, but does not completely characterize actual fuels. A better understanding of the compositions of fuels can be obtained by simultaneously using the FID with a UV detector. The two detectors provide complementary information. The FID produces an accurate indication of the amount of carbon eluting per unit time. However, the FID provides no insight into the chemical identity of compounds in individual peaks. The FID provides quantitation while the UV detector provides semi-qualitative information.

At low aromatics content, the UV response allows better estimation of integration parameters. The front end of the aromatics peak can be easily differentiated from saturated hydrocarbons in the UV trace but not in the FID trace.

Multiring aromatic compounds have much higher response factors than smaller compounds in the UV detector. Using the UV detector alone (as in LC), these compounds can appear to make up a large fraction of the total. Since different large ring structures can have dramatically different response factors, the UV detector is unreliable in quantifying an unknown mixture. The FID has an essentially constant response factor for all hydrocarbons. It clearly shows that these large ring structures actually represent only a tiny fraction of the mass in the sample, as shown in Figures 11.2a and 11.2b. The figures represent two diesels with similar aromatics content but with a different distribution of ring sizes.

Higher concentrations of larger rings are often associated with 'smokey' exhaust. The Canadian General Standards Board has proposed a method for determining ring numbers based on the use of several aromatic standards, as shown in Figure 11.3. The upper curves present both UV and FID response to a diesel fuel. The lower inverted curve indicates the relative elution position of six aromatic compounds from ethyltoluene to pyrene. Clearly, larger compounds elute later, even in 'group' separations. The intent of the analysis is to allow regulators to measure and control the quantity of large ring number compounds in fuels.

Because the approach is accurate at low aromatics content, it also can be used to determine aromatics in other fuels. ASTM committees are working on additional SFC standard methods. Aromatics analysis by SFC has been applied to jet fuels, naphthas, and gasolines as shown in Figure 11.4. In addition, a residual fuel oil, a blended vacuum oil, and a heavy distillate are shown from top to bottom in Figure 11.5.

Why SFC?

ASTM Method 5186–91 replaces the fluorescent indicator adsorption (FIA) method.[2,3] The FIA method is old and reliable, using open column chromatography. Fluorescent dyes are placed at the head of a silica column. A sample of diesel fuel is added, then eluted through the column with isopropyl alcohol. Saturates, olefins, and aromatics travel down the column at different speeds, separating into discrete bands. Under UV light, saturated hydrocarbons remain colourless, olefins turn yellow, aromatics turn blue, and the alcohol front turns red. The lengths of the bands are literally measured with a ruler. The length of each band corresponds to the volume percent of each group.

There were several reasons to develop a replacement technique for the FIA method. The Clean Air Act and other anti-pollution initiatives, particularly from California, require that the aromatics content in fuels be lowered. There is also a trend toward eliminating larger multi-ring aromatics from fuels. The FIA analysis has poor precision, typically ± 2–3%, but more importantly is not reliable below *ca.* 10% aromatics. Since regulators appear to be focusing on a **maximum** allowable aromatic content of 10%, the FIA method was inadequate as a test of legal compliance.

There have been instrumental replacement technologies for FIA, involving both GC and LC, available for many years but none were widely implemented until the SFC method. The most important reason was that these alternative methods gave an answer different from the FIA result.

The SFC method actually has no better chromatographic selectivity than earlier LC methods. However, the LC method is unreliable since it only uses a UV detector with a variable response factor.

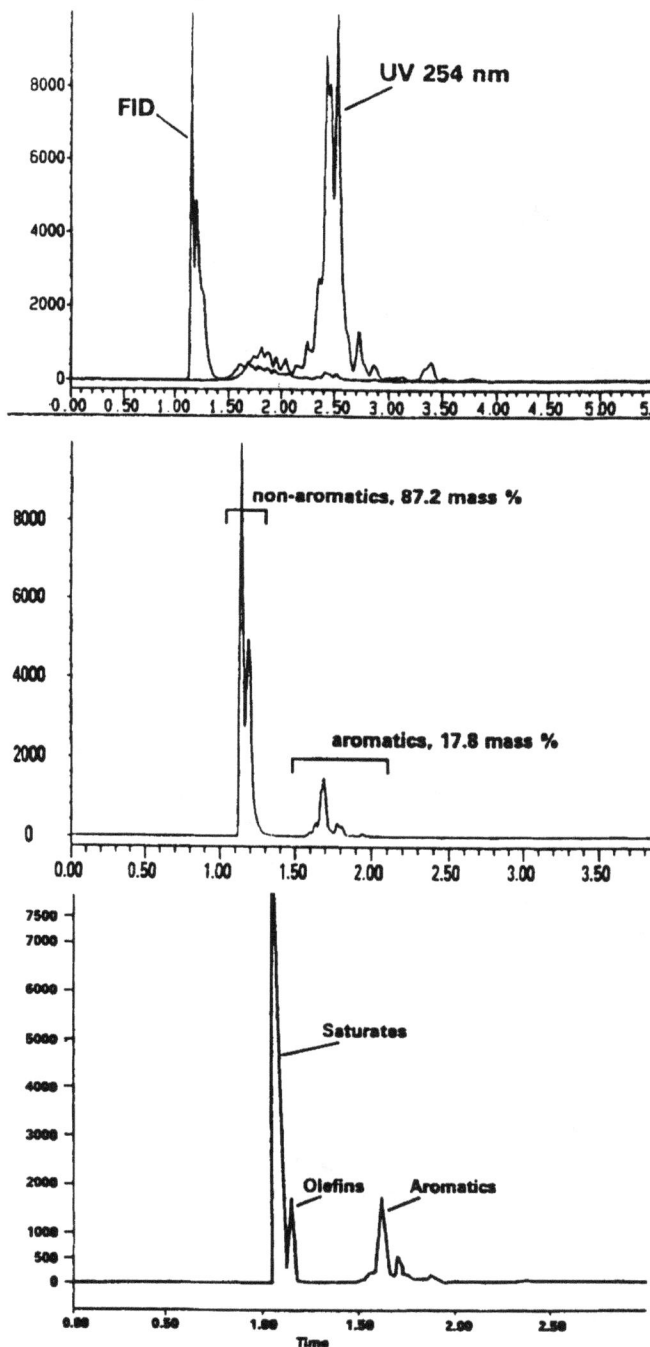

Figure 11.4 *The ASTM method is accurate to low aromatics concentrations making it usable with lighter fuels. Examples include: top, jet fuel; middle, naphtha; and bottom, gasoline*

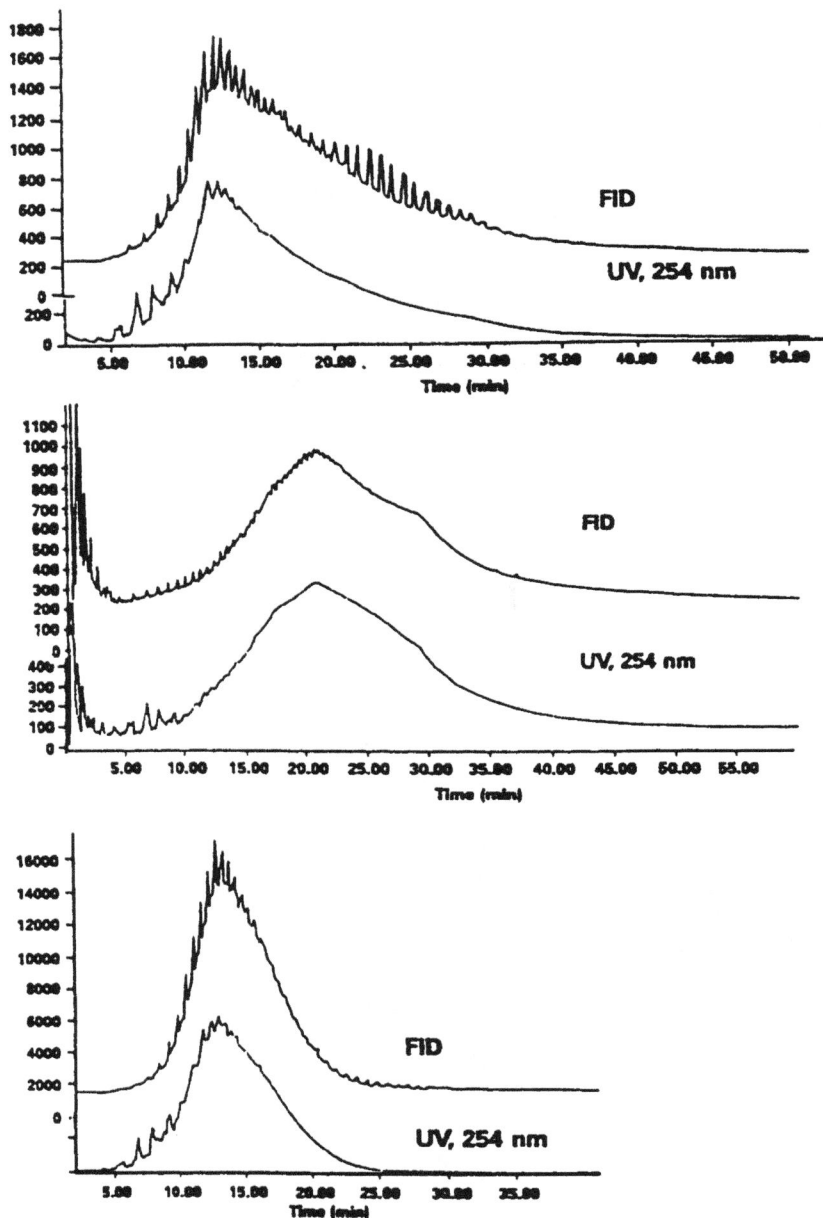

Figure 11.5 *SFC group separations have also been applied to heavier materials. Examples include: top, residual fuel oil; middle, blended vacuum oil; and bottom, heavy distillate*

LC with refractive index and dielectric constant detectors provide a nearly uniform response to all the groups. However, they exhibit poor sensitivity and drift due to temperature variations. The aromatics emerging from a column tail off and asymptotically approach the baseline. The drifting baseline of these detectors compounds the difficulty in quantifying such response. The SFC method gives the same information as LC methods, plus a direct measure of carbon distribution through the FID.

Other Aspects

Columns

There is a tremendous variation in the ability of different silicas to produce group separations. Some silicas produce almost no group separation whereas others show almost miraculous selectivities. It is unclear which attributes are responsible for these differences, although pure silicas, with lower metals content, appear to provide better group selectivity.[4]

Temperature and Pressure: Subcritical vs Supercritical

Resolution between groups improves as temperature decreases. The best resolution was attained at 28 °C, and 150 bar. At these conditions CO_2 is not supercritical $(T < T_c)$. Technically, the technique is an odd form of LC. Pressure had little effect on retention or selectivity. Since the temperature was low, the fluid density was always high, typically exceeding 0.9 g cm^{-3}. However, removing the back pressure regulator would result in decompression of the fluid into a gas.

There was no abrupt change in behaviour when the temperature was decreased to below T_c. The pressure drop across the column did not increase (near constant viscosity). Apparent efficiency did not change (D_M/μ constant). Retention times barely changed (marginal change in density). As in other fields, such as pharmaceuticals, and agricultural chemicals, the defined state of the fluid is irrelevant to the chromatography. The advantages of the technique are mostly practical.

Historical Development

The development of the SFC method required many years. Rawdon and Norris at Texaco[5] were probably the first to propose SFC for group separations for fuels. They used a modified HP 1082 LC with independent control of pressure and flow. This was combined with a stand-alone Gow-Mac FID.[6] Their results could not be precisely repeated, owing to the instability of silver loaded columns with which they tried to separate olefins from paraffins. The downstream controlled 1082 was discontinued in 1983.

Later workers abandoned the silver loaded column, and concentrated on the simpler task of separating the aromatics from the other hydrocarbons.

However, a controversy arose which caused considerable difficulty in improving the method for nearly a decade. A number of groups observed what appeared to be catastrophic losses in efficiency in packed column SFC at certain combinations of temperature and pressure. Some workers subsequently repeatedly asserted[7-11] that pressure drops must always be < 20 bar to avoid these problems. An unfortunate consequence was that developers of group separations tried to avoid pressure drops.

The preferred approach used a 50 cm long, 1 mm diameter column packed with 10 μm particles, operated at < 0.1 cm s^{-1}. Such operation produces broad, inefficient peaks with little resolution, and very long analysis times.

Surprisingly, pressure drops seldom actually cause efficiency losses.[12-14] Modern group separations use shorter columns with smaller particles. Analysis times are *ca.* 20 times faster with much better resolution and peak shapes.

The earlier bizarre acceptance of horrific performance to avoid a problem which didn't exist is difficult to understand today, except in the context of the hardware and concepts available at that time.

Upstream vs. *Downstream*

A recurring theme in this book has been the superiority of instruments with independent control of pressure and flow. This superiority is demonstrated with the historical development of group separations.

After many years of development, an upstream mode, FID based group separation was near adoption as an ASTM method.

However, despite years of effort the method was less than ideal. The upstream mode inherently controls density (pressure at fixed temperature) at the head of the column. To change μ at constant inlet density, the **restrictor** must be changed! To generate a Van Deemter plot with the same reference pressure, many different restrictors would be required. Changing fixed restrictors is time consuming and requires a significant level of physical dexterity. This procedure requires at least an hour per point in a Van Deemter curve.

Because of the difficulties in changing flow at constant pressure, most workers settled for changing the pressure to change flow and use one restrictor. Changing the head pressure changes density (and D_M), k', and μ, while removing any reference point between measurements. The p_o is unknown. Such results are equivalent to trying to solve one equation with multiple unknowns. The worker is left with a confusing hodgepodge of results that are in fact not directly comparable.

In contrast, direct control of the flow rate in the downstream mode allows complete, reproducible Van Deemter plots to be generated within a few hours. This made it immediately obvious that shorter columns, with smaller d_p at higher μ, dramatically shortened t_R while improving R_s. The configuration allows direct measurement of p_o. The point with the poorest solvent strength (p_o and T) is monitored and controlled. Flow and dP no longer changes with time. Over all, the technique ceases to be an art.

The downstream mode revealed the optimum conditions, cutting t_R nearly 10 times while improving R_s > 2 times. This allowed an almost immediate, dramatic improvement in **upstream** performance. The fact that these improvements occurred **only** because the downstream mode clarified the trade-offs has not been fully appreciated.

Once the trade-offs were apparent, either method can produce acceptable performance. However, systems relying on fixed restrictors still produce a slow drift in t_R as the restrictor plugs. With the back pressure regulator controlled system, no such drift occurs.

The fixed restrictor feeding the FID in the downstream mode does slowly plug. However, this restrictor does not control system mass flow. The conditions in the column (T_r, R_s, μ) are constant, since they depend on the pump control of mass flow rate, and the back pressure regulator controls p_o. Area measurement and the relative areas of the different groups are determined from the signal from one detector. Since the area measurement is relative, the very slowly changing restrictor flow has almost no impact on the ratio of one peak to another within one analysis. If restrictor flow drops, both the total hydrocarbons and aromatics decrease in the same manner.

2 High Temperature GC *vs.* SFC

SFC has other uses in the petroleum industry in addition to group separations. Previously unusable high boiling residues are becoming a significant source of raw material, but very high boiling, heterogeneous materials are difficult to analyse. Researchers appear to be divided into two camps on how to separate such materials.

Some workers promote the use of very high temperature GC (HTGC). Others believe that HTGC potentially alters samples, and note the poor stability of columns at T > *ca.* 350 °C. Some[15,16] have promoted the use of SFC as a replacement for HTGC. SFC has an advantage over LC in that CO_2 is compatible with the FID. Thus, the detector of choice can be used at low temperatures to separate larger hydrocarbons.

Polywaxes are alternate numbered hydrocarbons, used as molecular weight markers in simulated distillation. Chromatograms of several polywaxes were shown, previously, in Figure 6.7. The number (*e.g.* Polywax 500) represents the average molecular weight in the sample. Homologues at least up to C_{120} have been eluted from both packed and capillary columns.

3 High Efficiency Packed Column SFC

High efficiency packed column SFC applied to petroleum based samples has only been reported in passing.[12] Normally GC columns with 250 000 plates are compared with SFC columns with ≤ 20 000 plates. Such comparisons make SFC appear artificially inefficient. Packed SFC columns can produce efficiency comparable to 250 μm GC capillary columns with only a modest time penalty. There are several reasons why an analyst might be interested in such capability. The SFC separations will exhibit different chromatographic selectivity com-

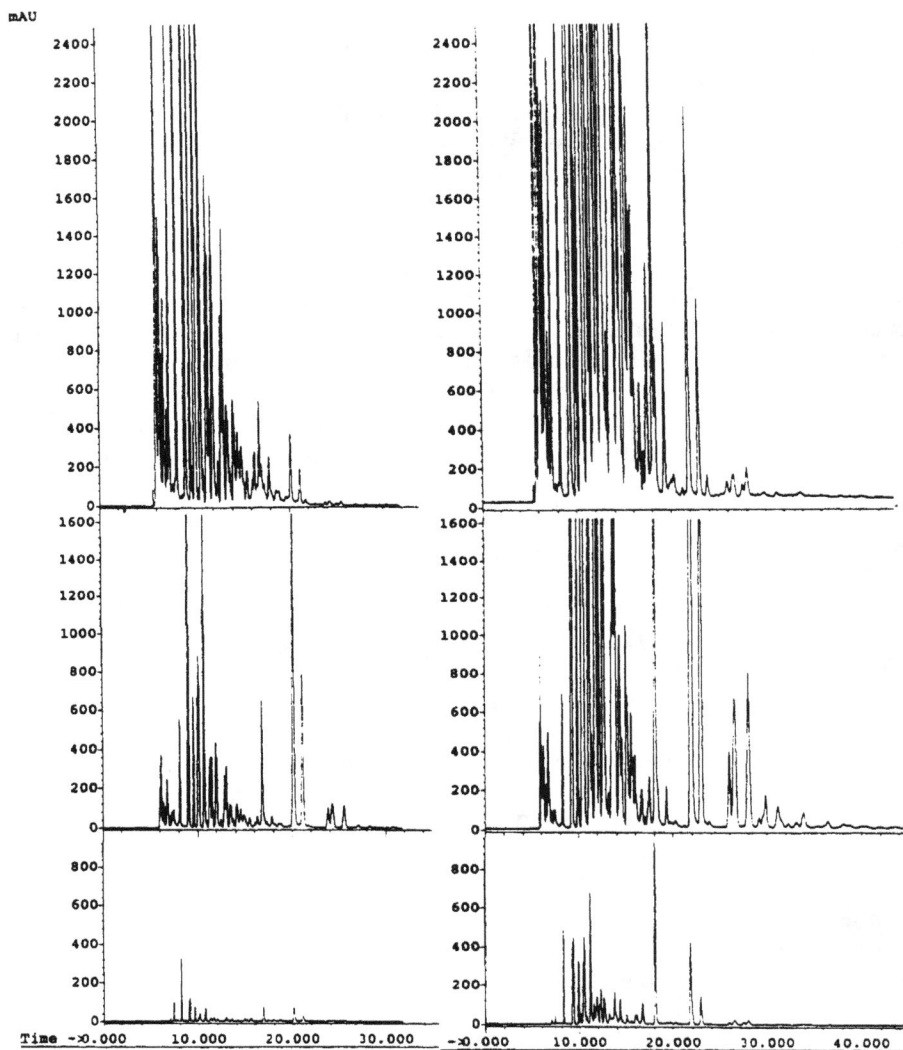

Figure 11.6 *High efficiency packed column SFC of two different gasolines (left and right) using UV detection at 190 (top), 210 (middle), and 254 (bottom) nm. The column was 1.4 m long, packed with 5 μm Hypersil silica. Flow was 2 ml min⁻¹ of CO_2, 60 °C, initially 70 bar*

pared with GC. In addition, other detectors can be used which are not very common in GC. These include the UV and fluorescence detectors from LC. Combinations of detectors provide unique insights. The UV or fluorescence detector can be used simultaneously with the FID, NPD, and/or the SCD to observe carbon mass, and/or the distribution of nitrogen, and/or sulfur in a high resolution chromatogram.

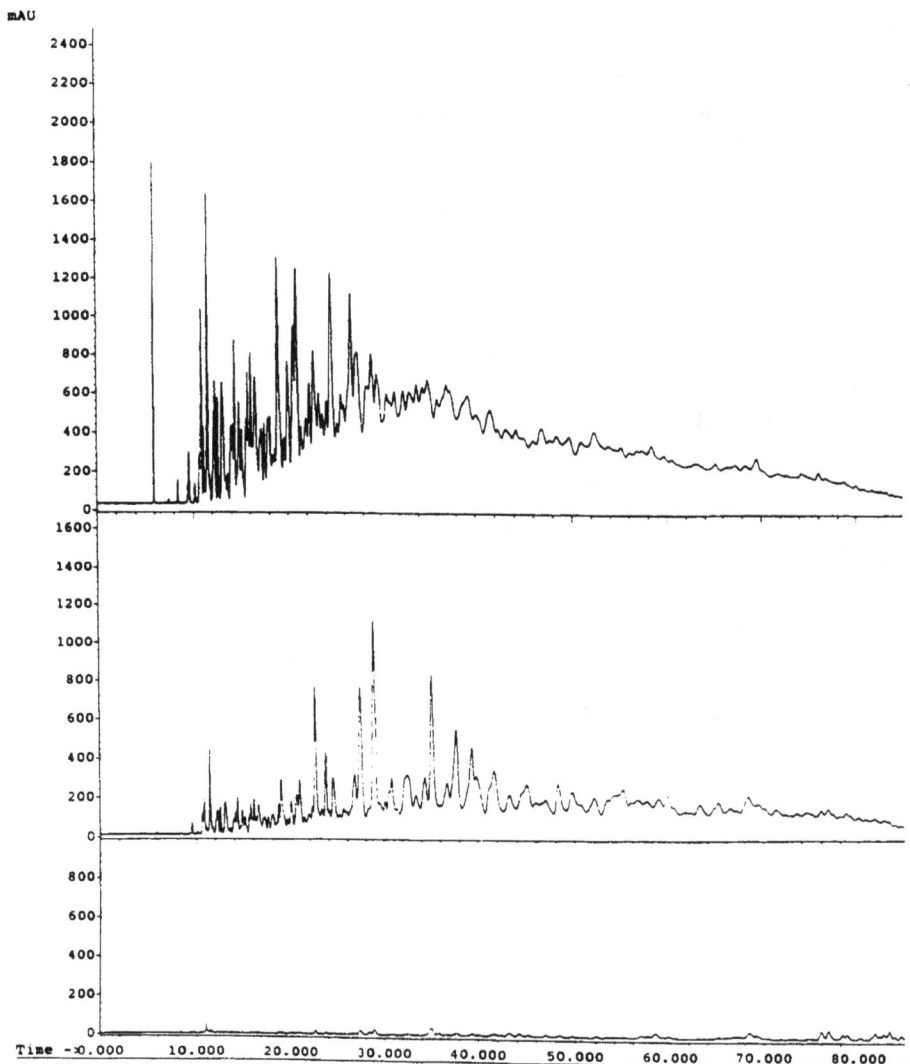

Figure 11.7   *High efficiency chromatogram of a diesel fuel collected under conditions similar to those of Figure 11.6*

Figures 11.6a and 11.6b are chromatograms of several gasolines collected using relatively long packed columns. Efficiency approaches that found in GC. Each injection was monitored at three UV wavelengths chosen to respond primarily to: aromatics (at 254 nm), aromatics plus olefins (at 220 nm), and aromatics, olefins, and saturates (at 190 nm). Of course response factors are dramatically different for the different classes at the different wavelengths.

Since the mobile phase was pure CO_2, the FID could be used simultaneously to produce a total carbon chromatogram to supplement the partial speciation possible with the UV detector. Chromatograms, like those found in **Figure 11.6**, have some role to play in the analysis of these sorts of samples but are likely to provide only supplemental or confirmatory information resulting from other detection options not normally available in LC.

Similar chromatograms can be collected for diesel fuels, as shown in Figure 11.7. Such chromatograms have not previously been published. With such a chromatogram it is clear that there are hundreds of different aromatic compounds present.

Gasoline and diesel chromatograms collected by GC are often based on boiling point. The separations using SFC can be based on other factors such as shape. This different selectivity may prove to be of value. The other aspect of the SFC chromatogram that should be of interest is the UV detector trace. Although UV spectra can be taken in GC, the process is far from ideal.

Another related separation that may be of interest involves polycyclic aromatic hydrocarbons (PAH). They can be separated with high efficiency[12], as shown in Figure 11.8. Compounds from the chimney of a wood burning stove have also been separated,[12] and indicate a wide range of aromatic compounds are present.

4 Several Instrumental Concerns Seldom Mentioned in Print

Characteristics of the FID

The full flow from a 1 mm i.d. column operated at μ_{opt} is *ca.* 50–100 ml min^{-1}, measured as an expanded gas. The flow from a 2 mm column is four times

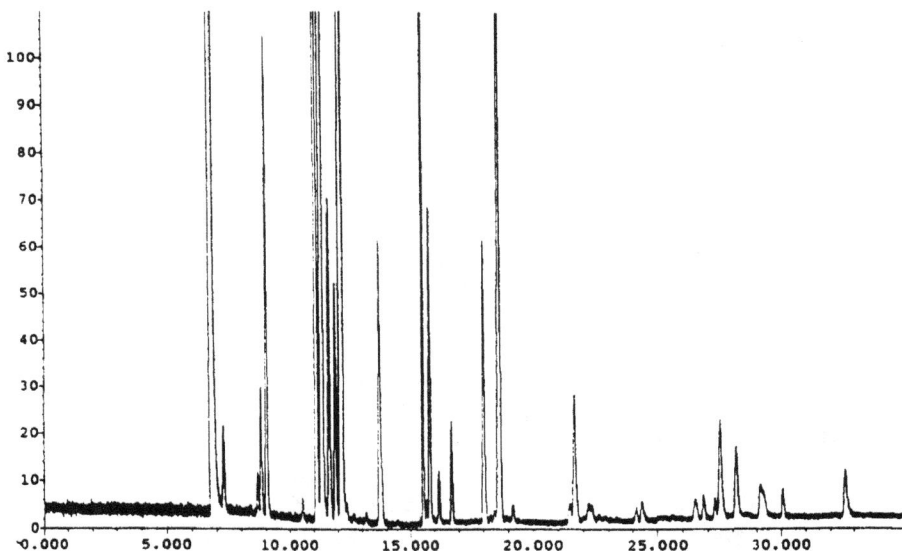

Figure 11.8 *High efficiency separation of selected polycyclic aromatic hydrocarbons by packed column SFC*

greater. Such flows are much larger than in capillary GC and require substantially different detector operating conditions.

The most obvious change is to increase the jet diameter. The jet orifice used in capillary GC is 0.28–0.46 mm (0.011–0.018 inches) in diameter. Packed column GC jets are up to *ca.* 0.76 mm (*ca.* 0.030 inches) in diameter. These larger jets also work well in SFC. With large carrier flows, the flow rates of the flame gases must also be increased. The presence of large volumes of CO_2 tends to dilute and expand the flame.

Although not of great concern in the analysis of finished fuels, which contain only relatively light hydrocarbons, both the restrictor type and detector heated zone design can result in discrimination against heavy components.[17,18]

Linear and, to a lesser extent, frit restrictors[19] drop the pressure over an extended distance. The density of the fluid, and its solvating power, drops along the length of the restrictor. For light hydrocarbons the increased μ caused by the fluid expansion tends to sweep components off the restrictor walls and into the flame. For very heavy hydrocarbons, the increased μ, caused by the fluid expansion, cannot make up for the loss of solvating power of the fluid. Some solutes may be lost on the restrictor walls. Integral restrictors[20] cause fewer problems with solute precipitation but are easier to plug. When working with very heavy solutes, the integral type should probably be preferred.

If the fluid is heated before it reaches the end of the restrictor, the local density drops. For compounds with extremely high boiling points, higher thermal energy may not make up for the loss in solvating power resulting from the drop in pressure and increase in temperature. Heavy molecules can drop out of solution and stick to the walls of the restrictor. A short heated zone minimizes such losses. Several manufacturers have commercially introduced such short zone detectors.

Without proper attention to details, the response factor of the FID to different groups of hydrocarbons can vary dramatically. By purposely using mismatched detector flows, the response factors to aromatics and aliphatics can be made to differ by as much as 30%. With appropriate matching of carrier and flame gases as well as geometry, the response factors for the different groups can be made virtually identical. The user should verify response factors when using a new method or instrument.

The Problem of Changing Split Ratio with Pressure Programming

Flow through fixed restrictors increases when pressure increases. During a pressure program, with the column effluent split between an FID and a UV detector, the total flow remains constant but the flow into the fixed restrictor in the FID increases. Under these conditions, the FID 'response factor' ceases to be a near constant. Apparent molecular weight distributions are distorted, and indicate that larger amounts of heavier components are present than is actually the case.

There are several ways to deal at least partially with this problem. One could calibrate response *vs.* pressure and multiply all the results by correction factors. One could program the detector temperature to try to maintain constant mass flow through the restrictor.[17] One could use a second high pressure source in a sheath flow restrictor[21] to try to maintain a constant split ratio. Finally, one could substitute a negative temperature ramp[22] for a pressure ramp to produce increasing density vs time. None of these approaches is particularly appealing.

It might appear that an operation mode with all the column flow going into the detector offers superior performance in terms of response factors. At least several FIDs are available that can accept the full column flow from a 1 or even 2 mm i.d. column without the flame blowing out. However, such operation is not without its problems. Back pressure regulators tend to cause serious band broadening if used in front of an FID. Fixed restrictors are irreproducible, no two being exactly the same. Two different restrictors in either the same or different instruments will give slightly different retention times. Retention also gradually changes over time, due to plugging. More importantly, pressure programming causes major changes in column flow, dramatically changing peak widths and efficiency, and potentially changing the detector response factor.

5 References

1. M.S. Klee and M.Z. Wang, Hewlett-Packard application note 228–226, 1993, see also M.S. Klee, M.Z. Wang, Hewlett-Packard application note 228–167, 1992; V. Giarrocco, M. Klee, Hewlett-Packard application note 228–231, 1993.
2. A.L. Conrad, *Anal. Chem.*, 1948, **20**, 725.
3. D.W. Criddle and R.L. LeTourneau, *Anal. Chem.*, 1951, **23**, 1620.
4. J. Nawrocki and B. Buszewski, *J. Chromatogr.*, 1988, **499**, 1.
5. T.A. Norris and M.G. Rawdon, *Anal. Chem.*, 1984, **56**, 1767.
6. M.G. Rawdon, *Anal. Chem.*, 1984, **56**, 831.
7. P.J. Schoenmakers and L.G.M. Uunk, *Chromatographia*, 1987, **24**, 51.
8. P.J. Schoenmakers, in 'Supercritical Fluid Chromatography', ed. R.M. Smith, RSC Chromatography Monographs, The Royal Society of Chemistry, London, 1988, Chapter 4.
9. H.G. Janssen, H.M. Snijders, J.A. Rijks, C.A. Cramers and P.J. Schoenmakers, *J. High Resolut. Chromatogr.*, 1991, **14**, 438.
10. P.A. Mourier, M.H. Caude and R.H. Rosset, *Chromatographia*, 1987, **23**, 21.
11. D.P. Poe and D.E. Martire, *J. Chromatogr.*, 1990, **517**, 3.
12. T.A. Berger and W.H. Wilson, *Anal.Chem.*, 1993, **65**, 1451.
13. T.A. Berger, *Chromatographia*, 1993, **37**, 645.
14. T.A. Berger and L.M. Blumberg, *Chromatographia*, 1994, **38**, 5.
15. H.E. Schwartz, J.W. Higgins, and R.G. Brownlee, *LC–GC*, 1986, **4**, 639.
16. H.E. Schwartz, R.G. Brownlee, M.M. Boduszynski, and F. Su, *Anal. Chem.*,1987, **59**, 1393.
17. T.A. Berger and C. Toney, *J. Chromatogr.*, 1989, **465**, 157.
18. T.A. Berger, *Anal. Chem.*, 1989, **61**, 356.
19. K.E. Markides, S.M. Fields, and M.L. Lee, *J. Chromatogr. Sci.*, 1986, **24**, 254.
20. E.J. Guthrie and H.E. Schwartz, *J. Chromatogr. Sci.*, 1986, **24**, 236.
21. D.E. Rainie, K.E. Markides, M.L. Lee, and S.R. Goates, *Anal. Chem.*, 1989, **61**, 1178.
22. M. Novotny, W. Bertsch, and A. Zlatkis, *J. Chromatogr.*, 1971, **61**, 17.

CHAPTER 12

Miscellaneous Applications of Packed Column SFC

1 Introduction

This chapter attempts to provide an overview of more traditional uses of SFC for those readers disappointed by the limited discussion of these topics in most of this book.

This chapter contains representative packed column SFC chromatograms. Much of the earliest work in SFC involved the separation of oligomers and light polymers, performed on packed columns, up until the early 1980s. However, most of this early work use the poorly packed, large and irregular particles typical in LC 20 years ago. It has only been during the last 15 years or so that modern packings have been widely used.

2 Specific Separations

Polystyrenes

Klesper, the first to use SFC, has published extensively on the separation of polystyrene oligomers, including some separations of > 100 000 molecular weight. Much of his early work used more polar modifiers, such as 1,4–dioxane, in light hydrocarbons, such as n-pentane, at elevated temperatures. A fairly late chromatogram[1] of molecular weight = 2200 is presented in Figure 12.1. The three traces represent pure pentane at the top, 5% MeOH in pentane in the middle, and 10% MeOH in pentane at the bottom. Pressure was programmed at 3.3 bar hr^{-1}, temperature was 220 °C. Note that the addition of modifier caused retention to increase. With CO_2 as the main fluid, retention decreases with increasing modifier concentration.

More recently, a micropacked column was used[2] to resolve up to 70 repeating units of polystyrene oligomers, from a higher molecular weight mixture, as shown in Figure 12.2. It is unclear from the chromatogram whether the largest oligomers actually eluted. The last 'bump' on the right hand side of the chromatogram involves polystyrene oligomers with an average molecular weight range of 9000. The 70-mer should have a molecular weight less than 8000.

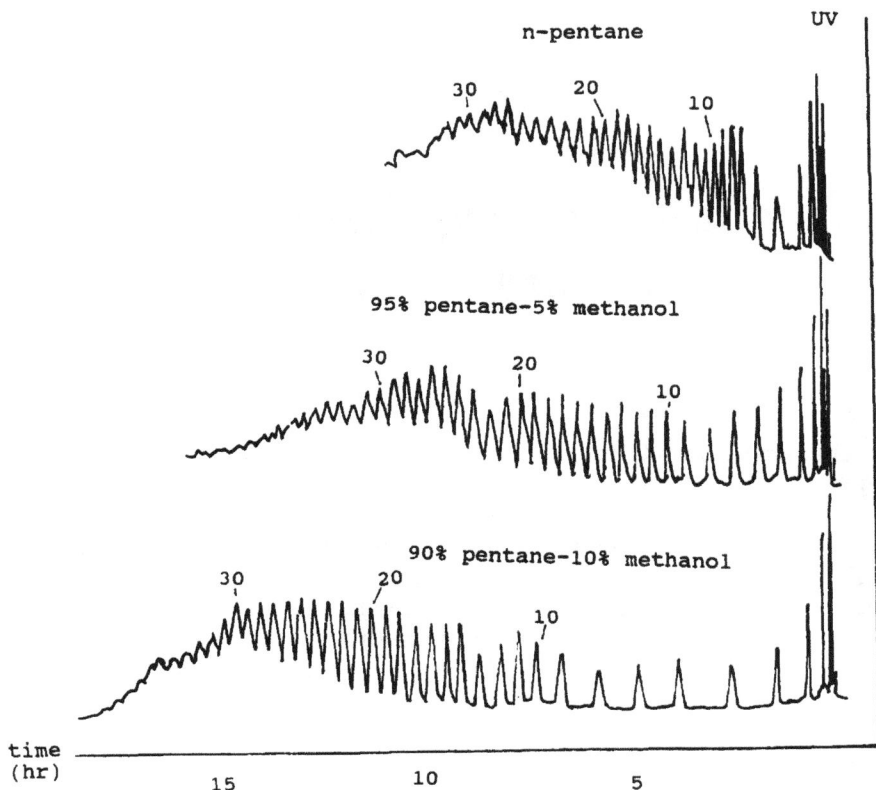

Figure 12.1 *Separation of Polystyrene 2200 showing the effect of methanol modifier when hexane is uses as the main fluid*

This author eluted polystyrenes with up to 3200 average molecular weight using MeOH/CO_2 mixtures at > 90 °C. Larger polystyrenes precipitated in the UV detector flow cell which was at near ambient temperature. The system overpressured and shut down. It was difficult to remove the precipitate from the flow cell and plumbing. This behaviour should not be surprising since MeOH is often used to precipitate polystyrenes from dioxane mixtures during polymerization.

Klesper has also presented chromatograms of other polymers similar to polystyrenes. Chromatograms of 2–vinylnaphthalene, collected using a 4.6 mm i.d. column packed with 10 μm Lichrosorb Si60 or Si100, and 1,4–dioxane/pentane mixtures, is presented[3] in Figure 12.3. The chromatogram indicates that significant progress was made in analysis time between 1978 and 1987.

Silicone Oils

Other work[4] also demonstrates the early influence of Klesper since it also used supercritical light hydrocarbons as the main part of the mobile phase. Several

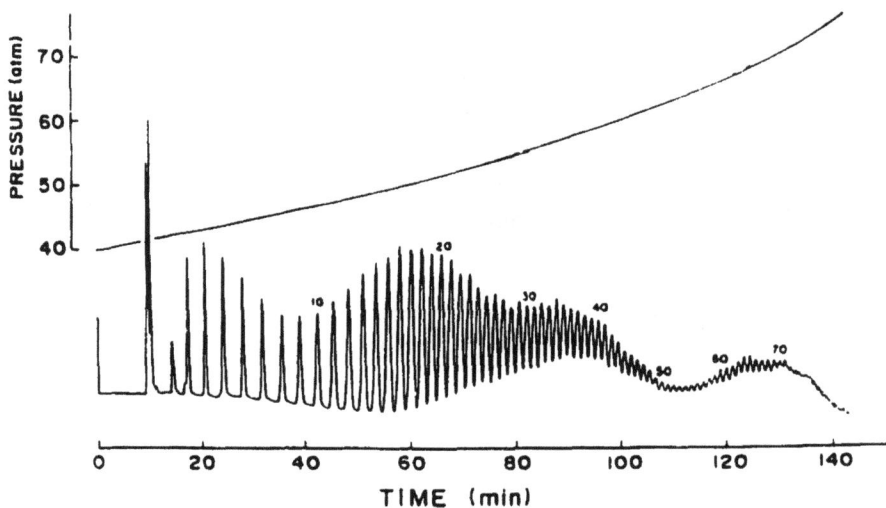

Figure 12.2 *A micropacked column separation of an artificial mixture of Polystyrenes including 580, 2100, 4250 and 9000 average molecular weight*

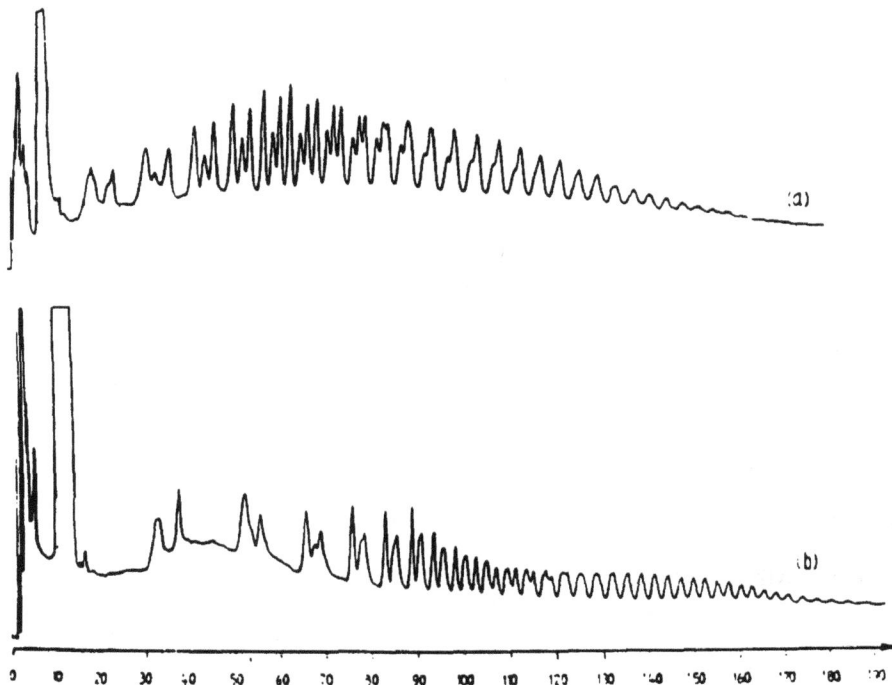

Figure 12.3 *Separations of 2–vinylnaphthalene oligomers*

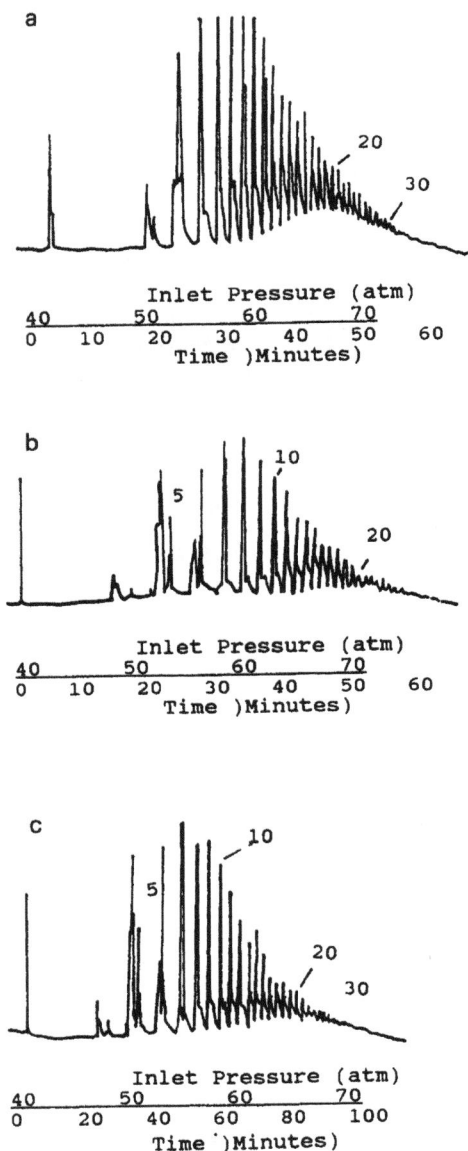

Figure 12.4 *Separations of the silicone oil DC-710*

polymethylphenylsiloxanes were separated on packed columns. The column was 200 μm i.d., 50 cm long, packed with Finesil C18–10. The mobile phase was 10% ethanol in hexane at 260 °C. The UV detector had a cell volume of 0.1 μl. Chromatograms of DC-710 are presented in Figure 12.4, and chromatograms of OV-17 in Figure 12.5.

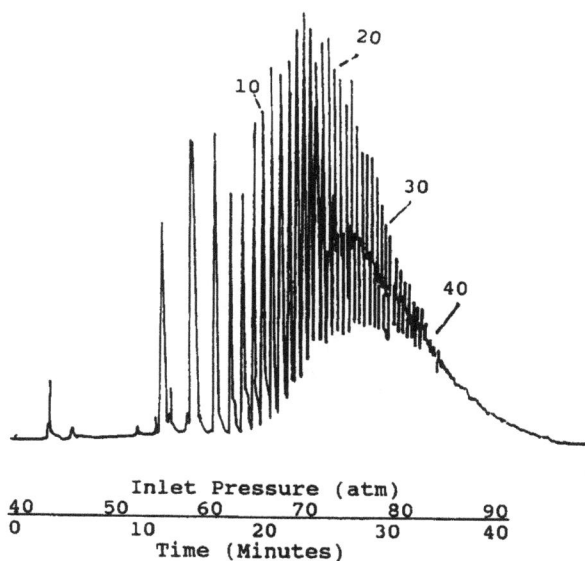

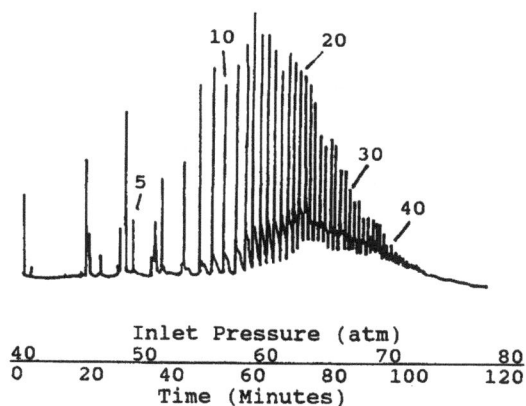

Figure 12.5 *Separations of OV-17 often used as a stationary phase in GC*

More recently, the trend away from the use of hydrocarbons at high temperature was demonstrated[5] with another, but much lighter, silicone oil. It is unclear what upper limit of molecular weight can be solubilized using these more benign fluids. DC-200 was separated on a 20 cm, 320 μm i.d., micro-packed column with 5 μm TVBS coated particles. The MP was pure CO_2, programmed from 15 to 32 MPa, over 120 minutes, at 80 °C, with FID detection. More than 60 peaks were apparent, as seen in Figure 12.6. The authors state; '...a result somewhat better than achieved by capillary column SFC'.

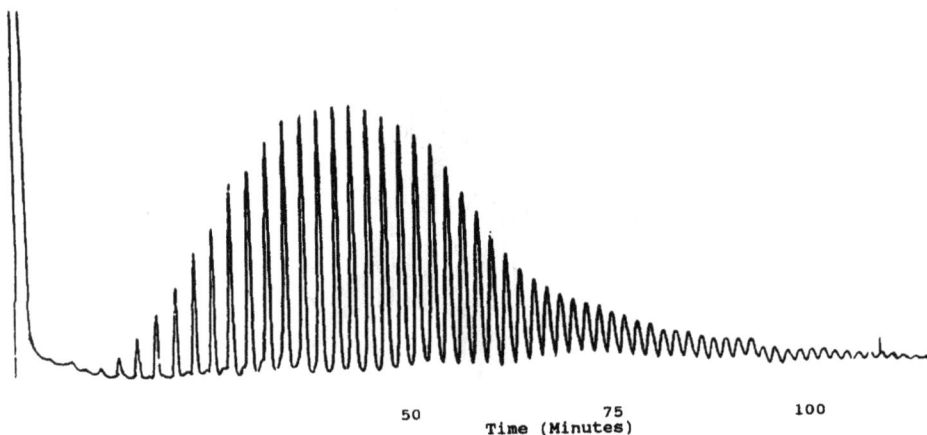

Figure 12.6   *Micropacked column separation of silicone DC-200*

Methyl Vinyl Silicone Stationary Phases–Peroxide Reaction Mixtures

Organic peroxides are often used in making polymers. Both dibenzoyl peroxide and dicumyl peroxide are chain reaction initiators for polymerization processes. SFC allows reaction mixtures to be studied to determine the decomposition product left after polymerization. Figure 12.7 (actually four chromatograms) shows that dicumyl peroxide leaves more benign reaction products than dibenzoyl peroxide in reactions with both SE-54 and SE-33.

Methyl Methacrylates

Methyl methacrylate oligomers were separated[7] and the fractions collected for further, off-line analysis. The eluent was ethanol in CO_2. Dichloromethane was apparently also added to the mobile phase. The column used was 1.7 × 250 mm, packed with 5 μm ODS. An unusual aspect of this separation was the use of a constant outlet pressure (155 kg cm^{-3}) and composition (25 μl min^{-1} EtOH, 300 μl min^{-1} CO_2), with a temperature program from 110 °C at 4 °C min^{-1}. It was not stated whether this involved a positive or negative temperature ramp but a positive ramp is implied.

A representative separation[7] of the MMA oligomers is shown in Figure 12.8. The collected fractions are underlined in the Figure. These fractions were subsequently dried by warming the collection vial filled with a trapping solvent on a hot plate while a stream of pure CO_2 was passed over the solvent. The dry fractions were then subjected to a wide range of instrumental analyses, such as NMR, MS, and FTIR.

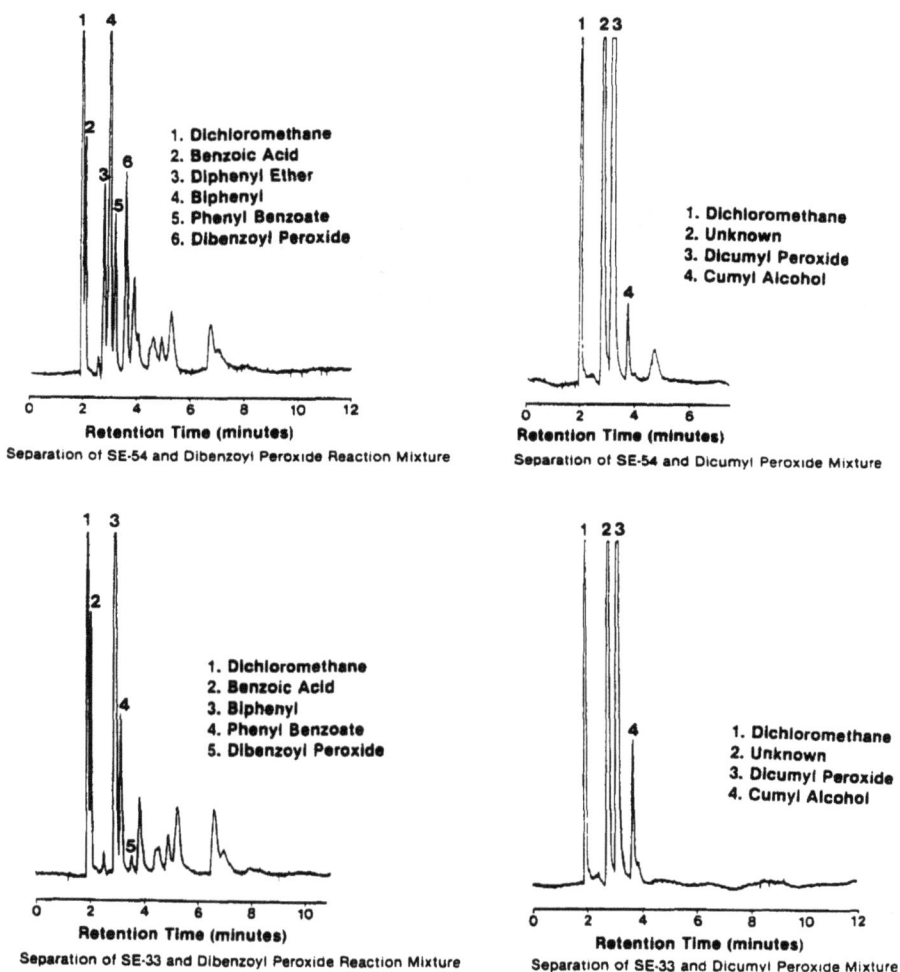

Figure 12.7 *Packed column chromatograms of reaction mixtures showing the reaction products from two different GC stationary phases (SE-54 and SE-33) with two different polymerization initiators*

Epoxy Resins

Ciba Geigy has been working with SFC for more than a decade, and is probably the most sophisticated industrial user. In the last few years they have begun to publish some of their work. In several papers they have reported the separation of epoxy resins by packed column SFC. An example[8] is shown in Figure 12.9. The column was 4 × 125 mm, packed with 3 μm Spherisorb CN. The MP was 0.2 ml min^{-1} MeOH in 2 ml min^{-1} CO_2. The temperature was 90 °C and the pressure program from 130 to 350 bar, 2.5 min hold. Detection was by UV at

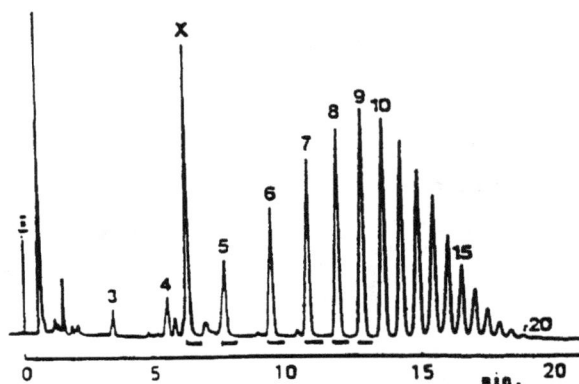

Figure 12.8 *Packed column chromatogram of the oligomers of methyl methacrylate. The underlined fractions were collected, dried and subjected to extensive external analysis with NMR, FTIR, etc.*

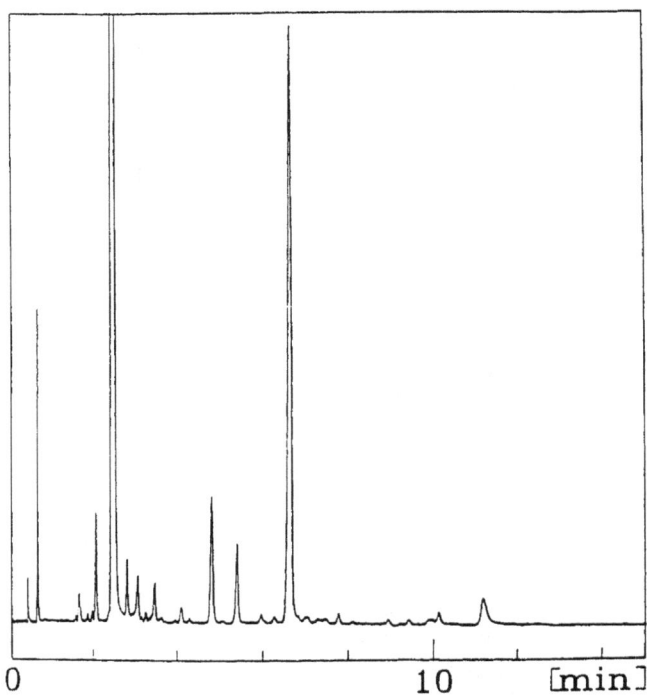

Figure 12.9 *Separation of an epoxy resin. The column was 4×125 mm, 3 μm Spherisorb CN, 0.2 ml min^{-1} MeOH, 2 ml min^{-1} CO$_2$, at 90°C. Pressure was programmed from 130 to 350 bar in 10 min. UV detection at 280 nm*

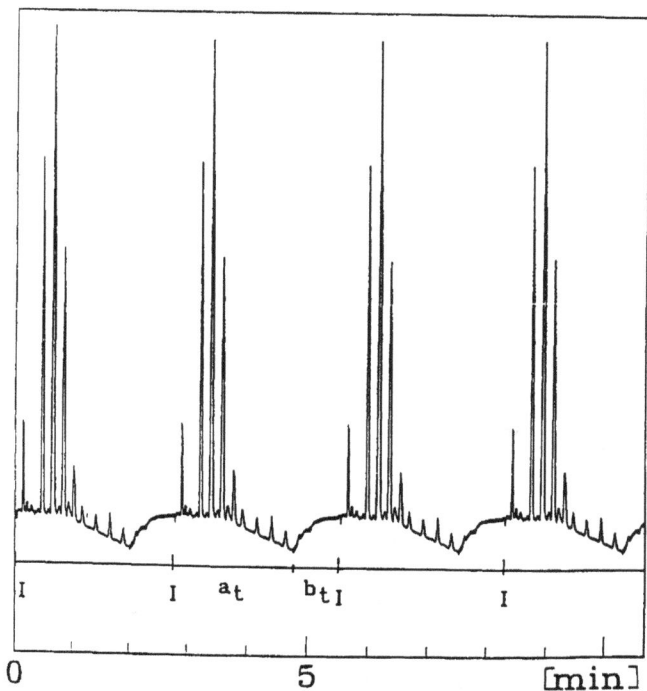

Figure 12.10 *Four consecutive high speed chromatograms monitoring an epoxy resin manufacturing process: see text*

280 nm. The authors state that the analysis was four times faster than their best HPLC analysis.

A second figure[8] shows sequential analyses from an actual epoxy resin manufacturing process (Figure 12.10). Turn around time was 165 seconds and the important components were resolved. This speed allows the process to be monitored in near real time.

Polymer Additives

In another application,[8] Ciba Geigy reported on the separation of polymer additives. Irganox 1425 is a calcium salt. In HPLC it cannot be eluted with the rest of the Irganox family. This requires two HPLC analyses on formulations containing this component. In addition the two analyses require a great deal of organic solvent.

Using pressure programmed, packed column SFC, all the components of formulations, including Irganox 1425, could be rapidly separated in a single analysis in less than 6 minutes, as shown in Figure 12.11. Two columns were used in series. The first column was 4×60 mm, packed with 3 μm Spherisorb C_8. The second column was 4×125 mm, packed with 3 μm Spherisorb CN. The MP was 0.24 ml min^{-1} of 0.15% citric acid in MeOH, 2 ml min^{-1} CO_2, at

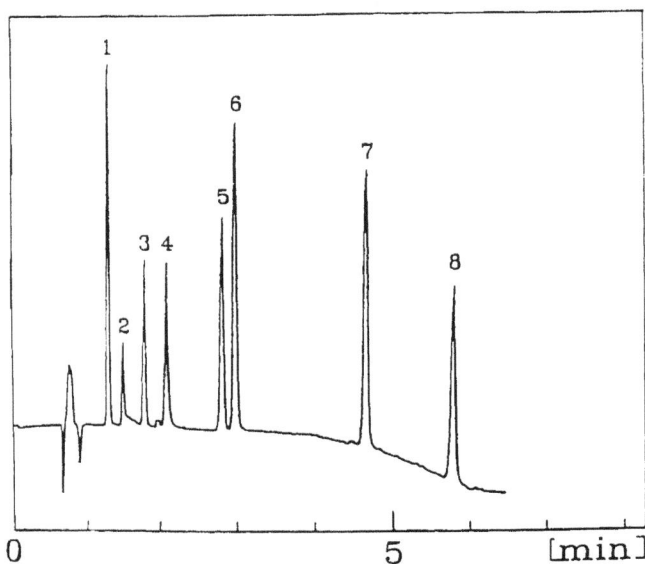

Figure 12.11   *Chromatogram of polymer additive. Irganox 1425 (peak 4) is a calcium salt which causes problems in LC, requiring a separate analysis. With packed column SFC, all the solutes were rapidly eluted in less than 6 minutes. Key: 1, Metilox; 2, Irgafos 168; 3, Irganox 1076; 5, Irganox 259; 6. Irganox 1330; 7, Irganox 1010; 8, Irganox 1098*

90 °C. Pressure was programmed from 130 to 200 bar in 4 minutes then 200 to 350 bar in 2 minutes. The detector was UV at 278 nm.

High Speed Chromatograms

Several mixtures of polar solutes were resolved[9] on 1.5 μm pellicular silica particles coated with a diol phase. The column was 3 cm long, and was operated at 5 ml min^{-1}. On the left of Figure 12.12, benzoic, phthalic, and trimesic acids were separated in 7 s. On the right, testosterone, estradiol, hydroxycortisone, and estriol were resolved in less than 9 s. The chromatograms were actually broadened by the 4 Hz bandwidth of the HP3390 integrator used to collect the chromatograms shown. Using a high speed X–Y plotter, they were baseline resolved (not shown) and the fastest peaks produced over 2700 plates s^{-1}.

Triton X-100

At one point, virtually every paper about SFC included an almost mandatory chromatogram of Triton X-100, a widely used surfactant. The first such chromatogram, by Gere,[10] resolved at least 20 of the oligomers in *ca.* 15 min, as shown in Figure 12.13. Separation was on a 4.1 × 150 mm, 10 μm PRP-1 polymer based column. The MP was 3.2 ml min^{-1} MeOH modified CO_2, programmed from 1 to 5%, at 60 °C, 307 bar.

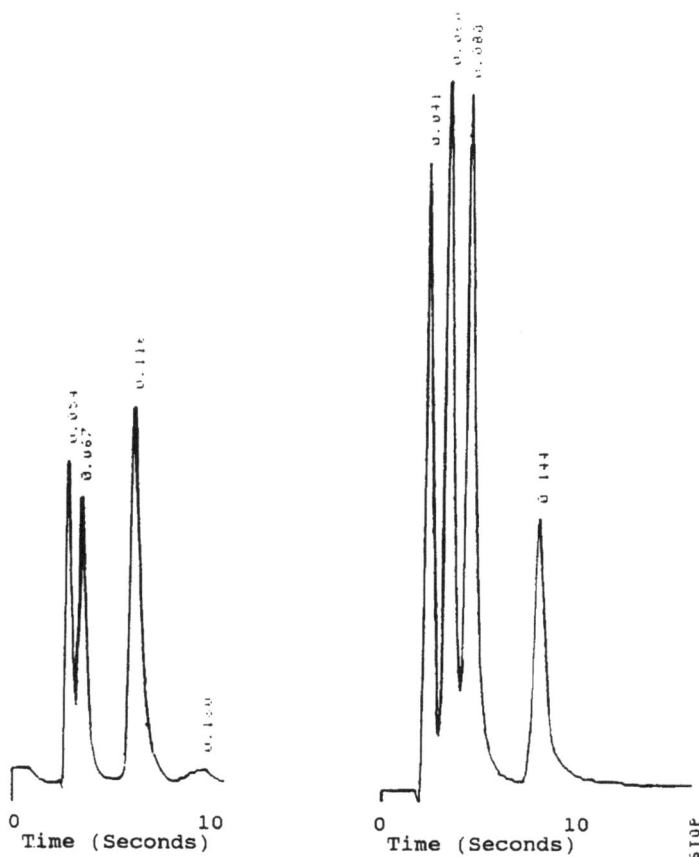

Figure 12.12 *High speed separation of (left) acids, (right) hydroxysteroids using a 4 × 40 mm, 1.5 μm pellicular diol column (provided by C. Horvath)*

A more recent chromatogram[8] of Triton X-165, by both packed and capillary SFC, is shown in Figure 12.14. The packed column chromatogram was obtained from a small bore, short silica column (2 × 100 mm, 3 μm Nucleosil C_{18}) with MeOH/CO_2 (0.15 ml min^{-1} MeOH, 2 ml min^{-1} CO_2) but at 170 °C. Pressure was programmed from 130 to 375 bar over 12 minutes. The last peak actually emerged before the pressure reached 300 bar.

In independent work in this laboratory,[11] it was found that Triton X-100 would not completely elute with the conditions specified above. Much higher modifier concentrations were required (> 15%) to elute the last oligomers. It is speculated that the chromatograph used to generate Figure 12.14 was a converted LC without an extended compressibility range. Consequently, the increasing pressure actually may have caused the CO_2 flow to drop. This would result in an inadvertent, uncontrolled (but reproducible) flow and composition program. A later commercial version of this chromatograph, modified for SFC, employs an extended compressibility adjustment range and does not suffer from this problem.

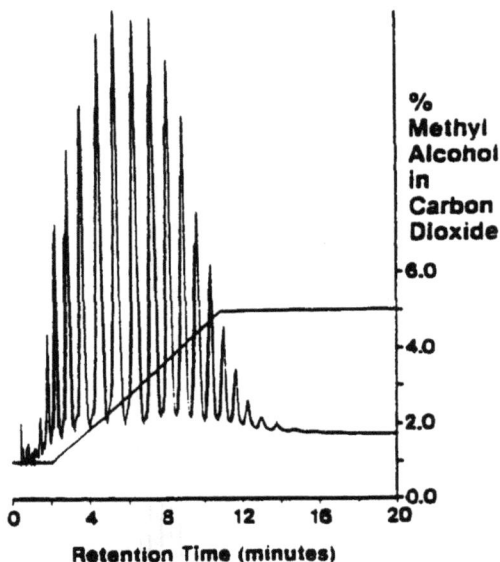

Figure 12.13 *The first Triton X-100 chromatogram that was so widely copied that it became a cliché. The separation was performed on a polymer based column because difficulties were encountered with silica*

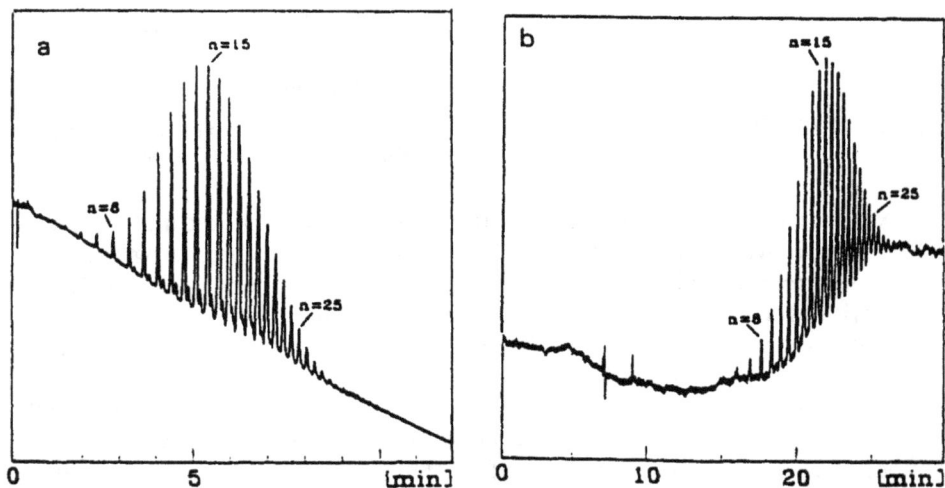

Figure 12.14 *Triton X-165 separated on* (a) *a packed column and* (b) *a capillary column*

Note the difference in the time scale between the packed and capillary chromatograms. This is a direct result of the larger diffusion distance in the capillary column.

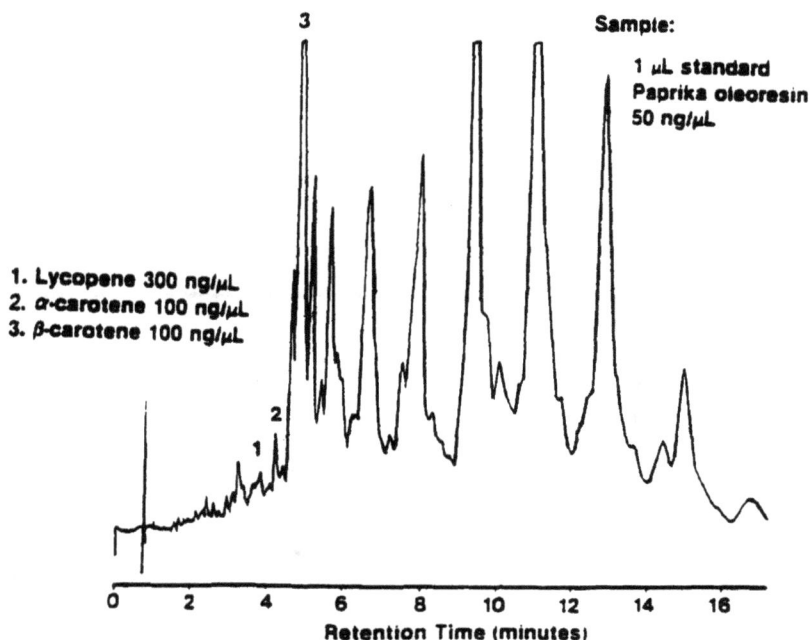

Figure 12.15 *Separation of an oil extracted from the paprika pepper showing large amounts of β-carotene*

Natural Products

Isoprenoids are often coloured molecules, which can also have pleasant fragrances. In addition, Vitamins A and K_2 are isoprenoids which have been separated by SFC. Lycopene and β-carotene supply the colour to tomatoes and carrots, respectively. Paprika oleoresin is an extract from a sweet red pepper. It contains small amounts of lycopene and α-carotene, and much larger amounts of β-carotene, as shown[12] in the chromatograms in Figure 12.15.

Tocopherols (Vitamin E) detected in a 2 μl injection of corn oil are shown separated[11] in the bottom of Figure 12.16. The column was a 4.6 × 200 Hypersil silica. The mobile phase was 2.5 ml min^{-1} of 2% MeOH in CO_2, at 30 °C and 110 bar. Detection was by UV at 295 nm. The three other chromatograms (above the corn oil) are α- (2.51 min), γ (3.44 min), and δ- (3.82 min) tocopherol standards.

Grapefruit oil was injected[13] onto a micropacked column 250 μm i.d. × 25 cm, packed with 5 μm Deltabond CN, as shown in Figure 12.17. Density was programmed from 0.145 g cm^{-3} (10 min hold) to 0.3 g cm^{-3}, at 0.015 g cm^{-3} min^{-1}, 100 °C. FID detection was used. Note the large number of compounds present in the FID trace.

Another citric oil was separated on a high efficiency packed column with UV

mAU

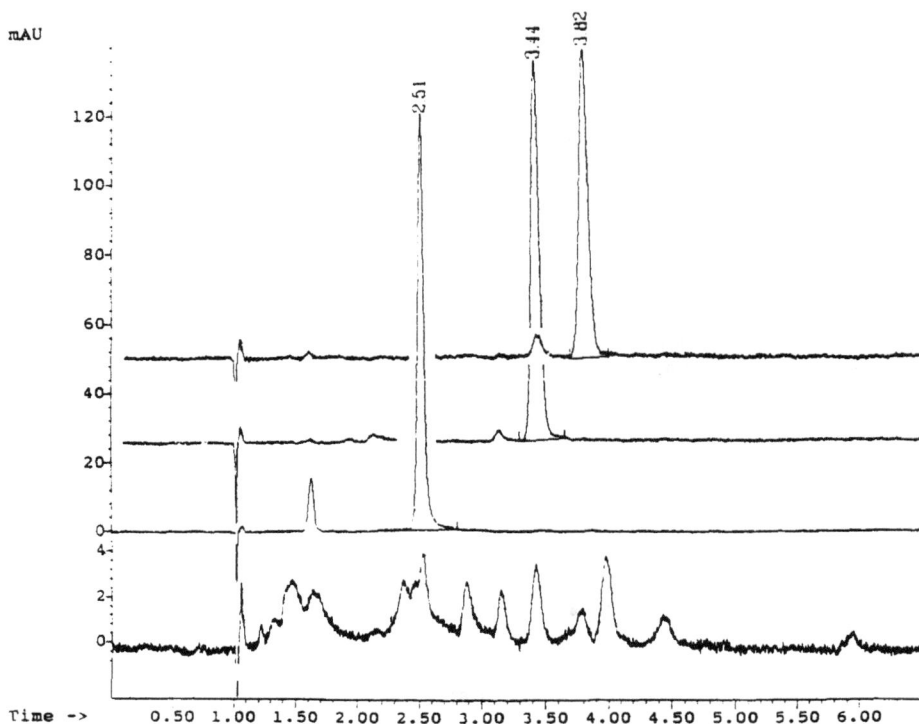

Figure 12.16   *Tocopherols detected in a direct injection of corn oil. The three upper traces are standards of α- (2.5 min), γ- (3.4 min), and δ- (3.8 min) tocopherol*

detection. A Brazilian Lemon oil was separated[14] on a 4.6 × 2000 mm, 5 μm Hypersil silica column with UV detection at 270 nm. The chromatogram was far simpler than the FID trace shown above. The compounds shown in Figure 12.18 are most likely aromatics. Spectra might allow semiqualitative analysis for aldehydes, ketones, *etc.* The conditions were: 2 ml min^{-1} 5% MeOH in CO_2, 60 °C, 150 bar outlet.

Ubiquinones in *Legionella*

Ubiquinones are isoprenoid quinones that are widely occurring in nature, including as constituents of bacterial plasma membranes. One ubiquinone (Q-10) has been widely used as a hypertensive drug in Asia. An earlier name for one ubiquinone was Coenzyme Q. Chromatograms[15] of two different bacterial extracts are presented in Figure 12.19, indicating the dramatic differences in content between subspecies of *Legionella*. Such differences could be used to supplement the more widely used cell wall fatty acid profiling for identification of bacteria.

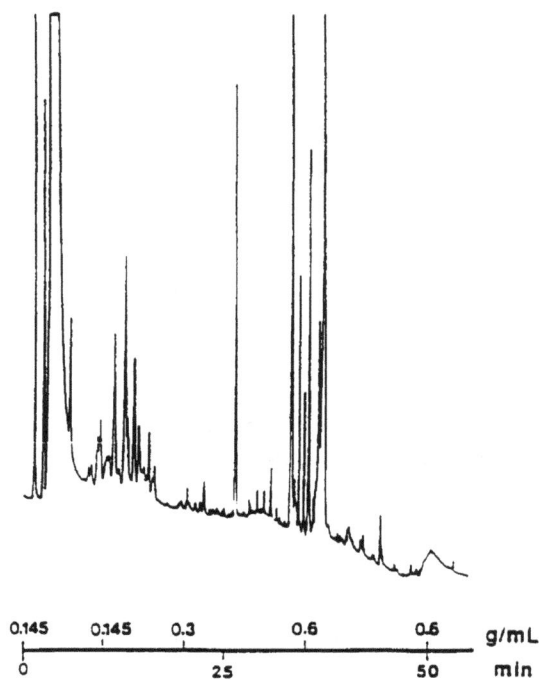

Figure 12.17 *Grapefruit oil separated on a micropacked column with FID detection*

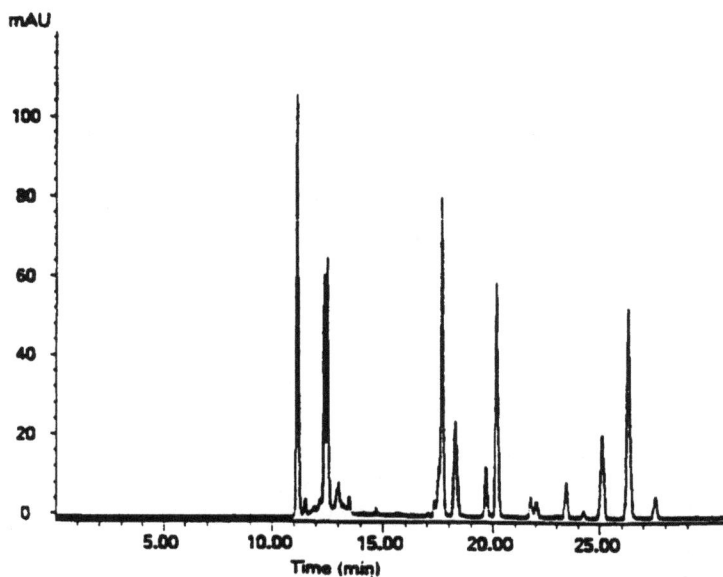

Figure 12.18 *A Lemon oil separated on a 2 m long packed column with UV detection at*
270 nm

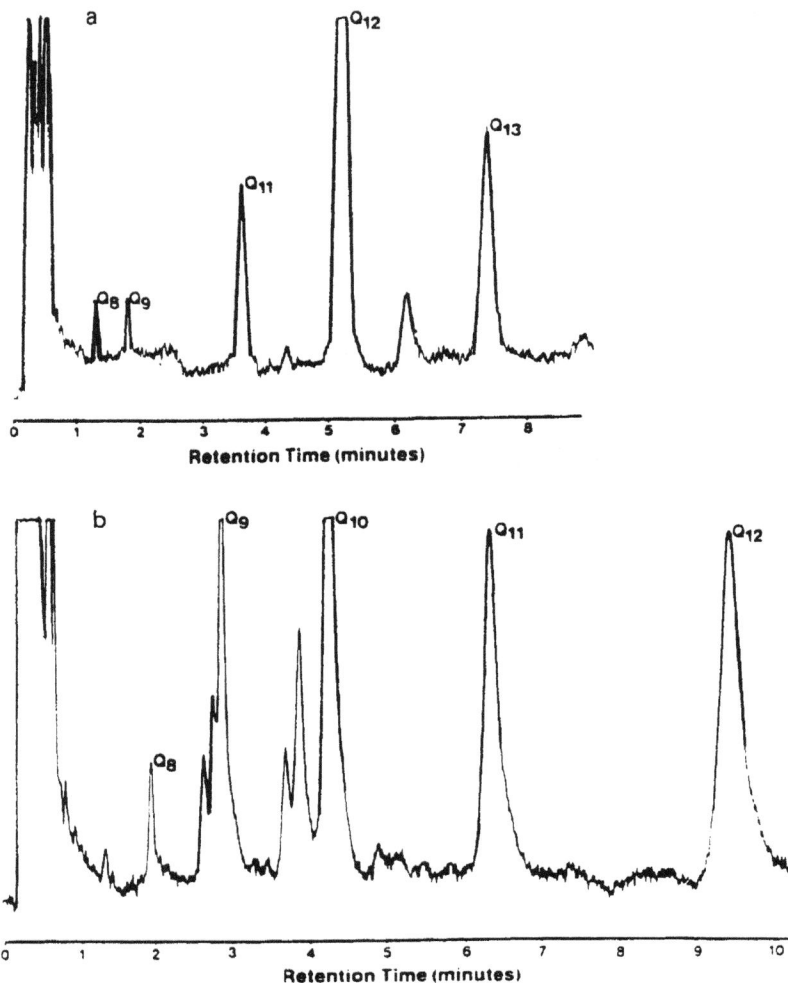

Figure 12.19 *Packed column SFC separation of the isoprenoid ubiquinones found in extracts from two sub-species of* Legionella *bacteria*

Underivatized Fatty Acids

An SFC chromatogram[16] of C_9 to C_{23} fatty acids, separated on a 0.75×150 mm column packed with PRN-300 polystyrene particles, is presented in Figure 12.20. The mobil phase was CO_2 at 170 °C. Pressure was programmed from 200 to 415 atm at 8 atm min^{-1}, after a 3 minute hold.

The checkout sample sent with the original Suprex SFC contained a number of hydrocarbons and underivatized fatty acids. The standard column was 1×100 mm, packed with 5 μm Deltabond C_{18}. A chromatogram[11] of this checkout sample, with FID detection is presented in Figure 12.21.

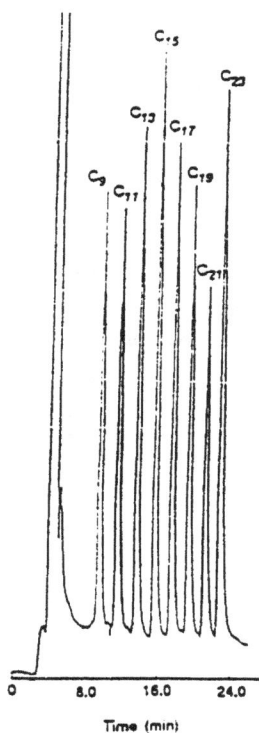

Figure 12.20 *Underivatized C₉–C₂₃ fatty acids separated on a polymer based column. Column: 0.75 × 150 mm, polystyrene PRN-300; CO₂ at 170 °C; pressure: 200–415 atm at 8 atm min⁻¹ after 3 min hold. Last peak emerged at ca. 360 bar*

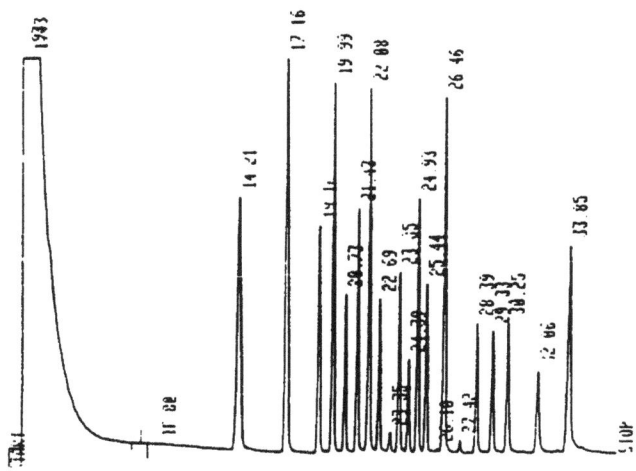

Figure 12.21 *Suprex SFC checkout sample containing hydrocarbons and underivatized fatty acids. Column 1 × 100 mm, Deltabond C₁₈ with pure CO₂, 80 °C, FID*

3 References

1. E. Klesper and W. Hartmann, *Euro.Polym. J.*, 1978, **14**, 77.
2. Y. Hirata and F. Nakata, *J. Chromatogr.*, 1984, **295**, 315.
3. F.P. Schmitz and E. Klesper, *J. Chromatogr.*, 1987, **388**, 3.
4. Y. Hirata, F. Nakata, and M. Kawasaki, *J. High Resolut. Chromatogr.*, 1986, **9**, 633.
5. Y. Schen, H. Huang, X. Liu, Q. Wang, and L. Zhou, *J. High Resolut. Chromatogr.*, 1994, **17**, 74.
6. D.R. Gere, T.J. Stark, and T.N. Tweeten, Hewlett-Packard Application Note AN 800–4, 1983 (out of print).
7. M. Takeuchi and T. Saito, *J. High Resolut. Chromatogr.*, 1991, **14**, 347.
8. A. Giorgetti, N. Pericles, H.M. Widmer, K. Anton, and P. Datwyler, *J. Chromatogr. Sci.*, 1989, **27**, 318.
9. T.A. Berger and J.F. Deye, unpublished results.
10. D.R. Gere, Hewlett Packard Application Note AN 800–3, 1983 (out of print).
11. T.A.Berger, unpublished results.
12. D.R. Gere, Hewlett Packard Application Note AN 800–5, 1983 (out of print).
13. L.T. Taylor and H.-C. K. Chang, *J. Chromatogr. Sci.*, 1990, **28**, 357.
14. T.A. Berger and W.H. Wilson, *Anal. Chem.*, 1993, **65**, 1451.
15. D.R. Gere, Hewlett-Packard Application Note AN 800–2, 1983 (out of print).
16. F.J. Yang, Symposium Workshop on SFC, Snowbird, UT, 1989.

Subject Index

Acebutolol 187

Acids 88, 125, 152, 155, 233, 236–237, 242–243

Active sites 56, 60, 75–76, 84–85, 88–89, 114–115

Additives 18, 84–94
  acidic 87, 88, 90, 92, 128, 173
  basic 118, 129, 152, 159, 163–167, 170, 180
  choosing 93–94

Adenine 74

Adiabatic heating 31

Adsorption 72
  pure carbon dioxide 77–78
  from binary mixtures 80–84

Agricultural chemicals 1–7, 99–101, 192–211
  by capillary SFC 7

Albendazole sulfoxide 187

Alcohols 124

Aldicarb 193

Aldicarb sulfone 193

Aldicarb sulfoxide 193

Alkaloids 159

Ally 201

Alprenolol 187

Amides 126, 187

Amines
  aliphatic 126, 152, 169
  aromatic 125, 152, 169

Amino acids 153

Amino acid derivatives 153, 187, 188

Amino acid esters 187

Amino alcohols 187

Aminoamides 187

Amino-, APS, aminopropyl stationary phase 60, 95, 118, 154, 163

Ammonia 29, 159

Amphetamines 169

Amphetamine sulfate 170

Anilines 125, 152

Anthracene 213

Anticonvulsants 160

Antidepressants 167

Antiparasitic agents 153

APS *see* amino-

Aqueous
  injections 36, 150, 171–2
  with SPE 3, 193, 199

Arginine 154

Aromatic acids 88

Aromatics in diesel 212

Ashai Pac amino column 95

Aspartic acid 154

ASTM D5186–91 212

Atenolol 187, 188

AZT 152

Baccatin 160

Baraphazoline 170

Barbiturates 158

Back pressure regulator (BPR) 24

Bases 12, 60, 62, 74, 85, 117, 125, 126, 129, 152, 169, 170

Baygon 195

Bendroflumethazide 189

Benzamide 74, 129

Benzodiazepam 184

Benzodiazepines 163

Benzoin 188

Benzphetamine 170

Benzylamines 60, 62, 74, 85, 117, 129

Betaxolol 187–189

Bile acids 155

Binary fluids 155
  pre-mixed 25, 27
  mixing 32, 33
  density range in 62, 63

Binders 160

β-Blockers 187, 188, 189

Buclizine 189

Butyl esters 187

C_8-, MOS, octyl-stationary phase 85, 88, 118, 125, 235

C_{18}-, ODS, octadecyl-stationary phase 57, 77, 82, 86, 125, 130, 162, 230, 232, 243
Caffeine 160
Camazepam and metabolites 163, 184, 189
Canadian General Standards Board 216
Capillary electrophoresis 19–20
Capillary LC 8
Capillary SFC 5–7, 12
  figures of merit 18–19
Capsules 156
Carbamates 2–3, 99–101, 193–199
Carbaryl 195
Carbofuran 195
Carbohydrates 157
Carbon dioxide 1
  cylinder pressure 26
  as mobile phase with packed columns 108–109
carotenes, α, β 239
Cephalomannin 160
Cephalosporins 154
Chenodeoxycholic acid 155
Chilled pump 23, 29, 30
Chiral 1, 48, 176–191
Chirapak AD, OD 189
Chirocel AD, OBH, OD, ODH 179, 180, 181–185, 188, 189
Chirex 184
Chloroprothixene 105, 164
Cholesterol 15
Cicloprolol HCl 189
Classic 201
Clustering 53
c_{max}, maximum solute concentration in peak 11
CN- *see* cyano-
c_0, solute concentration in sample 11
Cocaine 169
Codeine 159
Co-enzyme Q 240
Columns 3, 5, 40
  polarity 57, 88, 89
Column diameter 5
Column impedance 13
Composition
  control 23
  programming 6

Compressibility
  adjustment 30, 237
  factor (Z) 24, 31
Corn oil 153
Cortisone 155
Coupled columns 4, 183, 184, 185, 186, 221
CPS- *see* cyano-
Critical point 5, 66
Critical pressure (P_c) 5
Critical temperature (T_c) 5
Crotamitron 159
Cryptopine 159
Cyano-, cyanopropyl-, CN, CPS
  stationary phase 84, 104, 105, 118, 157, 162, 163, 171, 173, 188, 200, 233, 235, 238–241
Cylinder
  storage 26
  location 26
  padded 27
  pre-mixed 27
Cyclobond I 188–189
Cyclodextrin, β, stationary phase 157
Cysteine 154

d_c, inner diameter of a capillary 11, 12, 13
Deltabond
  C_8 85, 118
  C_{18} 243
  CN 84, 118, 239–243
Density 25, 56–59, 109, 110
  of $MeOH/CO_2$ 59, 62–63
Deserpidine 105, 164
Detectors 12, 15, 37, 148, 193
  interfacing GC detectors 39
  selective 12, 15
  universal 12, 15
d_f, film thickness 75
Diamines 126
Dibenzoyl peroxide 232
Dicumyl peroxide 232
Dielectric constant 52
Diffusion coefficients (D_M) 1, 5, 10, 12, 43, 44
  in gases 10
  in liquids 10, 45
  in supercritical fluids 10, 45
  in near critical fluids 47

Diffusivity 1
Diode array 4, 209
Diol stationary phase 2, 60, 74, 117, 118, 125, 129, 152, 161, 172, 199, 201
Dip tube 26
Distribution coefficient (K_D) 73, 74, 113
Diuretics 173
Diuron 199
DNA 152
Downstream control 23
d_p, particle diameter 11
dP, column pressure drop 45, 73

Ecdysteroids 156
Efficiency 9
    effect of pressure 106–107
    effect of temperature 107
    effect of modifier concentration 107
    effect of flow rate 108
Electron capture detector (ECD) 4, 6, 15, 154, 204
Elution strength 55, 56
Elutrophic series 55, 56
Enantiomers 176–191
Environmental Protection Agency (EPA) Method 531.1 2, 99–101, 193
Ephedrine 170
Epoxy resins 233–235
Erythromycine A 154
Estradiol 154
Estratriol, estriol 154
Estrone 15
Ethosuximide 189
Ethylmorphine 159
Excipients 156, 159, 160
Extensive *vs.* intensive interactions 51, 60, 73
Extra column effects 5

F-13 84
Fat soluble vitamines 153
Fatty acids 242
Figures of Merit 9
    column impedance 13
    efficiency, definition 9
    peak capacity 10
    resolution, definition 9
    sample capacity 11

selectivity 10
sensitivity 11
speed 10
Finesil C18–10 230
Fish oil 153
Flame Ionization Detector (FID) 2, 6, 7, 12, 15, 213–218, 224–225, 243
Flow control 23
Flow sensor 34
Fluorescence 3
Flurbiprofen 189
Fluroform (CHF_3) 1
Flux 12
Fructose 157
Functional groups, effect on retention 126–127

Games, D.E. 154, 157
Gasoline 217
GC, gas chromatography
    comparison to SFC 6, 7
    figures of merit 14–15
GC detectors 193
    ECD 4, 6, 15, 154, 201, 204
    ELCD 15
    FID 2, 6, 7, 12, 15, 40, 213–215, 224–225
    heated zone design 40
    interfacing to packed columns 39
    NPD 2, 3, 15, 162, 193, 197, 199–201, 204
    PID 15
    MS (mass spectrometry) 15, 154, 157
Giddings, J.C. 51, 55, 56, 59
Glean 201
Grapefruit oil 239
Glucose 157
Glutamic acid 154
Golay equation 44
Group separations 76, 212
Guanidine 152
Guidelines
    solute characteristics that affect retention 137
    phase selection 137
    instrumental strategy 137

H, plate height 9
Hannay 51
Heavy distillate 217

High efficiency SFC 4, 221–224
High speed 236–237
Hildebrand solubility parameter 51, 55,
    56, 59, 122
Hydrocortisone 155, 159
Hydroxyacids 152
Hydroxyamphetamine 170
3-Hydroxycarbofuran 195
2-Hydroxyhippuric acid 154
Hydroxysteroids 85–87, 117, 236–237
Hydroxyzine 188, 189
Hypersil
    APS 118
    CPS 118, 157
    silica 110, 221–224, 239, 240, 241

Ibuprofen 154, 182, 183, 189
Imidazole derivatives 159
Indapamide 189
Independent flow control 6
Inertness 28
Infinite dilution 60
Injection 11, 35, 149–150
Intermolecular interactions 1, 5
in vivo metabolites 163
Ionization suppression 86, 90–91
Irgafoz 235
Irganox 235–236

Jet fuel 217

Ketoprofen 189

L, column length 9
Lactones 187
LC, liquid chromatography
    comparison to SFC 1, 3, 8
    figures of merit 15
Leak
    compensation 32
    through check valves 34
Legionella 240
Lemon oil 239
Lichrospher
    APS 163
    CN 105, 163, 171
    diol 2, 99, 161
    silica Si-60 155
Light scattering detector 154, 159
Linear velocity (μ) 14, 44, 79–81

Linuron 199
Liquid–vapour equilibrium 63, 67, 68
Londax 201
Lotion 159
Lorazepam 189
Lormetazepam 184
Lycopene 239

Maltose 157
Mass spectrometry 15, 154, 157
    negative ion CI 154
Mefloquine 154
Melezitose 157
Mercaptopurine 152
Methamphetamine 170
Method development 133–150,
    178–182
Methomyl 195
Methotrimeprazine 105, 164
Methyl methacralates 232–234
Methylparabene 159
Methyltestosterone 155
Metobarbital 189
Mitomycin C_{22} 154
Metilox 235
Metoprolol 184, 187, 188
Mixing 32, 33
Mobile phase characteristics 43
Molecular weight 128–130
Molindone 105, 164
Morphine 159

N, efficiency 9
Nadolol 189
Naphtha 217
Naphthalene, 2-vinyl 228
1-Naphthol 195
Naphthylamine 188
Narcotine 159
Near critical *see* subcritical
Nile Red 52, 61, 92
Nitrogen–phosphorus detector (NPD)
    2–4, 15, 160, 162, 193, 197–201,
    204
Nitrous oxide (N_2O) 1, 28, 159
Non-linear retention *vs.* % modifier
    61–62
Noproxen 154
Norcamazepam 163
Normal phase 114, 156

Nucleosil
  diol 74, 118, 152
  SA 58, 60, 117, 118
Nylidrine 170

Oligosaccharides 157
Opium alkaloids 159
Organochlorine pesticides 209
Organophosphorus pesticides 209
Oust 201
Over-compressed fluid 30
Oxamyl 195
Oxazepam 163
Oxprenolol 187, 188
Oxyphenylcyclimine 188

P', solvent strength scale 52
Packings, column 40
  materials 72–73
  pore diameter 72
  surface area 73
Padded tanks 27, 29, 30
PAH's 59, 224
Papaverine 159
Paprika oil 239
Partition ratio (k') 9, 74, 75, 113
P_c, critical pressure 5
Peak capacity (n_c) 10
Peltier 30
Pentane 28, 227
Permethrin 152
Peroxides, initiators 232–233
Perphenazine 105, 164
Pesticides 2–4, 192–211
Petroleum 212–226
Pharmaceuticals 151–191
Phase diagrams 63–69
  methanol in carbon dioxide 66–69
  pure carbon dioxide 64
Phase ratio (β) 73, 74, 75, 113
  of adsorbed film 78
Phase selection guide 139
Phenmetrazine 170
Phenobarbitone 158
Phenols 61, 152
Phenothiazines 160
Phenyl stationary phase 87, 88, 117, 172
Phenylalaninol 188
Phenylephrine 170

2-Phenylethanol 159
Phenylpropanolamine 170
Phenylureas 3, 199
Phosphine oxides 187
Pindolol 187, 188
Plant extracts 156
Plasticizers 28
Plumbing 26
Polarity window 113
Polymer additives 235
Polymer stationary phases/supports 76, 94–95, 163, 231, 236, 238, 242, 243
Polystyrene 227
Polywax 129–130
Poppy straw extracts 159
Preservative 159
Pressure
  control 23, 34, 109
  drop 3, 45, 73
  outlet (p_0) 24
  programming 6, 7, 52, 55
Progesterone 155
Promazine 105, 164
Propafenone 189
Propranolol 180, 181, 187, 188
Propylparabene 159
PRN-399 243
PRP-1 236, 238
PS-DVB 163
Pseudoephedrine 152
PTH amino acids 153
Pumps 22, 29
  flow source 24
  pressure source 25
Purines 152
Pyrimidines 152

Q-10 (ubiquinone) 240
Quality control 193
Quality, of fluids 29

Raffinose 157
Reciprocating pumps 29
Reserpine 105, 164
Residual fuel oil 217
Resolution (R_s) 9, 102
Restrictors
  fixed 25, 34, 35, 225
  linear 35
  frits 35

integral 35
Retention 10, 25
  effect of density 56–57
  effect of modifiers 54, 57–60, 98–99
  effect of pressure 49–50, 99–100
  effect of temperature 50, 100–101
  effect of flow rate 102
Ribose 157
RNA 152

Sacchrides, mono-, di-, penta- 153, 158
Safety 28
Salicylic acid, glycine conjugate 154
Sample capacity 11
Saunders 55
Scabicide 159
Scorbic acid 159
Screening, for pesticides 206
Selectivity 6, 9, 10, 76
  effect of modifier concentration 103
  effect of pressure 103–104
  effect of temperature 104
  effect of flow rate 106
Sensitivity 11, 12
SFC
  definition 1, 47
  figures of merit 17
Silica as stationary phase 82, 119, 154,
    155, 221–224, 239–241
Silicone oils 228–232
Solid phase extraction (SPE) 154
Solubility
  'infinite dilution' 60–61
  with modifiers 59
Solvatochromic dyes 52, 61, 62, 92,
    121
  Reichardt 122
  Kamlet and Taft 122
  Carr 122
Solvent strength 51
  non-linearity 53–55
  p' 52–53, 120–121, 123
  Rohrschneider 120
  Saunders 55, 119
  Snyder 55, 119
  of tertiary fluids 92
Speed 10, 11, 236–237
Spherisorb C_8 235; CN 233, 235, 238
Spidergrams 14–19
Split ratio 225

Stationary phase characteristics 72
  polarity 89, 117, 128
  relative retentivity 60, 117, 130
Statistically designed experiments 147
Steric hindrance 127, 129
Steroids 85–87, 117, 154–155, 156,
    236–237
Stimulants 169
Subcritical fluid 47–48, 67
  GC 49–50
  LC 50
  vs. supercritical 219
Sucrose 157
Sugar 157
Sulconazole 189
Sulfadimethoxine 173
Sulfa drugs 104
Sulfaguanidine 173
Sulfamerazine 173
Sulfamethazine 74
Sulfamethazole 173
Sulfamethoxazole 173
Sulfamethoxypyridazine 104, 172
Sulfaquinoxaline 173
Sulfisomidine 104, 172
Sulfonamides 74, 104, 171–3
Sulfonic acid (SA) stationary phase 58,
    60, 117, 118
Sulfonylureas 3, 201
Sulfur hexafluoride 159
Supercritical fluid definition 5
Surface coverage
  by additives 89
  by carbon dioxide 77
  by modifiers 82
Suxazole 189
Swelling 79
Syringe pumps 29

Tablets 156, 160
Tailing 75, 76
Taxol 160
T_c, critical temperature 5
Temazepam 163
Testosterone 155
Tertiary fluids 6
Tetrahydrozaline 172
TFA, 152
Thebaine 159
Theobromine 160

Theophyline 160
Thiothixene 105, 164
Thermally labile compounds 14, 156, 193–199, 200
Thymine 74
t_0, column transit time 9, 74
Tocopherols 152, 239
Toluene 213
Toluidines 125
t_R, solute retention time 9, 74
T_R, reduced temperature (T/T_c) 45
Triazines 3, 204
Tricyclic antidepressants 167
Trifluoroacetic acid 152
2,2,2-Trifluoro-1-(9-anthryl)ethanol 188
Triflupromazine 105, 164
Trifluridine 152
Trimethoprim 152
Triprofenic acid 184
Triprolidine 152
Triple point 64, 66
Triton X-100 and other Tritons 236–238

Ubiquinones 240–242
Ultra-high pressures 59
Under-compressed fluid 30
Upstream control 24
Urine 154, 183

Ursodeoxycholic acid 155
UV detector 2, 4, 6, 152, 153, 155, 158, 161–164, 168, 172–173, 180–185, 194, 198–203, 205, 209, 210, 213–215, 217–218, 222–224, 228–231, 233–239, 242
  cell 37
  sensitivity 38
  spectra 38
  temperature 37

Van Deemter
  equation 79
  plots 80–81
Viscosity 1, 5, 45–47
  in near critical fluids 47–48
Vitamins A, E, D_2, K_1, D_3, tocopherols 153
V_r, retention volume 11
V_s, sample injection volume 11

Warfarin 189
W_h, peak width at half height 9

Xylose 157

Z, compressibility factor 24
Zidovudine (AZT) 152
Zorbax CN 104

www.ingramcontent.com/pod-product-compliance
Lightning Source LLC
Chambersburg PA
CBHW050524190326
41458CB00005B/1658